Climate Change Risks in Brazil

Carlos A. Nobre • Jose A. Marengo
Wagner R. Soares
Editors

Climate Change Risks in Brazil

Springer

Editors
Carlos A. Nobre
Institute for Advanced Studies
University of São Paulo
São Paulo, Brazil

Jose A. Marengo
National Center for Monitoring and Early
Warning of Natural Disasters
São Paulo, Brazil

Wagner R. Soares
National Center for Monitoring and Early
Warning of Natural Disasters
São Paulo, Brazil

ISBN 978-3-030-06538-6 ISBN 978-3-319-92881-4 (eBook)
https://doi.org/10.1007/978-3-319-92881-4

Cover Illustration: Germano Ribeiro Neto, CEMADEN, São Paulo, Brazil

This Springer imprint is published by the registered company Springer Nature Switzerland AG
The registered company address is: Gewerbestrasse 11, 6330 Cham, Switzerland

This book is dedicated to my wife Ana Amelia for her endless love, support, wise advice, and encouragement. It is also dedicated to my 6-year-old grandson Rafael, hoping that his generation will have the wisdom and determination that lacked to my generation to put our Planet on a sustainable pathway.

Carlos A. Nobre

This book is dedicated to my beloved wife Angela Cristina and son Jose Antonio, without whose love, encouragement, support, and inspiration I would never have made it this far. I also wish to dedicate this book to my mother and brothers for their continuous love and support all the way since the beginning.

Jose A. Marengo

This book is dedicated to my father Antônio Flávio de Almeida Soares (In Memoriam), my mother Lígia Soares, my family, and work colleagues for the support received in the elaboration of this scientific material.

Wagner R. Soares

Foreword

Political decision makers have to decide how much effort to spend on countering the impacts of climate change. That decision must be informed by a full risk assessment, and that assessment takes three different approaches. We need first to establish what we human beings are doing to our climate. Second, we must establish how the climate is likely to change in response to our actions, and how these changes will affect us and our essential habitat. Third, we need to examine carefully what we, in one country and also between countries, might do to each other.

We should not be in any doubt that climate change poses a great risk to all societies, and to our efforts at poverty eradication and economic prosperity. Eventually, this will threaten global civilization. But we need to get beyond that level of understanding, to look at specific risks that are possible threats to us. In doing this, we need to start with an approach not commonly used by climate scientists.

If you own or manage a house or other building, you would take out an insurance policy to guard against the risks of burglary or fire. Even though these are low probability events, the impact for you would be very great if one of these events did happen. Who are the experts at evaluating these low probability but high impact events? It is the actuaries, the statisticians working in the insurance and re-insurance industries, who use whatever scientific, social, and historical information available to evaluate the likelihood of these events. Of course, with climate change the likelihood of high impact events happening is dependent on time. As the average global temperature and the sea level continue to rise, so will the impact of extreme weather events, such as storms at sea flooding further inland or hot summer days producing greater heat stress to populations. People have limited toleration for combinations of high temperature and high humidity. Coastal cities, particularly those at the mouths of large rivers, will have thresholds in the rate of sea level rise that they can deal with. These threats, viewed holistically, have unprecedented implications for humanity. Where will the 100 million people who live around the Ganges estuary, in Calcutta and Bangladesh, relocate to when these areas become unlivable?

In this book, the authors take this risk analysis approach to analyze in particular the impacts of climate change for health, energy, and biodiversity in Brazil, and regional risks to the Amazon and North Eastern Brazil. Its importance cannot be underestimated. The editors have brought together a full range of appropriate experts to evaluate the risks that Brazil faces unless actions are taken quickly across the whole world to manage the challenges.

All countries need to engage actively in the business of defossilizing their economies, and where possible not only in avoiding deforestation, but in reforestation to create more carbon sinks. The era of fossil fuels has provided a massive boost to global economies since the beginning of the industrial revolution. Coal, oil, and gas have been an abundant source of the energy needed in those revolutionary developments. And now we know that, together with deforestation, burning fossil fuels is the major cause of the increased content of greenhouse gases in the atmosphere. This increase captures more of the energy from the earth that would otherwise escape into space, causing global warming. The new era, the era of clean fossil-fuel-free energy, is the way forward, with the co-benefit of providing clean air for all to breathe, free of carbon particulate matter, NO_x and SO_x. Renewable energy is already competitive in the market place for most countries in the world. Energy is the largest global industry: the clean energy sector provides a new era for wealth creation too.

The pledge of the Paris Agreement, to keep the average global temperature rise to less than 2°C and to aim for no more than 1.5°C rise, is now a common commitment of 195 signatory countries. No other issue has brought such universal agreement for action. But this agreement is not matched by universal actions to meet the challenge. This book underlines the importance for all of us, as individuals, as urban, regional and national citizens, as city mayors and regional governors, and as politicians and heads of government, to act now to help to avoid the severe risks set out here.

Cambridge, UK Sir David King

Preface

In a high greenhouse gas emissions scenario, the country has a high likelihood (over 70%) of suffering a greater than 4°C temperature rise before the end of the century. For high degrees of global warming exceeding 4°C in relation to the pre-industrial era, climatic conditions in large areas of Brazil may become dangerous for the population: mean temperatures may reach 30°C—double the current global average—increasing the risk of dying from heat stress, particularly for infants and the elderly. Maximum temperatures above the human body's ability to adapt will reduce labor productivity in sectors like agriculture and civil construction. In some regions, heat waves and changes in the rainfall regime may increase the spatial-temporal incidence of diseases.

Aside from the serious impacts on human health, higher warming rates may increase the risk of extreme or even catastrophic events like species extinction, reduce the availability of water and hydropower, and severely impact food production, limiting the crop area available for major agricultural crops in Brazil, severely limiting the country's role as a key provider of global food security.

The aim of this book is to explain how changes in the physical climate can lead to multiple complex changes in human systems, many of which are hard to predict, and which tend to have adverse consequences when the changes fall far outside the normal range of variability.

This study aimed to map extreme temperature increase scenarios in Brazil and their impacts on four key sectors: agriculture, health, biodiversity, and energy. It also examines in-depth likely consequences of high degrees of warming for two critical regions in Brazil, the Amazon and Northeast Brazil. The assessments of climate risks for high degrees of warming were accomplished through a careful review of the literature and climate projections, including relative risk estimates. This synthesis, which summarizes the state of the art of knowledge on the subject, provides decision-makers with risk analysis tools: events that are unlikely to occur but could have significant and even catastrophic consequences need to be understood and incorporated to public policy planning.

São Paulo, Brazil

Carlos A. Nobre
Jose A. Marengo

Acknowledgments

The book *Climate Change Risks in Brazil* is a result of contributions from a wide range of experts from across Brazil. We thank everyone who contributed to its richness and multidisciplinary outlook.

It has been written by a team from the National Center for Monitoring and Early Warning of Natural Disasters CEMADEN, the Brazilian Agricultural Research Corporation EMBRAPA, the Federal University of Rio de Janeiro UFRJ, the Brazilian Foundation for Sustainable Development FBDS, the National Public Health School at the Oswaldo Cruz Foundation FIOCRUZ, and the National Institute for Space Research, including Carlos A. Nobre, Jose A. Marengo, Wagner R. Soares (Editors), Eduardo Assad, Sandra Hacon, Fabio Scarano, Roberto Schaeffer, and Gilvan Sampaio, leading authors of the various chapters.

Insightful comments were provided by scientific peer reviewers. We thank Dan Bernie, Stoecio Maia, Jurandir Zullo Junior, Andrea Sobral, Eliane Ignotti, Donald Wilhite, Rita Vieira, Jose Maria Cardoso da Silva, Rafael Loyola, Chris Jones, Sergio Margulis, and Luis Alberto da Cunha Saporta for their review of early versions of the various book chapters.

The report was commissioned by the United Kingdom Embassy in Brasilia, Brazil, that was funding the research project Climate Change Risks - Limits to Adaptation. Results presented in this book were derived from research funded by the National Institute of Science and Technology (INCT) for Climate Change Phase 1 under the Brazilian National Council for Scientific and Technological Development (CNPq) Grant 573797/2008-0 and the São Paulo Research Support Foundation (FAPESP) Grant 2008/57719-9; and Phase 2 under CNPq Grant 465501/2014-1, and FAPESP Grants 2014/50848-9; the National Coordination for High Level Education and Training (CAPES) Grant 16/2014. We are also grateful to the Brazilian Platform on Biodiversity and Ecosystem Services (BPBES) for support

(CNPq; project: 405593/2015-5). The authors also would like to thank the Brazilian Network of Climate Change (Rede CLIMA) for its collaboration in performing studies and analyses that helped in the elaboration of the chapters.

Carlos A. Nobre, Jose A. Marengo, and Wagner R. Soares provided valuable guidance and oversight. We would like to thank Elisangela Rodrigues de Sousa and Ana Paula Soares for outreach efforts to partners, the scientific community, and the media.

São Paulo, Brazil

Carlos A. Nobre
Jose A. Marengo
Wagner R. Soares

Contents

List of Contributors

Felipe Ferreira Alexandre Earth System Science Center/National Institute for Space Research (CCST/INPE), São Paulo, Brazil

Regina C. S. Alvalá National Center for Monitoring and Early Warning of Natural Disasters (CEMADEN), São Paulo, Brazil

Lincoln M. Alves National Institute for Space Research (INPE), São Paulo, Brazil

Eduardo Delgado Assad Embrapa, São Paulo, Brazil

Laura S. Borma Earth System Science Center/National Institute for Space Research (CCST/INPE), São Paulo, Brazil

Manoel Cardoso Earth System Science Center/National Institute for Space Research (CCST/INPE), São Paulo, Brazil

Isabella V. L. Costa COPPE/Federal University of Rio de Janeiro (COPPE/UFRJ), Rio de Janeiro, Brazil

Luz A. Cuartas National Center for Monitoring and Early Warning of Natural Disasters (CEMADEN), São Paulo, Brazil

Ana Paula Cunha National Center for Monitoring and Early Warning of Natural Disasters (CEMADEN), São Paulo, Brazil

Sheila S. de Barros Brito National Center for Monitoring and Early Warning of Natural Disasters (CEMADEN), São Paulo, Brazil

Beatriz Fátima Alves de Oliveira National Public Health School at the Oswaldo Cruz Foundation (FIOCRUZ), Rio de Janeiro, Brazil

Sandra de Souza Hacon National Public Health School at the Oswaldo Cruz Foundation (FIOCRUZ), Rio de Janeiro, Brazil

Vanessa Huback COPPE/Federal University of Rio de Janeiro (COPPE/UFRJ), Rio de Janeiro, Brazil

Karinne Leal National Center for Monitoring and Early Warning of Natural Disasters (CEMADEN), São Paulo, Brazil

André F. P. Lucena COPPE/Federal University of Rio de Janeiro (COPPE/UFRJ), Rio de Janeiro, Brazil

Antonio R. Magalhaes Center for Strategic Studies, CGEE, Brasília, Brazil

Jose A. Marengo National Center for Monitoring and Early Warning of Natural Disasters (CEMADEN), São Paulo, Brazil

Alan Massaru Nakai Embrapa, São Paulo, Brazil

Carlos A. Nobre Institute for Advanced Studies, University of São Paulo, São Paulo, Brazil

Germano Ribeiro Neto National Center for Monitoring and Early Warning of Natural Disasters (CEMADEN), São Paulo, Brazil

Rodrigo Rudge Ramos Ribeiro Embrapa, São Paulo, Brazil

Daniel Andrés Rodriguez Federal University of Rio de Janeiro (UFRJ), Rio de Janeiro, Brazil

Gilvan Sampaio Center for Weather Forecasting and Climate Studies/National Institute for Space Research (CPTEC/INPE), São Paulo, Brazil

Fabio Rubio Scarano Federal University of Rio de Janeiro (UFRJ) and Brazilian Foundation for Sustainable Development (FBDS), Rio de Janeiro, Brazil

Roberto Schaeffer COPPE/Federal University of Rio de Janeiro (COPPE/UFRJ), Rio de Janeiro, Brazil

Ismael Silveira National Public Health School at the Oswaldo Cruz Foundation (FIOCRUZ), Rio de Janeiro, Brazil

Institute of Social Medicine at the University of the State of Rio de Janeiro, Rio de Janeiro, Brazil

Ana Paula Soares Center for Environmental Studies and Research State University of Campinas (UNICAMP), São Paulo, Brazil

Wagner R. Soares National Center for Monitoring and Early Warning of Natural Disasters (CEMADEN), São Paulo, Brazil

Roger R. Torres National Center for Monitoring and Early Warning of Natural Disasters (CEMADEN), São Paulo, Brazil

Eveline Vásquez COPPE/Federal University of Rio de Janeiro (COPPE/UFRJ), Rio de Janeiro, Brazil

Cindy Viviescas COPPE/Federal University of Rio de Janeiro (COPPE/UFRJ), Rio de Janeiro, Brazil

Celso von Randow Earth System Science Center/National Institute for Space Research (CCST/INPE), São Paulo, Brazil

List of Acronyms

ANA	National Water Agency
ANEEL	National Agency for Electric Energy
AR4	IPCC Fourth Assessment Report
AR5	IPCC Fifth Assessment Report
AVHRR	Advanced Very High Resolution Radiometer
BT	Brightness temperature
CC	Climate Change
CDD	Consecutive Dry Days
CEMADEN	National Center for Monitoring and Early Warning of Natural Disasters
CGEE	Center for Strategic Studies
CMIP5	Coupled Model Intercomparison Project Phase 5
CO_2	Carbon Dioxide
CO_2eq	Carbon Dioxide Equivalent
CONAB	National Company for Food Supply
COP	Conference of the Parties
CPTEC	Center for Weather Forecasts and Climate Studies
CRU	Climate Research Unit
EbA	Ecosystem Based Adaptation
EMBRAPA	Brazilian Agricultural Research Corporation
ETA	Regional Model with the Eta vertical coordinate
FAO	Food and Agriculture Organization
FBDS	Brazilian Foundation for Sustainable Development
FBMC	Brazilian Forum on Climate Change
FIOCRUZ	National Public Health School at the Oswaldo Cruz Foundation
GCM	General Circulation Model
GHG	Greenhouse gases
HI	Heat Index
HLF	Habitat Fragmentation
IBGE	Brazilian Institute for Geography and Statistics
INDC	Intended Nationally Determined Contributions

INMET	National Institute of Meteorology
INPE	National Institute for Space Research
IPCC	Intergovernmental Panel on Climate Change
MME	Ministry of Mines and Energy
NDVI	Normalized Difference Vegetation Index
NEB	Northeast Brazil
OECD	Organisation for Economic Co-operation and Development
ONS	National Operator of the Electric System
PBMC	Brazilian Panel on Climate Change
RCP	Representative Concentration Pathway
SAE	Secretariat of Strategic Affairs of Brazil
SPI	Standardized Precipitation Index
SPEI	Standardized Precipitation Evapotranspiration Index
SRES IPCC	Special Report on Emissions Scenarios
TCI	Temperature Condition Index
UNIFEI	Federal University of Itajuba
UNFCCC	United National Framework Convention on Climate Change
UTCI	Universal Thermal Climate Index
VCI	Vegetation Condition Index
VHI	Vegetation Health Index
WBGT	Wet-Bulb Global temperature
WEF	World Economic Forum
WGI	IPCC Working Group I
WGII	IPCC Working Group II
WGIII	IPCC Working Group III
WHO	World Health Organization
WSDI	Warm Spell Duration Index
WMO	World Meteorological Organization

Chapter 1
Introduction

Carlos A. Nobre, Jose A. Marengo, Wagner R. Soares, and Ana Paula Soares

The risks of global climate change go from high to very high with an average increase in temperature of 4 °C or more. This includes serious and generalised impacts related to the extinction of endangered species, great risks to global and regional food security, in addition to climate risks associated to alterations in extreme events like heat waves, extreme rainfall and coastal flooding, which are already moderate to high with 1 °C increase in temperature. The consequences of a dangerous climate change due to 4 °C or above warming are suggested to be devastating: the inundation of coastal cities; increasing risks for food production potentially leading to higher malnutrition rates; many dry regions becoming dryer, wet regions, wetter; unprecedented heat waves in many regions, especially in the tropics; substantially exacerbated water scarcity and drought in many regions; increased frequency of high-intensity tropical cyclones; and irreversible loss of biodiversity, including coral reef systems (World Bank, 2012).

And most importantly, a 4 °C world is so different from the current one that it comes with high uncertainty and new risks that threaten our ability to anticipate and plan for future adaptation needs. The lack of action on climate change not only risks putting prosperity out of reach of millions of people in the developing world, it threatens to roll back decades of sustainable development. It is clear that we already know a great deal about the threat before us. The science is unequivocal that humans are the cause of global warming, and major changes are already being observed: the globally averaged combined land and ocean surface temperature data as calculated

C. A. Nobre (✉)
Institute for Advanced Studies, University of São Paulo, São Paulo, Brazil

J. A. Marengo · W. R. Soares
National Center for Monitoring and Early Warning of Natural Disasters (CEMADEN), São Paulo, Brazil

A. P. Soares
Center for Environmental Studies and Research State University of Campinas (UNICAMP), São Paulo, Brazil

C. A. Nobre et al. (eds.), *Climate Change Risks in Brazil*,
https://doi.org/10.1007/978-3-319-92881-4_1

by a linear trend, show a warming of 0.85 [0.65–1.06] °C, over the period 1880–2012; the total increase between the average of the 1850–1900 period and the 2003–2012 period is 0.78 [0.72–0.85] °C; ocean warming dominates the increase in energy stored in the climate system, and the upper 75 m warmed by 0.11 [0.09–0.13] °C per decade over the period 1971–2010, and are acidifying; sea levels rose over the period 1901–2010, global mean sea level rose by 0.19 [0.17–0.21] m; an exceptional number of extreme heat waves occurred in the last decade; major food crop growing areas are increasingly affected by drought (IPCC, 2014; Clarke et al., 2014; Collins et al., 2013).

Despite the global community's best intentions to keep global warming below a 2 °C increase above pre-industrial climate, as decided in the 21st Conference of the Parties – COP 21 in Paris in 2015, higher levels of warming are increasingly likely and the consensus reached in there among countries were to include a temperature limit increase of 2 °C, aiming at 1.5 °C. Scientists agree that countries' current United Nations Framework Convention on Climate Change emission pledges and commitments would most likely result in 3.5–4 °C warming. And the longer those pledges remain unmet, the more likely a 4 °C world becomes. With 1.5 °C, the risk or the irreversible melting of Greenland's ice layer would be, for example, much lower than if we surpassed 2 °C. Avoiding a dangerous anthropogenic interference in the climate system is the final goal of the COP-21.

Understanding how an increase in global temperature may turn into risks for sectors of society and ecosystems is critical in facing the scale of impacts, which may be worse in different regions of the planet. This conclusion may be drawn from scientific studies compiled in Working Group I (WGI) Report of IPCC-AR5 – which addresses climate change science – and Working Group II (WGII) – which assesses impacts, adaptation and vulnerability -, published in 2013 and 2014 respectively, and for Brazil from the results of the First National Assessment Report of the Brazilian Panel on Climate Change (PBMC), published in 2016.

The IPCC AR5 used four scenarios-referred to as 'Representative Concentration Pathways' (RCPs) to show how GHG may evolve this century. For example, in 2100 the RCP 4.5 intermediate scenario projects that global temperature will reach around 1.4–3.1 °C above pre-industrial levels (Collins et al., 2013). In the IPCC AR5 most extreme scenario (RCP 8.5) – where a business as usual rate of emissions is considered, in other words, as seen in the last decades – global temperature increase may exceed 4°C by 2100. In fact, unless emissions stop during this century, temperatures will continue to rise much beyond the end of the century. In RCP 8.5, projections for 2150 show an increase of 6 °C. Now, if it is already difficult to project what the world will be like 2 °C hotter, it is even more complex foreseeing what will happen with an increase of 4 or 6 °C in temperature. In addition, the report suggests that impacts resulting from a high rise in temperature are not only relatively unknown, but have not been studied much. Nonetheless, the risk of triggering big changes in the climate system is clear. Furthermore, some physical systems or ecosystems may be at risk of abrupt and irreversible changes (IPCC, 2014). Events like storms, floods, landslides and heat waves had stronger effects particularly in poorer countries. For example, an increase in sea level may force up the migration of about 760 million people – the equivalent to Europe's population – this century. According to a study, this will happen if there is a 4 °C increase in the global temperature (Kreft, Eckstein, & Melchior, 2017).

In the last 15 years, Brazil has witnessed climate extremes, some of the characterised as 'events of the century': droughts in 2005, 2010 and 2015–16 and floods in 2009, 2013, 2014 and 2017 in the Amazon; drought in the semiarid region of Northeast Brazil since 2012; the 2014–2015 drought in Brazil's Southeast, which led to the worst water crisis in the city of São Paulo since 1960. Heat waves have also affected the whole of Brazil and in the winter of 2015, temperatures reached 3–4 °C above normal. An increase in extreme rainfall has also been observed in Southeast and South, affecting vulnerable areas in towns and cities in these regions. This has added to the risk of flooding and landslides in rural and urban areas, just like drought in the Amazon and the Central-west have increased the chance of forest fires. The possibility of these events becoming more frequent and/or intense in a warmer climate in the future has been shown by climate model predictions in the IPCC-AR5 reports (IPCC, 2014).

According to the PBMC (2016), the most vulnerable coastal areas in Brazil are: the coast of the state of Santa Catarina, the cities of Santos and Guarujá in the state of São Paulo, Recife and Rio de Janeiro, as well as São Paulo's metropolitan region. According to Strauss, Kulp, and Levermann (2015), Brazil is among the countries most affected by the advance of waters. In its coastal zone, 16 million people would lose their homes before 2100. Based on the study, Rio de Janeiro will be among the twenty most affected metropolises on the planet. The drought and water crisis that has affected the São Paulo Metropolitan Region in 2014–15 shows the vulnerability of this megacity to drought, and is a sign of serious problems Brazil will have to face in the future, as a consequence of the deficient rainfall, particularly in warmer scenarios (Marengo, Torres, & Alves, 2016; Nobre, Marengo, Seluchi, Cuartas, & Alves, 2016). From 2012 to 2016, Northeast Brazil has experienced the longest continuous drought, affecting cities and population (Alvalá et al., 2017; Marengo et al., 2017).

Climate risks may be divided into two categories: risk of extreme climate events and risks related to long term changes to average conditions. Assessments aimed at informing disaster risk reduction policies will focus on the risk of extreme climate events. As extreme climate events already represent a danger in the present and may be predicted a few days or season earlier, it might be reasonable for such risk assessment to have a relatively short-term focus. However, a risk assessment aimed at informing society's response in relation to climate change as a whole should take into account all of the time scale, which will be affected by our current decisions on greenhouse effect gases emissions. Hence, more than 2 °C increase in temperature through the next centuries may not be discarded. In addition, extreme events are part of the global climate change scale. Therefore, assessments need to consider the 'worst case' scenarios, not just in terms of individual events like climate extremes, but also in terms of long term changes, in other words: the risk that average climate change may reach extreme values.

Climate impacts are not distributed evenly throughout the world and in Brazil, and local ability to adapt to such impacts may differ significantly between regions. This is why until now, no strong political consensus has been reached in relation to what level of climate change should be considered 'dangerous' and, thus, what level of warming may be considered as one to be clearly avoided. Therefore, limiting analyses on global warming below 2 °C in relation to pre-industrial levels may disguise regional impacts. Hence, other levels of warming should be looked at, to derive a set of 'vital signs' that allow for local risks to be studied (King, Schrag, et al., 2015). Climate targets may be expressed in a variety of ways. However, there are uncertainties associated to how the Earth's system responds to anthropogenic pressure. Therefore, temperature targets are usually expressed in terms of a specific temperature limit and a minimum likelihood of this limit being met. While the international community uses 2 °C as the basic principle of 'dangerous' warming, some climate impacts are already included, particularly for low altitude regions and island countries. Nevertheless, contrary to this optimistic scenario, GHG emissions have continued to increase. What happens if we exceed the 2 °C target established to limit global warming? The size of the task increases every year. Limiting warming in under 2 °C already implies great global challenges (Clarke et al., 2014; Rogelj et al., 2015; and Riahi et al., 2015). According to a World Bank study (World Bank, 2012), 'the world is heading towards an average of over 3 °C in global warming' and current climate models still suggest a 20% chance of a 4 °C increase in relation to the pre-industrial period.

Assessing how large increases in temperature translate into risks to natural and human systems is critical to face the scale of impacts, as well as proposing adaptation measures that may stand out more in different regions. Therefore, this study presents an assessment of sectorial impacts and risks for temperature increases of 4 °C and warmer for Brazil. An extensive literature review was carried out on sectorial risks, based on temperature increase projections. In addition, situations and scenarios where regional warming reaches over 4 °C were assessed, as well as its impacts in the sectors considered key in this study: agriculture, human health, biodiversity and energy. Some case studies are also assessed for two regions vulnerable to the impacts of extreme climate variability and climate change: the Northeast Brazil and the Amazon regions. On this book, we review the impacts of regional warming above 4 °C, because the sectorial impacts may be higher under a regional intense warming.

References

Alvalá, R. C., Cunha, A. P. M. A., Brito, S. B., Seluchi, M. E., Marengo, J. A., Moraes, O. L. L., et al. (2017). Drought monitoring in the Brazilian semiarid region, In Press, Annais da Academia Brasilieira de Ciencias do Brasil.

Clarke, et al. (2014). In O. Edenhofer et al. (Eds.), *Climate change 2014: Mitigation of climate change, Contribution of working group III to the Fifth assessment report of the intergovernmental panel on climate change* (pp. 413–510). Cambridge, UK: Cambridge University Press.

Collins, M., Knutti, R., Arblaster, J., Dufresne, J.-L., Fichefet, T., Friedlingstein, P., et al. (2013). Long-term climate change: Projections, commitments and irreversibility. In T. F. Stocker, D. Qin, G.-K. Plattner, M. Tignor, S. K. Allen, J. Boschung, A. Nauels, Y. Xia, V. Bex, & P. M. Midgley (Eds.), *Climate change 2013: The physical science basis. Contribution of working group I to the fifth assessment report of the intergovernmental panel on climate change*. Cambridge, UK and New York: Cambridge University Press.

IPCC. (2014). Climate Change 2014: Synthesis Report. Contribution of Working Groups I, II and III to the Fifth Assessment Report of the Intergovernmental Panel on Climate Change [Core Writing Team, R.K. Pachauri and L.A. Meyer (eds.)]. Geneva: IPCC, p. 151

King, D., Schrag, D., et al. (2015). Climate change: a risk assessment. Report commissioned by the Foreign and Commonwealth Office, available online at http://www.csap.cam.ac. uk/media/uploads/files/1/climatechange–a-risk-assessment-v9-spreads.pdf

Kreft, S., Eckstein, D., & Melchior, I. (2017). Global Climate Risk Index. Who suffers most from extreme weather events? Weather-related loss events in 2015 and 1996 to 2015.

Marengo, J. A., Alves, L. M., Alvalá, R. C., Cunha, A. P., Brito, S. S., & Moraes, O. L. L. (2017). Climatic characteristics of the 2010–2016 drought in the semiarid Northeast Brazil region, Anais da Academia Brasileira de Ciências (2017) (Annals of the Brazilian Academy of Sciences) Printed version ISSN 0001–3765 / Online version ISSN 1678–2690 http://dx.doi. org/DOI.

Marengo, J. A., Torres, R. R., & Alves, L. M. (2016). Drougth in northeast Brasil – Past, present and future. *Theoretical and Applied Climatology, 20*, 1–12.

Nobre, C. A., Marengo, J. A., Seluchi, M. E., Cuartas, L. A., & Alves, L. M. (2016). Some characteristics and impacts of the drought and water crisis in southeastern Brazil during 2014 and 2015. *Journal of Water Resource and Protection, 8*, 252–262.

PBMC. (2016). Mudanças Climáticas e Cidades. Relatório Especial do Painel Brasileiro de Mudanças Climáticas. In S. K. Ribeiro & A. S. Santos (Eds.), *PBMC, COPPE – UFRJ* (p. 116). Rio de Janeiro, Brasil. isbn:978-85-285-0344-9.

Riahi, K., et al. (2015). Locked into Copenhagen pledges—Implication of short-term emission targets for the cost and feasibility of long-term climate goals Technol. *Forecast Social Change, 90A*, 8–23.

Rogelj, J., Luderer, G., Pietzcker, R. C., Kriegler, E., Schaeffer, M., Krey, V., & Riahi, K. (2015). Energy system transformations for limiting end-of-century warming to below 1.5 [deg] C. *Nature Climate Change, 5*(6), 519–527.

Strauss, B. H., Kulp, S., & Levermann, A. (2015). *Mapping choices: Carbon, climate, and rising seas — Our global legacy*. Princeton, NJ: Climate Central.

World Bank. (2012). *4C: Turn down the heat*. A report for the World Bank by the Potsdam Institute for Climate Impact Research and Climate Analytics, November 2012, Washington, DC 20433.

Collins, M., Knutti R., Arblaster, J., Dufresne, J.-L., Fichefet, T., Friedlingstein, P., et al. (2013). Long-term climate change: Projections, commitments and irreversibility. In T. F. Stocker, D. Qin, G.-K. Plattner, M. Tignor, S. K. Allen, J. Boschung, A. Nauels, Y. Xia, V. Bex & P. M. Midgley (Eds.), *Climate change 2013: The physical science basis. Contribution of working group I to the fifth assessment report of the intergovernmental panel on climate change*. Cambridge, UK and New York: Cambridge University Press.

IPCC. (2014). Climate Change 2014: Synthesis report. Contribution of Working Groups I, II and III to the Fifth Assessment Report of the Intergovernmental Panel on Climate Change [Core Writing Team, R. K. Pachauri and L. A. Meyer (Eds.)]. [illegible]

[illegible]

[illegible]

[illegible]

[illegible]

[illegible]

[illegible]

[illegible]

[illegible]

[illegible]

Chapter 2
Assessment of Warming Projections and Probabilities for Brazil

Wagner R. Soares, Jose A. Marengo, and Carlos A. Nobre

2.1 Introduction

Climate targets can be expressed in different ways. Temperature targets are usually expressed in terms of a specific temperature limit and a minimum probability to reach such limit. Climate models considering several greenhouse gas emissions from socio-economic development scenarios generate future temperature change projections. Targets are generally expressed with a minimum probability of *avoiding* a limit rather than reaching it. Although targets tend to arise from political decisions, technically the timescale is also important in specifying a target. 2100 is generally assumed, but in reality avoidance over longer time scales as well need to be assessed. The 2 °C target for global warming had been under severe scrutiny in the run-up to the climate negotiations in Paris in 2015 (COP21). Clearly, with a remaining carbon budget of 470–1020 $GtCO_2eq$ from 2015 onwards for a 66% probability of stabilizing at concentration levels consistent with remaining below 2 °C warming at the end of the twenty-first century and yearly emissions of about 40 $GtCO_2$ per year, not much room is left for further postponing action (Fuss, 2017).

The SRES emission scenarios (*Special* Report on Emissions *Scenario,* Nakicenovic & Swart, 2000*),* used in the 2001 IPCC TAR and 2007 AR4 reports (the third and fourth IPCC reports), are based on a set of coherent and physically consistent assumptions on their forcing, such as demography, socio-economic development and technological changes. The 2013 IPCC AR5 introduced the new RCP scenarios (Representative Concentration Pathways, Moss et al., 2010), which

W. R. Soares (✉) · J. A. Marengo
National Center for Monitoring and Early Warning of Natural Disasters (CEMADEN), São Paulo, Brazil
e-mail: wagner.soares@cemaden.gov.br

C. A. Nobre
Institute for Advanced Studies, University of São Paulo, São Paulo, Brazil

C. A. Nobre et al. (eds.), *Climate Change Risks in Brazil*,
https://doi.org/10.1007/978-3-319-92881-4_2

adopt a more complete system and take into account emission impacts, i.e., how much the radiation balance will change in the Earth system.

RCPs are identified by their total radiative forcing (expressed in W/m^2) to be reached during (or by the end of) the twenty-first century: RCP 2.6 (mitigation scenario, leading to a very low forcing level), RCP 4.5 and RCP 6.0 (two intermediates scenarios), and RCP 8.5 (scenario with very high GHG emissions) (IPCC, 2014). Each RCP provides spatially distributed datasets on land-use changes and sectorial emissions of air pollutants, and specifies annual GHG concentrations and anthropogenic emissions up to 2100 (Burkett et al., 2014). RCP 8.5, projections for 2150 show an increase of 6 °C in mean global surface temperature. Now, if it is already difficult to project what the world will be like 2 °C hotter, it is even more complex foreseeing what will happen with an increase of 4 °C to 7 °C in temperature. In addition, the report suggests that impacts resulting from a high rise in temperature are not only relatively unknown, but have not been studied much. Nonetheless, the risk of triggering large changes in the climate system is clear. Furthermore, some physical systems or ecosystems may be at risk of abrupt and irreversible changes (IPCC, 2014). Events like storms, floods, landslides and heat waves had stronger effects particularly in poorer countries (Kreft, Eckstein, & Melchior, 2016). For example, an increase in sea level may force up the migration of about 760 million people – the equivalent to Europe's population – this century. According to a study, this will happen if there is a 4 °C increase in the global temperature (Strauss, Kulp, & Levermann, 2015).

The radiation balance is the ratio between the amount of solar radiation entering and leaving the Earth. By measuring GHG and aerosol concentrations, we can find out how much energy is stored in the Earth system. Each RCP was developed by a set of integrated assessment models (IAMs), and for each scenario synthetic datasets were researched and created from available representative studies, which were repeatedly reviewed by different stakeholders. The RCP projections are part of the set of CMIP5 - Coupled Model Intercomparison Project Phase 5 simulations and projections (Taylor, Stouffer, & Meehl, 2012).

2.2 Climate Modelling and Climate Scenarios Used

In order to make climate projections, global or regional climate models are used so as to try to represent nature as realistically as possible according to existing knowledge levels and computing tools. According to Jones et al. (2004), climate scenarios are plausible representations of future emissions of substances that are active from a radiative perspective (greenhouse gases), or that have the characteristic of affecting other components that are active from the same radiative perspective (such as sulphur dioxide, which forms sulphate aerosols).

The primary characteristics of each RCP are described below. Table 2.1 presents the main features, such as radiative forcing, CO_2 equivalent concentration, behaviour throughout the twenty-first century, temperature variation and sea level for the four scenarios.

Table 2.1 Characteristics of CMIP5 RCPs

RCP	Radiative Forcing:	Concentration (ppm) CO_2-equiv.	Behaviour:	Temp. Rise in the Planet in 2100	Sea Level Rise	Source
2.6	Peak of <3 W/m^2 in 2100, and then declines to 2.6	~ 490	Rise with peak in 2040, and declines	Between 0.3 °C and 1.7 °C	Between 26 and 55 cm	Van Vuuren et al. (2007)
4.5	Additional storage of ~4.5 W/m^2 stabilises after 2100	~650	Rise until 2060 and stabilises	Between 1.1 °C and 2.6 °C	Between 32 and 63 cm	Clarke et al. (2007), Smith and Wigley (2006) and Wise et al. (2009)
6.0	Additional storage of ~6W/m^2	~850	Rise until 2100 and stabilises	Between 1.4 °C and 3.1 °C	Between 33 and 63 cm	Fujino, Nair, Kainuma, Masui, and Matsuoka (2006) and Hijioka, Matsuoka, Nishimoto, Masui, and Kainuma (2008)
8.5	Additional storage around 8.5 W/m^2 in 2100	>1370 in 2100	Rise until 2100	Between 2.6 °C and 4.8 °C	Between 45 and 82 cm	Riahi, Grübler, and Nakicenovic (2007)

RCP 2.6 was developed by the IMAGE modelling team of the Netherlands Environmental Assessment Agency (PBL). This emission pathway is representative for scenarios in the literature leading to very low greenhouse gas concentration levels. It is a more optimistic 'peak-and-decline' scenario. Initially, its radiative forcing level first reaches a value around 3.1 W/m^2 mid-century, returning to 2.6 W/m^2 by 2100. In order to reach such radiative forcing levels, greenhouse gas emissions (and, indirectly, emissions of air pollutants) are substantially reduced over time (van Vuuren et al., 2007).

RCP 4.5 was developed by the modelling team at the US Pacific Northwest National Laboratory's Joint Global Change Research Institute (JGCRI). It is a stabilisation scenario where total radiative forcing is stabilised shortly after 2100, without surpassing the long-term high radiative forcing level (Clarke et al., 2007; Smith & Wigley, 2006; Wise et al., 2009).

RCP 6.0 was developed by the AIM modelling team at the National Institute for Environmental Studies (NIES), Japan. It is a stabilisation scenario where total radiative forcing is stabilised shortly after 2100 by employment of a range of technologies and strategies to reduce GHG emissions (Fujino et al., 2006; Hijioka et al., 2008).

RCP 8.5 was developed using the MESSAGE model and IIASA's integrated assessment framework (IIASA is Austria's International Institute for Applied Systems Analysis). This RCP is a pessimistic scenario characterised by increasing GHG emissions over time, representative for scenarios in the literature leading to high concentration levels. (Riahi et al., 2007).

It's important make sure that climate models don't consider the socio-economics other than that the emissions scenarios are a consequence of them, they use the concentrations of GHG (IPCC, 2014).

RCP projections go up to the year 2100; after 2100, they are called ECPs (Extended Concentration Pathway), which are mere extensions of the RCPs (Meinshausen et al., 2011; Van Vuuren et al., 2011).

2.3 Region Analysed, Models and Data Used

The region analysed is the whole Brazilian territory (green area in Fig. 2.1) for four RCPs. The Brazil analysis serves two purposes: a) To think about plausible adaptation strategies that require local risk knowledge; and b) To focus on average temperature changes which is crucial for a cost-benefit analysis of mitigation policies such as warming limit targets established by the international community to reduce the risks of impacts and damages caused by climate change.

The models used in this study, which integrate the CMIP5's data archive, are shown in Table 2.2. The simulations and projections were used in IPCC AR5's reports from 2013 and 2014, as well as in several publications related to climate change. As seen in Table 2.2, RCPs 2.6, 4.5, 6.0 and 8.5 use, respectively, 32, 42, 25 and 39 simulations. For RCP 8.5's ECP, 12 simulations were considered.

The reliability of a climate model (global or regional) is based on three main aspects: (a) whether it is based on well-established laws of physics; (b) whether it is able to simulate important aspects of the current climate (such assessment is conducted by comparing simulation results with observations of the atmosphere, the oceans, land surface etc., or by comparing the results of multiple climate models); and (c) whether it is able to reproduce past climate change and climate characteristics (Randall et al., 2007). When climate models fulfil these three requisites, they are considered capable of providing consistent quantitative estimates of future climate change based on different GHG emission projections (IPCC, 2014).

According to Burke, Dykema, Lobell, Miguel, and Satyanath (2015), it is necessary to bear in mind that the projections made by climate models contain uncertainties due to imperfect knowledge and/or mathematical representation of: all relevant physics processes; the model's sensitivity to the initial conditions adopted; future values of variables that may affect the climate system (notably GHG emissions); and how changes in these variables translate into climate change.

Such uncertainties are more significant in the case of regional models RCMs, since they are highly dependent on the boundary conditions provided by the global

Fig. 2.1 Brazilian territory considered in this study

models GCMS. During the downscaling process, large scale GCM climate uncertainties are absorbed by RCMs, and they can also be amplified due to their wider spatial resolution (Pielke Sr. & Wilby, 2012). In addition, if an RCM is nested on very few GCMs, the high resolution scenarios cannot cover the whole range of projected changes, thus increasing the uncertainty of the results obtained (Hewitson et al., 2014).

CMIP5 data on annual average temperatures close to the surface were obtained from the KNMI-Climate Explorer website - http://climexp.knmi.nl/. The simulations available on the website were used, i.e., CMIP5's Atlas subset. For RCP 6.0, the analyses go up to 2100, as only two outputs have ECP (data after 2100) available for this scenario. By convention, all projections up to 2100 are called RCPs, and after that they are called ECPs. Actually, the RCPs are built bottom up from internally consistent description of socio-economics and interactive models of the energy system, agriculture, trade, policy, international cooperation etc. The ECP are effectively just scaling of the end of the RCPs to extend their duration.

In Table 2.2, the line below the name of the model refers to the simulation, which can range from 1 to 3. There are three versions (different simulations) for some models that vary according to how aerosols and atmospheric chemistry are treated:

Table 2.2 Atlas subset of CMIP5 simulations used in this study. Simulations in boldface were used in the ECP (post 2100) for the RCP 8.5 scenario: (Source: KNMI-Climate Explorer 2017)

Simulations	RCP 2.6	RCP 4.5	RCP 6.0	RCP 8.5
1	bcc-csm1–1 r1i1p1	ACCESS1–0 r1i1p1	bcc-csm1–1 r1i1p1	ACCESS1–0 r1i1p1
2	bcc-csm1–1 m r1i1p1	ACCESS1–3 r1i1p1	bcc-csm1–1 m r1i1p1	ACCESS1–3 r1i1p1
3	BNU-ESM r1i1p1	bcc-csm1–1 r1i1p1	CCSM4 r1i1p1	**bcc-csm1–1 r1i1p1**
4	CanESM2 r1i1p1	bcc-csm1–1 m r1i1p1	CESM1-CAM5 r1i1p1	BNU-ESM r1i1p1
5	CCSM4 r1i1p1	BNU-ESM r1i1p1	6–0_r1i1p1	CanESM2 r1i1p1
6	CESM1-CAM5 r1i1p1	CanESM2 r1i1p1	FIO-ESM r1i1p1	**CCSM4 r1i1p1**
7	CNRM-CM5 r1i1p1	CCSM4 r1i1p1	GFDL-CM3 r1i1p1	CESM1-BGC r1i1p1
8	CSIRO-Mk3–6-0 r1i1p1	CESM1-BGC r1i1p1	GFDL-ESM2G r1i1p1	CESM1-CAM5 r1i1p1
9	EC-EARTH r8i1p1	CESM1-CAM5 r1i1p1	GFDL-ESM2M r1i1p1	CMCC-CM r1i1p1
10	FGOALS-g2 r1i1p1	CMCC-CM r1i1p1	GISS-E2-H r1i1p1	CMCC-CMS r1i1p1
11	FIO-ESM r1i1p1	CMCC-CMS r1i1p1	GISS-E2-H r1i1p2	**CNRM-CM5 r1i1p1**
12	GFDL-CM3 r1i1p1	CNRM-CM5 r1i1p1	GISS-E2-H r1i1p3	**CSIRO-Mk3–6-0 r1i1p1**
13	GFDL-ESM2G r1i1p1	CSIRO-Mk3–6-0 r1i1p1	GISS-E2-R r1i1p1	EC-EARTH r8i1p1
14	GFDL-ESM2M r1i1p1	EC-EARTH r8i1p1	GISS-E2-R r1i1p2	FGOALS-g2 r1i1p1
15	GISS-E2-H r1i1p1	FGOALS-g2 r1i1p1	GISS-E2-R r1i1p3	FIO-ESM r1i1p1
16	GISS-E2-H r1i1p2	FIO-ESM r1i1p1	HadGEM2-AO r1i1p1	GFDL-CM3 r1i1p1
17	GISS-E2-H r1i1p3	GFDL-CM3 r1i1p1	HadGEM2-ES r2i1p1	GFDL-ESM2G r1i1p1
18	GISS-E2-R r1i1p1	GFDL-ESM2G r1i1p1	IPSL-CM5A-LR r1i1p1	GFDL-ESM2M r1i1p1
19	GISS-E2-R r1i1p2	GFDL-ESM2M r1i1p1	IPSL-CM5A-MR r1i1p1	**GISS-E2-H r1i1p1**
20	GISS-E2-R r1i1p3	GISS-E2-H r1i1p1	MIROC5 r1i1p1	**GISS-E2-H r1i1p2**
21	HadGEM2-AO r1i1p1	GISS-E2-H r1i1p2	MIROC-ESM r1i1p1	**GISS-E2-H r1i1p3**

(continued)

Table 2.2 (continued)

Simulations	RCP 2.6	RCP 4.5	RCP 6.0	RCP 8.5
22	HadGEM2-ES r2i1p1	GISS-E2-H r1i1p3	MIROC-ESM CHEM r1i1p1	**GISS-E2-R r1i1p1**
23	IPSL-CM5A-LR r1i1p1	GISS-E2-H r1i1p1	MRI-CGCM3 r1i1p1	**GISS-E2-R r1i1p2**
24	IPSL-CM5A-MR r1i1p1	GISS-E2-R r1i1p1	NorESM1-M r1i1p1	**GISS-E2-R r1i1p3**
25	MIROC5 r1i1p1	GISS-E2-R r1i1p2	NorESM1-ME r1i1p1	HadGEM2-AO r1i1p1
26	MIROC-ESM r1i1p1	GISS-E2-R r1i1p3		HadGEM2-CC r1i1p1
27	MIROC-ESM CHEM r1i1p1	GISS-E2-R-CC r1i1p1		HadGEM2-ES r2i1p1
28	MPI-ESM-LR r1i1p1	HadGEM2-AO r1i1p1		inmcm4 r1i1p1
29	MPI-ESM-MR r1i1p1	HadGEM2-CC r1i1p1		**IPSL-CM5A-LR r1i1p1**
30	MRI-CGCM3 r1i1p1	HadGEM2-ES r2i1p1		IPSL-CM5A-MR r1i1p1
31	NorESM1-M r1i1p1	inmcm4 r1i1p1		**IPSL-CM5B-LR r1i1p1**
32	NorESM1-ME r1i1p1	IPSL-CM5A-LR r1i1p1		MIROC5 r1i1p1
33		IPSL-CM5A-MR r1i1p1		MIROC-ESM r1i1p1
34		IPSL-CM5B-LR r1i1p1		MIROC-ESM CHEM r1i1p1
35		MIROC5 r1i1p1		MPI-ESM-LR r1i1p1
36		MIROC-ESM r1i1p1		MPI-ESM-MR r1i1p1
37		MIROC-ESM CHEM r1i1p1		MRI-CGCM3 r1i1p1
38		MPI-ESM-LR r1i1p1		NorESM1-M r1i1p1
39		MPI-ESM-MR r1i1p1		NorESM1-ME r1i1p1
40		MRI-CGCM3 r1i1p1		
41		NorESM1-M r1i1p1		
42		NorESM1-ME r1i1p1		

Physics_version = 1, aerosols and ozone are read through precomputed transient aerosol and ozone fields, and the indirect effect of aerosols is parametrised (Lee, Pierce, & Adams, 2013; Nazarenko et al., 2012; Romanou et al., 2015).
Physics_version = 2, aerosols and atmospheric chemistry are calculated online as a function of atmospheric status and transient emission levels. The indirect effect of aerosols is also parametrised (Lee et al., 2013; Nazarenko et al., 2012; Romanou et al., 2015).
Physics_version = 3, the atmospheric composition is calculated as in version 2, but the impacts of aerosol on clouds are calculated (Lee et al., 2013; Nazarenko et al., 2012; Romanou et al., 2015). These three versions are shown in Table 2.2 below the names of the models, and are: r1i1p1, r1i1p2 and r1i1p3. Thus, for these 'variations on a theme', the physics version is represented by an integer (or whole number): pN. The digit N = 1, 2, 3 is the physics version, as in Table 2.2.

The results are shown in absolute values for global annual average temperature (average of simulation) nearby surface (2 m) (Fig. 2.2), and calculated anomalies for these temperatures. The anomalies (Fig. 2.3) were calculated in relation to the pre-industrial period (1861–1890), and the average temperature in that period in Brazil is 24.41 °C.

In the results, 2030s represents the time slice mean between 2021 and 2040, 2060s is the time slice mean between 2051 and 2070, and 2090s is the time slice mean between 2081 and 2100.

The climate response for any future emissions or concentration scenarios must be expressed as a probability distribution or interval due to the level of uncertainty regarding the link between GHG emission changes/concentrations and the climate response. It is important to understand how these probabilities change with time, and look beyond 2100, particularly for the higher emission scenarios such as RCP 8.5, since in some cases the impacts and the probability of disruptions in natural physics systems will continue to increase.

Thus, the probabilities for a range of different warming levels were obtained, exceeding by 2 °C, 3 °C, 4 °C, and 7 °C for all four RCPs (Figs. 2.4, 2.5 and 2.6). The probabilities were obtained in a simple form as a function of the value for each output's individual anomaly. In other words, it depends on the number of simulations, where each simulation has an X% of weight which, when added, will result in a 100% chance of a given warming level being exceeded. For example: RCP 4.5, for a warming above 4 °C: if no simulation exceeds 4 °C of annual average temperature anomaly in relation to the pre-industrial period, the probability of the annual average temperature exceeding 4 °C will be zero. If all simulations project a warming above 4 °C, the probability will be 100%, and the same will apply to all RCPs and temperature limits (from 2 °C to 7 °C).

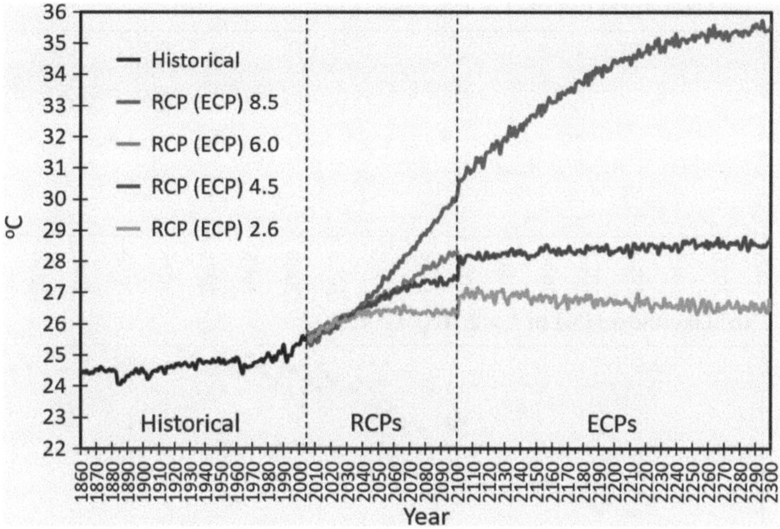

Fig. 2.2 Temporal series of annual average temperatures close to the surface from 1860 to 2300. The region analysed is Brazil. Each line represents the average of the simulations outputs listed in Table 2.2 for each RCP. For the period from 1860 to 2005, we have the historical period (*black line*), then the RCPs up to 2100, and then the ECPs up to 2300. The red line represents RCP 8.5, the green one represents RCP 6.0, the purple one, RCP 4.5, and the blue one, RCP 2.6. The average temperature for the period 1861–1890 (pre-industrial period) for Brazil is 24.41 °C. The unit is °C

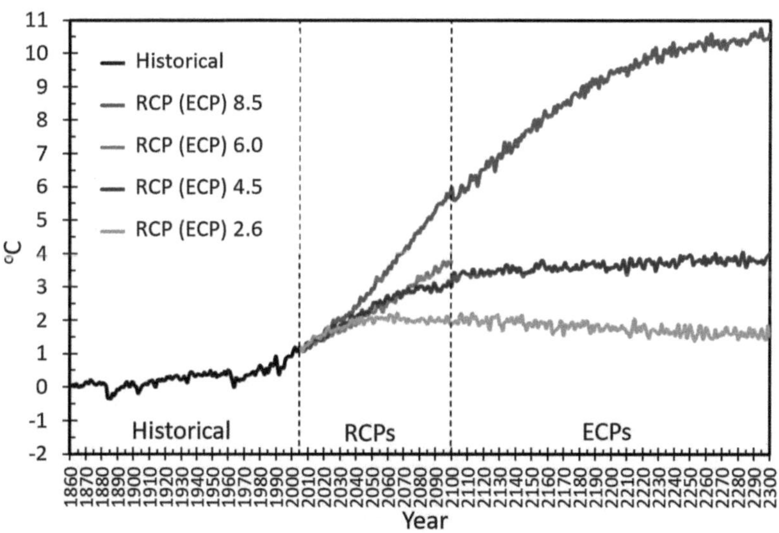

Fig. 2.3 Temporal series of annual average temperature anomalies close to the surface from 1861 to 2300 in relation to the pre-industrial period. The region analysed is Brazil. Each line represents the average of the simulations outputs listed in Table 2.2 for each RCP used in this study. For the period from 1860 to 2005, we have the historical period (black line), then the RCPs up to 2100, then the ECPs up to 2300. The red line represents RCP 8.5, the green one represents RCP 6.0, the purple one, RCP 4.5, and the blue one, RCP 2.6. The unit is °C

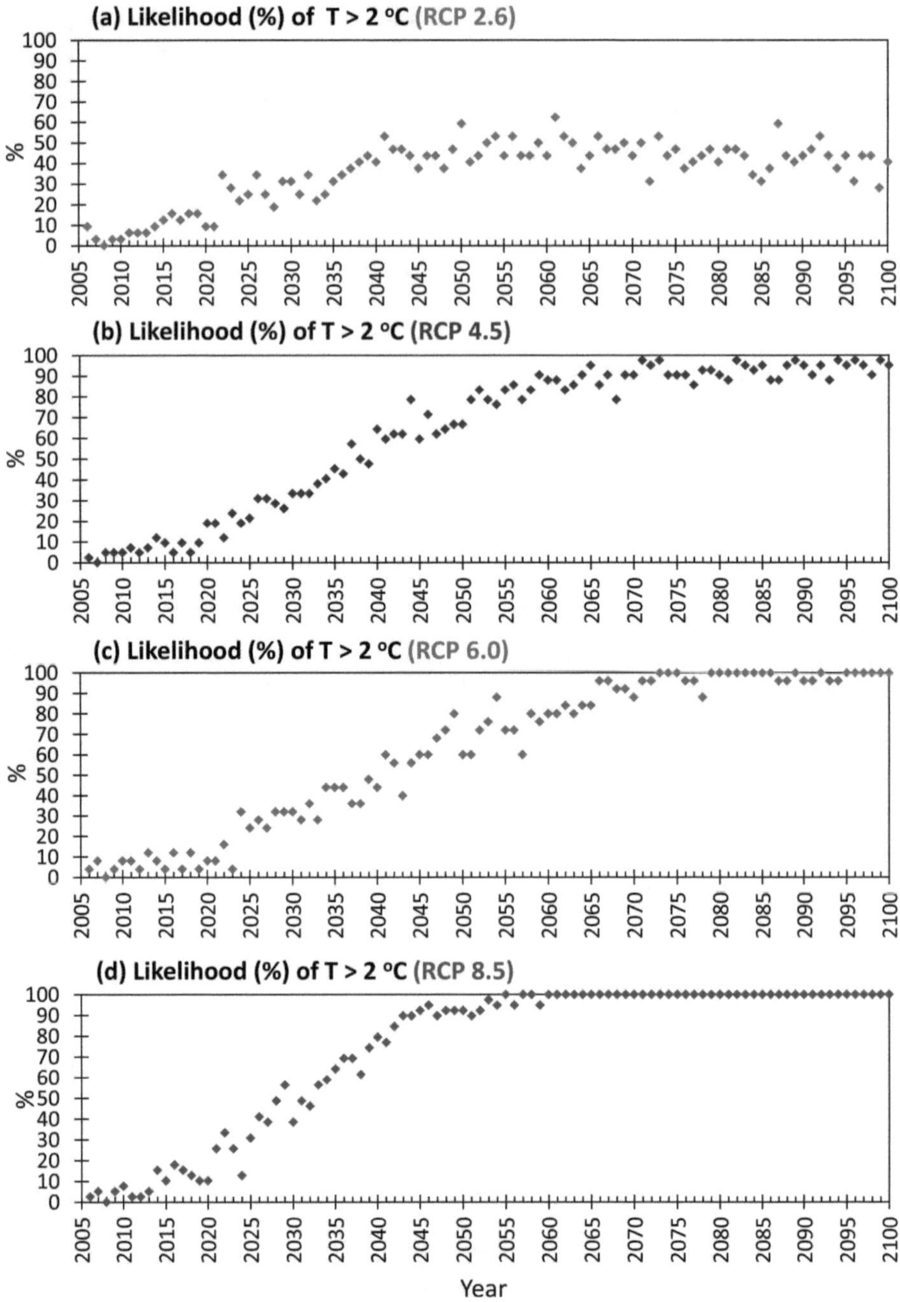

Fig. 2.4 Probability of annual average temperature exceeding 2 °C (**a**) RCP 2.6, (**b**) RCP 4.5, (**c**) RCP 6.0, and (**d**) RCP 8.5. The warming probabilities above 2 °C were obtained from the anomaly values of the simulations outputs listed in Table 2.2. The unit is %

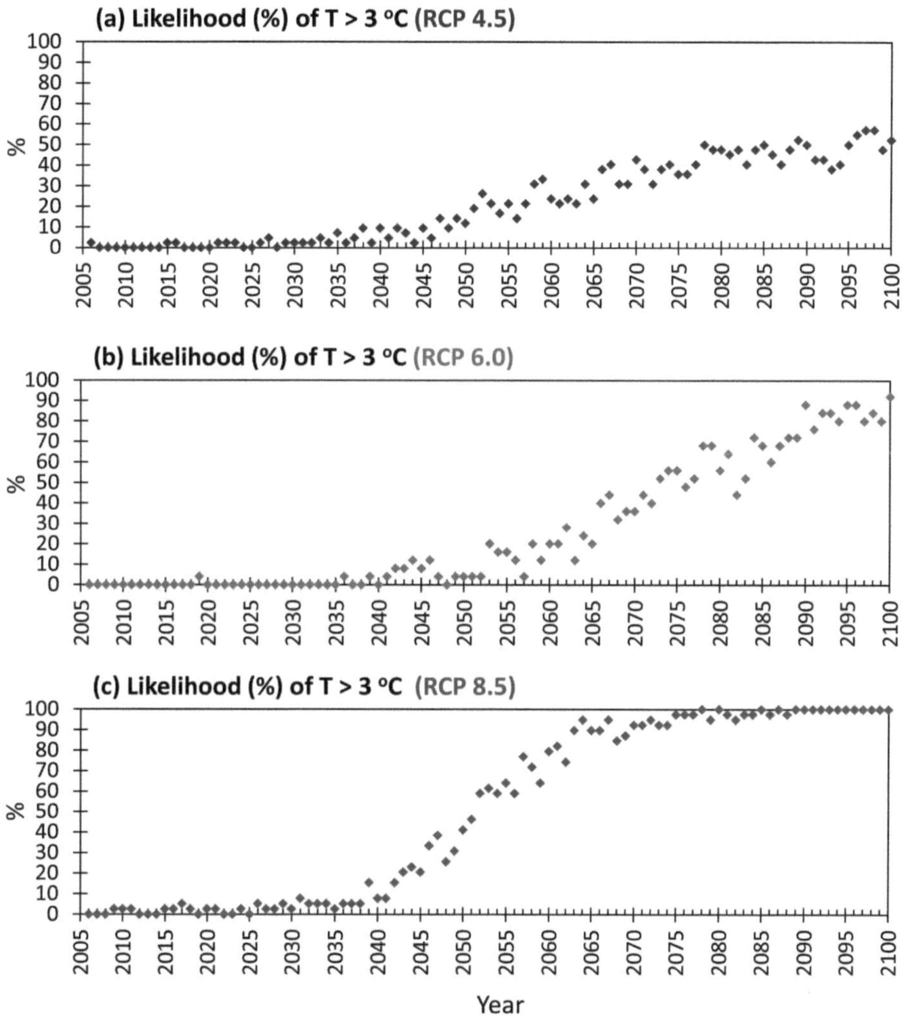

Fig. 2.5 Probability of annual average temperature exceeding 3 °C (**a**) RCP 4.5, (**b**) RCP 6.0, and (**c**) RCP 8.5. The warming probabilities above 3 °C were obtained from the anomaly values of the simulations outputs listed in Table 2.2. The unit is %

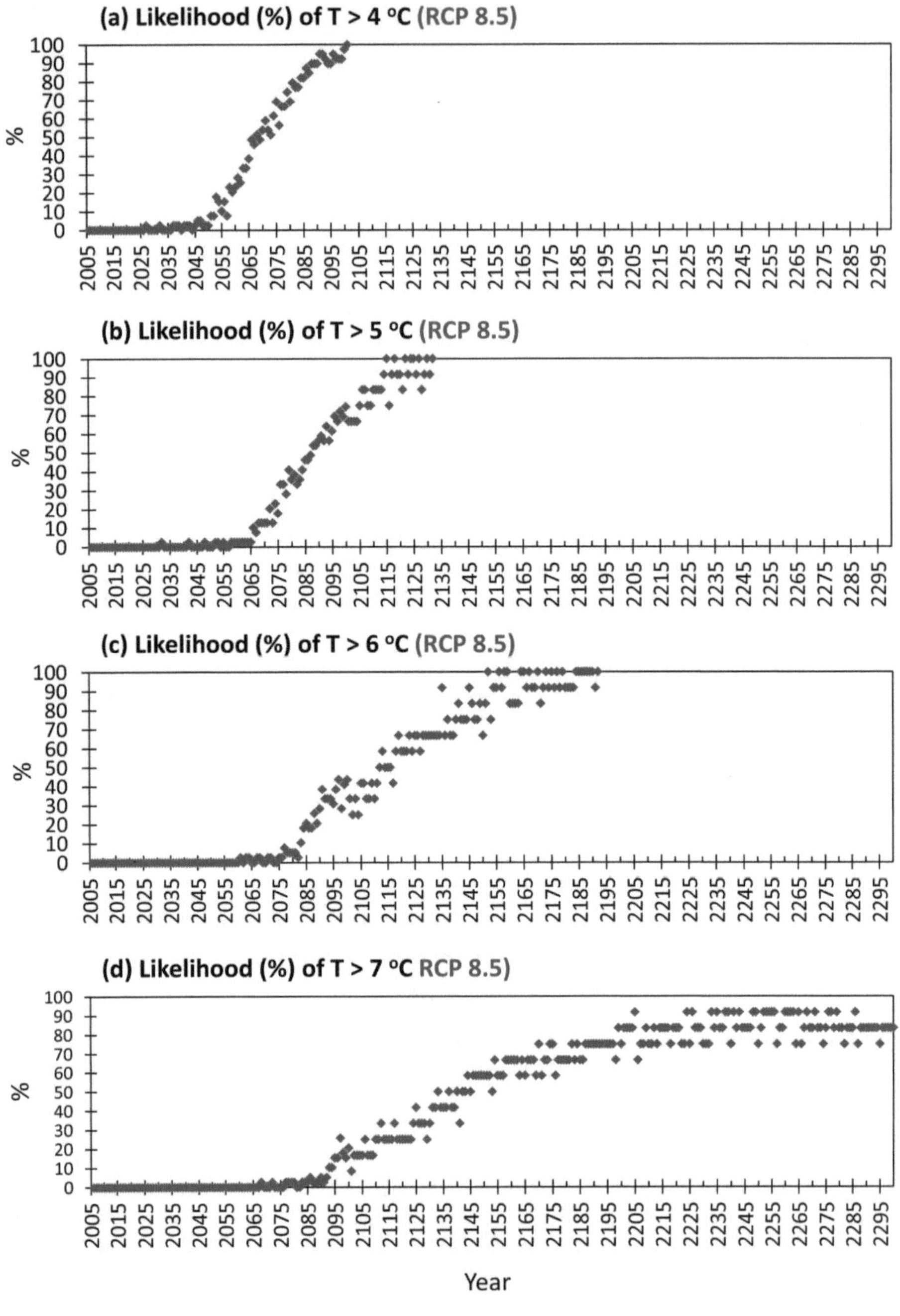

Fig. 2.6 Probabilities of annual average temperature projected by RCP 8.5 exceeding: (**a**) 4 °C, (**b**) 5 °C, (**c**) 6 °C, and (**d**) 7 °C. The warming probabilities were obtained from the values of temperature anomalies values of the simulations outputs listed in Table 2.2. Probabilities of Warming above 4 °C were not found in RCPs 2.6, 4.5 and 6.0. The unit is %

2.4 Long-Term Projections for Brazil: Analysis of Air Temperature Projections

Figure 2.2 shows the temporal series (average for the simulations contained in Table 2.2) of annual average temperatures near surface (2 m). The black line represents the 'historical' period, ranging from 1860 to 2005, where a slight temperature increase is observed.

With regards to future projections: the average behaviour of RCP 2.6 (blue line) projects temperatures around 25.8 °C in 2020 (near future), increasing up to 26.4 °C in 2050. At the end of the century, the temperature would be around 26.8 °C, slowly decreasing to approximately 26.4 °C at the end of the projection (2300). RCP 4.5 (purple line) projects 25.8 °C in 2020, then reaches 28 °C in 2100, and a maximum of 28.8 °C at the end of the series. RCP 6.0 (green line) projects 26 °C in 2020 and increases the temperature to 26.8 °C in 2050, reaching 28.3 °C around 2100. No intervals are presented for the projections beyond 2100 for this RCP (6.0) because only two models are available. RCP 8.5 (red line) projects 25.7 °C in 2020, reaching 27.1 °C in 2050, 30.2 °C in 2100, and 35 °C at the end of the series in 2300, as observed in Fig. 2.2.

Figure 2.3 shows the annual average temperature anomalies in relation to the pre-industrial period (1861–1900). RCP 2.6 (blue line) has a temperature anomaly of 1.5 °C in 2020, reaching 2 °C around 2050; after that, there is a decline, and in 2100 the temperature anomaly is 1.8 °C; then, at the end of the series (2300) it is 1.5 °C. This RCP is the most optimistic of the top four (2.6, 4.5, 6.0 and 8.5) RCPs used by the IPCC with regards to temperature increases, as shown in Fig. 2.3. RCP 4.5 (purple line) projects an anomaly of 1.5 °C in 2020, 2.3 °C in 2050, then it increases to around 3 °C at the end of the century, and reaches a maximum of 4°C at the end of the series. RCP 6.0 (green line) projects a temperature rise of 1.3 °C in 2020 and 2.3 °C in 2050, reaching 3.7 °C in 2100, the end of the series for this RCP. The most pessimistic scenario, RCP 8.5 (red line) projects a 1.5 °C temperature rise in 2020 (below the 2 °C limit), but in 2050 the temperature rise is 2.9 °C. In 2100, the anomaly is around 6 °C, and the maximum warming value of 10 °C is observed at the end of the series.

These degrees of warming may have strong impacts in the water cycle. For instance, while investigating changes in water stress in 112 major basins in our planet, Fung, Lopez, and New (2011) demonstrate that a 4 °C warming can have bigger runoff impacts than a 2 °C one, with dry areas becoming dryer, and humid areas becoming more humid. In fact, as we move from a 2 °C to a 4 °C world, there is also a small but growing number of basins that may end up facing less water stress, since they are located in regions where rainfall is projected to increase. Using population growth scenarios for the 2030s and 2060s, in a 2 °C warmer world, water stress tends to be dominated by population changes. However, when moving towards a 4 °C temperature rise, the signs of climate change become stronger and these alterations may play a more dominant role in determining water stress in water basins. The authors have also found that runoff seasonality may be more pronounced in a 4 °C warmer world than in a 2 °C one; thus, even if the annual average runoff growth levels increase, dry seasons may become more stressed. This could mean that more sophisticated adaptation strategies may be needed to cope with the impacts of warming above 4 °C than in a 2 °C warmer world, so as to avoid floods and droughts.

IPCC (2014) results show a small subset of climate models reaching global average temperature rises 10 °C or above relative pre-industrial levels. This could also affect the timing when probabilities will exceed a certain degree of warming (for a few years in the estimate, or for over a decade in more extreme cases). Globally, RCP 8.5's increased warming levels have no precedent in observational registers, even for around 1000 AD (through proxy measures) (Masson-Delmotte et al., 2013). For a longer period, IPCC AR5 concludes that, during the mid-Pliocene (3.3 to 3.0 million years ago), temperatures were 1.9 °C to 3.6 °C higher than pre-industrial levels. During the early Eocene (52 to 48 million years ago), the global average temperature was 9 °C to 14 °C higher than pre-industrial levels for a CO_2 atmospheric concentration around 1000 ppm, which is slightly higher than the RCP 8.5 concentration in 2100. When considering these distant historical periods, it is important to bear in mind both the uncertainty of the records and whether if the past period really represents an adequate analogue for the Anthropocene.

According to IPCC (2014), in the RCP 8.5 scenario the projections for 2150 show a 6 °C rise. This is an average worst-case scenario, and while it is very difficult to know how a 2 °C warmer world would affect natural and human systems, much less is our knowledge of impacts of warming of 4°C or 6 °C. However, the IPCC AR5 report points out that climate change has major consequences, and some physical systems may be at high risk of abrupt and irreversible changes. A potentially 4 °C warmer world may make it very difficult, or even impossible, for poorer countries to adapt. Thus, in the first instance, a 4 °C warmer world might seriously harm poverty reduction in many regions. In addition, the effects of a 4 °C warming have not been fully assessed, as they may potentially be more dramatic than projected. Linearity (as the idea that the impacts of a 4 °C warming would be twice as serious as those of a 2 °C warming) in the effects is not expected, nor is it in the regional changes (Good et al., 2012; Good et al., 2016). Different effects might combine in unexpected ways (scale and overlap of risks). For example, the non-linear effects of temperature on crops may be extremely pertinent in a 2 °C warmer world, but most existing harvest models still do not take this effect into full consideration, nor do they consider the potential impacts of increased variability, for example, extreme temperatures, new diseases and pests, sharp changes in critical climate factors which have major impacts on the yield and/or quality of grains.

2.5 Probabilities of Warming Exceeding the Limits of 2 °C to 4 °C

Figure 2.4 shows the probabilities of warming above 2 °C. For such warming, the projections go up to 2100. After that, warming above 4°C shall be analysed. For RCP 2.6 (Fig. 2.4a), the probability of a warming (annual average temperature) above 2 °C is around 50% between 2040 and 2060, decreasing to around 40% at the end of the series. Thus, even a more optimistic RCP presents probabilities of warming, albeit low, with the annual average temperature exceeding 2 °C in relation to the pre-industrial period. For RCP 4.5 (Fig. 2.4b), higher probabilities appear after 2035, reaching 100% around 2100. RCP 6.0 (Fig. 2.4c) presents probabilities above 50% after 2040, a little after RCP 4.5, but reaches a 100% of probability of $T > 2$ °C in 2080. RCP 8.5 (Fig. 2.4d) shows probabilities above 50% after 2030 and reaches 100% of probability of the annual average temperature exceeding 2 °C in mid-2055; in comparison with the three other RCPs, this is the one that reaches higher warming probabilities sooner, as observed in Fig. 2.4. According to IPCC 2014, the risks of aggregate impacts as a consequence of additional warming between 1 °C and 2 °C are moderate, with impacts on biodiversity and the economy as a whole (medium confidence); an additional warming around 3 °C (high confidence) would lead to extensive loss of biodiversity and associated impacts on ecosystem services, thus resulting in higher risks than expected.

Figure 2.5 shows projections for warming above 3 °C for RCPs 4.5, 6.0 and 8.5. RCP 4.5 shows a warming probability of around 50% at the end of the series as shown in Fig. 2.5a. RCP 6.0 (Fig. 2.5b) shows probabilities above 50% for warming higher than 3 °C around 2075, rising to near 100% in 2100. RCP 8.5 (Fig. 2.5c) begins to show probabilities above 50% for warming higher than 3 °C from mid-2053 (earlier than in the other RCPs), reaching 100% in 2075. RCP 2.6 did not show a probability above 50% for a warming higher than 3 °C at any point in the series, and therefore it is not included in the figure. According to Piontek et al. (2014), there is a significant risk of changes in the ecosystem in regions where temperatures rise between 3 °C and 4 °C. According to IPCC (2014), approximately 20 to 30% of all animal and vegetable species face a growing risk of extinction when average temperature exceeds 2 °C to 3 °C above pre-industrial levels.

2.6 Probability of Warming Exceeding 4°C or More (RCP 8.5)

Warming higher than 4 °C and beyond 2100 are assessed only for RCP 8.5, as the three other RCPs analysed do not project significant probabilities for warming above 4 °C in Brazil. Figure 2.6 shows probabilities higher than 50% from mid-2070, reaching 100% in 2100. For warming of 4.5 °C using RCP 8.5, Piontek et al. (2014) observed falling trends in water availability in some regions in South

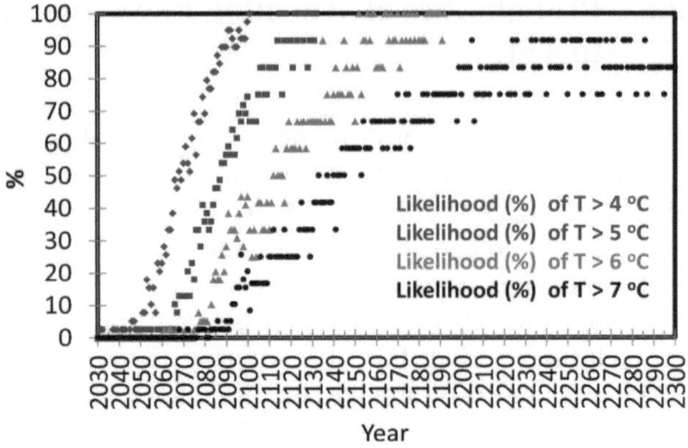

Fig. 2.7 Probabilities of annual average temperature projected by RCP 8.5 exceeding: 4 °C (*blue*), 5 °C (*brown*), 6 °C (*green*) and 7 °C (*black*). The warming probabilities were obtained from the values of temperature anomalies of the simulations outputs listed in Table 2.2. The unit is %

America, especially in the south of the Amazon basin. The probability is higher than 50% from 2088 onwards; in mid-2130, the probability of warming above 5 °C is 100%. Probabilities higher than 50% for warming above 6 °C start occurring in 2112 and reach 100% around 2184 (Fig. 2.6c). According to RCP 8.5, from 2142 onwards there is a 50% of probability for a warming higher than 7 °C; after 2200, these figures exceed 80%, and remain as such until the end of the series in 2300 (Fig. 2.6d).

Figure 2.7 shows the RCP 8.5 probabilities for warming of 4 °C, 5 °C, 6 °C and 7 °C. IPCC AR5 does not consider only high probability results; it also includes those with very low probabilities, but with more serious consequences. It also characterises climate alterations as a challenge for risk management, which opens the doors to a wide range of possible solutions.

Rahmstorf (2007), based in future warming scenarios of the IPCC, shows a projected sea-level rise in 2100 of 0.5 to 1.4 meters above the 1990 level, with a proportionality constant of 3.4 mm/year per °C.

Changes in wind and sea currents due to global warming and other factors will also regionally affect the rise in sea levels and the impacts of winds and storms (Brierley & Kingsford, 2009).

According to the World Bank (2012), warming above 4 °C will shift temperatures in the tropics more than 6 standard deviations for all months in the year. In particular, countries in tropical South America, Central Africa, and all tropical islands in the Pacific will see unprecedented extreme temperatures become the new norm all year long. According to IPCC AR5, more warming will take place on land, varying from 4 °C to 10 °C (in the Amazon, the projected temperature change in 2100 reaches 6 °C), with a dramatic increase in the intensity and frequency of high temperature extremes and drought periods. The association between less rain and higher temperatures makes the forest more vulnerable, with a higher tree mortality rates and risks of forest fires, in addition to the impact on climate regulation and carbon storage by means of a collapse of the forest.

Reductions in deforestation and mitigation options to reduce GHG emission and project warming may reduce the possibility of Amazon's forest collapse. For

Gemenne (2011), warming above 4 °C would bring unprecedented changes to the environment, which may affect human mobility in different ways. The relationship between environmental changes and migrations is highly complex and depends on many specific contexts and variables. It cannot be reduced to a direct causal relationship. Thus, the effects associated with a 4 °C warming may affect not only the magnitude of population movements induced by climate change, but could also particularly affect their nature of this migration.

Impacts on the sea level have been assessed by Horton, Rahmstorf, Engelhart, and Kemp (2014), who estimated a 17% of probability for a sea level rise above 2.0 m in 2100 for a temperature scenario higher than 4 °C in RCP 8.5. In 2300, in a more optimistic lower warming scenario (RCP 3, similar to RCP 2.6), the range is between 0.6 m and 1 m. Schaeffer, Hare, Rahmstorf, and Vermeer (2012) estimated a 2.0 m global rise for the same temperature scenario. IPCC AR5 projects 0.41 m to 0.85 m for RCP 2.6 in 2300. For the most pessimistic scenario (RCP 8.5) the 'probable' rise range would be between 2 m and 3 m (IPCC, 2014), with a possibility of a huge rise of up to 14.0 m (Horton et al., 2014). IPCC AR5 shows estimates from 0.92 m to 3.59 m of average sea level rise for 2300 under RCP 8.5. According to Horton et al. (2014), the estimates illustrate the risk that an increased temperature caused by non-mitigated emissions may harm coastal populations in the long term; there may be catastrophes over many coastal cities and lower lands if we do not succeed in reducing emissions. For Thomas and Lin (2015), RCP 8.5 impacts also increase with higher sea level thresholds; the peaks in RCP 4.5 and RCP 6.0 are close to the centre of the band, as expected, but are also systematically higher than RCP 2.6, even for lower sea level thresholds.

In general, higher probabilities result from more pessimistic RCPs, such as RCP 8.5, and they are also reached sooner in future projections. The United Nations Framework Convention on Climate Change aims at preventing potentially dangerous climate alterations, and has adopted a long-term target to maintain global average warming below 2 °C in relation to pre-industrial levels. The choice of an adequate level is a subjective political choice informed by estimates of future climate change impacts and their relative adaptation difficulties. IPCC AR5 has concluded that global 'climate change risks' are high to very high, with a global average temperature rise of 4°C or more above pre-industrial levels. It suggests that a large scale biodiversity loss is likely to take place with temperature rises of 4 °C and high CO_2 concentration, which would lead to a transition of Earth ecosystems. Thus, we may expect a dramatic reduction in the ecosystem services on which society depends (Stanton, Shoemaker, Pearson, & Akçakaya, 2015).

Reasons for concern include serious and generalised impacts on unique and threatened systems, substantial extinction of species, as well as major risks for world and regional food safety; in addition, the combination of high temperature and humidity might compromise normal human activities, including agriculture or open-air work in some areas of the planet (IPCC, 2014). The exact levels of climate alterations that would be sufficient to cause disruption points (thresholds for abrupt and irreversible changes) remain uncertain, but the risks associated with crossing several disruption points in the Earth system or in interconnected natural and human systems increase when temperature rises (medium confidence) (IPCC, 2014). In terms of regional changes, the models agree that a more pronounced warming (above 4 °C) is likely to take place on land. In Brazil, communities, cities and different regions are likely to experience serious disturbances, damages and displacement of people, and many of these risks will be unevenly distributed. The poor are likely to suffer more, and local communities might become more fragmented and unequal than they are today. For warming of 7 °C or more, there are still few studies on their impacts, but these limited studies suggest the possibility of even stronger impacts if temperatures rise above such level. Such impacts may potentially increase due to extreme heat events and the physiological tolerance of human beings in some regions (Sherwood & Huber, 2010).

2.7 Warming Risks for Brazil during the Twenty-First Century

Figure 2.8 shows the warming probabilities for Brazil in three time slices: 2030s (mean for the period from 2021 to 2040), 2060s (mean for the period from 2051 to 2070), and 2090s (mean for the period from 2081 to 2100). For the 2030s time slice (near future), warming risks of 2 °C are observed for all four RCPs. The values are between 30 and 50%. As temperature rises, risks probability decreases (Fig. 2.8a). For the 2060s time slice (Fig. 2.8b), the risks observed for a 2 °C warming are 50% for RCP 2.6, in line with Fig. 2.4a. RCPs 4.5 and 6.0 reach an 80% risk, and RCP 8.5 shows a 100% risk for a 2 °C warming. For a 3 °C warming, this RCP shows an approximate value of 90%, which decreases as additional warming increases, e.g.: for a 3.5 °C warming, the risk is 50%, whereas for an additional 4 °C warming, the risk in the period is 28%.

Figure 2.8c shows that for RCP 2.6 (blue line), there is a 40% of probability of temperatures exceeding 2 °C of additional warming in the 2090s time slice. Probabilities for warming above 3 °C are very low for RCP 2.6 (Fig. 2.8c). RCP 4.5 shows a value around 93% for a warming risk of 2 °C in the 2090s time slice. For $T > 3$ °C, this RCP shows a risk around 50% and low probabilities (lower risk) for warming above 3.5 °C or more. Just as in RCP 4.5, RCP 6.0 shows high-risk levels, with a 100% of probability for a 2 °C warming, and a 90% chance of warming above 3 °C at the end of the century. RCP 8.5 shows high risks of additional warming surpassing the limits of 2 °C, 3 °C and 4 °C in the 2090s time slice, as shown in Fig. 2.8c. There is approximately a 75% risk of Brazil exceeding 4.5 °C of additional warming according to this RCP. The probability of reaching a 5 °C limit or more is less than 55%. As temperature rises, the risks decrease reaching values under 10% for a 7 °C additional warming.

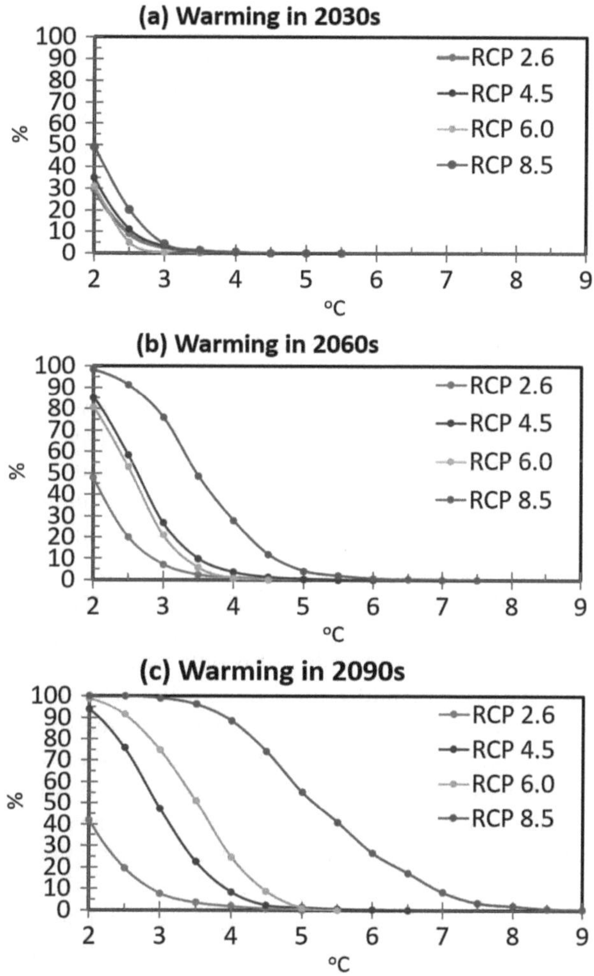

Fig. 2.8 Warming probability in the 2030s, 2060s and 2090s time slices. The warming probabilities were obtained from the anomaly values of the simulations outputs listed in Table 2.2 in relation to the pre-industrial period. The unit is % and the region is Brazil

2.8 Final Remarks

During the last decade, Brazil has experienced and continues to be affected by extreme climate events classified as 'events of the century', with significant impacts on the economy and on ecosystems all over the country. Munich-Re, the oldest reinsurance company in Germany, considered the drought in the South-east of the country in 2014–2015 the fifth costliest natural disaster in the world in 2014, with estimated losses around 5 billion US dollars. This shows that Brazil is vulnerable to extreme climate events at present, and that this may get worse in the future due to global warming, which has different regional impacts.

Over the next decades, global warming and climate change are likely to significantly alter human and natural systems, pushing the limits of variability and vulnerability beyond historical values and leading to significant changes in relation to typical conditions. Identifying the places, periods and features of such impacts for any given level of global warming enables the development of appropriate adaptation strategies, or may motivate strategies of adaptation and decisions to mitigate climate change. However, these impacts may be extremely important, as they may modify effects, restrict response options, and lead to indirect impacts on other regions, strongly increasing adaptation challenges.

Over time, any emissions pathway will present a wide range of possible global temperature rises with their regional differences, as well as sea level rises. In a high emissions pathway, such as RCP 8.5, where the most probable global temperature rise is estimated at 5 °C by 2100, any warming between 3 °C and 7 °C might be possible.

Apparently small changes in the climate may have significant effects, especially if important thresholds are surpassed. Crops have little tolerance to high temperatures, and as the climate gets warmer, these limits may be exceeded more and more often. This is one of the reasons why a temperature rise of 4°C or more might represent severe risks for global food safety and affect food producing countries like Brazil.

At some point between 5 °C and 7 °C of temperature rise, warm areas are likely to experience fatal conditions even for people lying in the shade. Population growth may well double the number of people living below the threshold of extreme water shortage in the middle of the century. Climate change may cause even more extreme water shortage in some regions, while increasing the risks of floods in others. Coastal cities probably have limits regarding the rate and extension of sea level rise that they can deal with, but we have little idea of where these thresholds are.

There are two possible ways to address climate change in a critical manner: mitigation, and finding a way to deal as effectively as possible with climate change that can no longer be avoided. IPCC AR5 considers the whole range of possible results, including not only high probability results, but also those with very low probabilities, but whose consequences are stronger; it also characterises climate change as a threat in risk management, thus opening the doors to a wide range of possible solutions and stressing the point that climate change is a challenge in risk management that can be substantially reduced through mitigation. Given the strong link between emissions, energy use and economic activity, de-carbonisation efforts inevitably affect global economy. Relevant economic indicators of the mitigation challenges include carbon prices, which quantify the marginal costs of emission reductions in the short and long terms.

A 4 °C warmer world will be different from today's world. This factor involves a high level of uncertainty and new risks that threaten our ability to foresee and plan future adaptation needs. Not only may inaction (on climate change) turn prosperity into something unachievable by millions of people in the developing world, but it also threatens decades of sustainable development, affecting local quality of life and health.

A limit is reached when adaptation efforts are unable to provide an acceptable level of safety against the risks for existing values and objectives, nor prevent the loss of key attributes, components or services of an ecosystem. A warming above 4 °C may have irreversible effects on human and natural systems, particularly in regions and sectors that are already exposed and that show high risks of climate change, variability and extreme events. In particular, if afforestation is included in the mitigation options, the global Earth biosphere may act as a natural sink for emissions. However, there are doubts about whether such capture would indeed be permanent (Rogelj et al., 2015).

As a result of the climate agreement signed in Paris during COP-21, it was determined that countries should do all in their power to maintain temperature rises well below 2 °C, and have 1.5 °C as a target for 2100. In order to do that, countries need to establish targets to reduce their GHG emissions, known as INDC (Intended Nationally Determined Contributions). Brazil has the target of reducing absolute emissions by 43% by 2030, aiming at the 2 °C target. Now, this target must be reviewed in order for the country to reach an appropriate contribution to global emissions reductions consistent with the 1.5 °C limit. In 2012, Brazil had already reduced its emissions by 41% thanks to a reduction in deforestation after 2005. Over the past few years, Brazilian emissions have grown driven by thermoelectric plants and transports. The problem is that in order to limit warming to 1.5–2 °C, global GHG emissions have to reach a peak in 2020 and then decrease to zero by 2050. The sum of current emission targets proposed by all countries at COP-21 does not allow us to reach that target. The objective of limiting temperature to 1.5–2 °C in relation to pre-industrial levels is strongly driven by island countries, which face the highest threats by sea level rise as a consequence of climate change. In order to limit temperature rises to 2 °C, governments must significantly strengthen their objectives, and jointly reduce global CO_2 equivalent emissions between 12 and 15 gigatonnes by 2015, and between 17 and 21 gigatonnes by 2030. According to the UN, current GHG emissions are around 50 gigatonnes.

For adaptation and mitigation planning, stakeholders need reliable information about regional precipitation changes under different emissions scenarios and for different time periods. A significant amount of current planning effort assumes that each K of global warming produces roughly the same regional climate change (Good et al., 2016).

The decisions taken at COP-21 represent a universal legal agreement on the long term fight against climate change aiming at ensuring that our planet's temperature does not increase more than 2 °C by the end of this century in comparison with pre-industrial levels. We hope this may help to prevent the serious sectoral impacts presented in this report in key sectors of the Brazilian economy and environment. These mitigation measures may help to reduce vulnerability and facilitate adaptation measures, which otherwise would be almost impossible to reach with a global warming above 4 °C.

References

Brierley, A. S., & Kingsford, M. J. (2009). Impacts of climate change on marine organisms and ecosystems. *Current Biology, 19*, 602–614.

Burke, M., Dykema, J., Lobell, D. B., Miguel, E., & Satyanath, S. (2015). Incorporating climate uncertainty into estimates of climate change impacts. *Review of Economics and Statistics, 97*(2), 461–471.

Burkett, V. R., et al. (2014). Point of departure. In *Climate change 2014: Impacts, adaptation, and vulnerability. Part A: Global and sectoral aspects. Contribution of working group II to the Fifth assessment report of the IPCC* (pp. 169–194). Cambridge, UK: Cambridge University Press.

Clarke, L., Edmonds, J., Jacoby, H., Pitcher, H., Reilly, J., & Richels, R. (2007). Scenarios of greenhouse gas emissions and atmospheric concentrations. *US Department of Energy Publications, 6*.

Fujino, J., Nair, R., Kainuma, M., Masui, T., & Matsuoka, Y. (2006). Multi-gas mitigation analysis on stabilization scenarios using AIM global model. *The Energy Journal*, 343–353.

Fung, F., Lopez, A., & New, M. (2011). Water availability in+ 2°C and +4°C worlds. *Philosophical Transactions of the Royal Society of London A: Mathematical, Physical and Engineering Sciences, 369*(1934), 99–116.

Fuss, S. (2017). The 1.5°C target, Political implications, and the role of BECCS. Oxford University Press, http://climatescience.oxfordre.com/view/10.1093/acrefore/9780190228620.001.0001/acrefore-9780190228620-e-585 (Accessed March 16, 2017). https://doi.org/10.1093/acrefore/9780190228620.013.585.

Gemenne, F. (2011). Climate-induced population displacements in a 4 C+ world. *Philosophical Transactions of the Royal Society of London A: Mathematical, Physical and Engineering Sciences, 369*(1934), 182–195.

Good, P., et al. (2012). A step-response approach for predicting and understanding non-linear precipitation changes. *Climate Dynamics, 39*, 2789–2803.

Good, P., et al. (2016). Large differences in regional precipitation change between a first and second 2 K of global warming. *Nature Communications, 7*, 13667.

Hewitson, B., et al. (2014). Regional context. In *Climate change 2014: Impacts, adaptation, and vulnerability. Part B: Regional aspects. Contribution of working group II to the Fifth assessment report of the IPCC* (pp. 1133–1197). Cambridge, UK: Cambridge University Press.

Hijioka, Y., Matsuoka, Y., Nishimoto, H., Masui, M., & Kainuma, M. (2008). Global GHG emissions scenarios under GHG concentration stabilization targets. *Journal of Global Environmental Engineering, 13*, 97–108.

Horton, B. P., Rahmstorf, S., Engelhart, S. E., & Kemp, A. C. (2014). Expert assessment of sea-level rise by AD 2100 and AD 2300. *Quaternary Science Reviews, 84*, 1–6.

Intergovernmental Panel on Climate Change-IPCC. (2014). Summary for policymakers. In C. B. Field, V. R. Barros, D. J. Dokken, K. J. Mach, M. D. Mastrandrea, T. E. Bilir, M. Chatterjee, K. L. Ebi, Y. O. Estrada, R. C. Genova, B. Girma, E. S. Kissel, A. N. Levy, S. MacCracken, P. R. Mastrandrea, & L. L. White (Eds.), *Climate change 2014: Impacts, adaptation, and vulnerability. Part A: Global and sectoral aspects. Contribution of working group II to the fifth assessment report of the intergovernmental panel on climate change* (pp. 1–32). Cambridge, UK and New York: Cambridge University Press.

Jones, R. G., Noguer, M., Hassell, D. C., Hudson, D., Wilson, S. S., Jenkins, G. J., et al. (2004). *Generating high resolution climate change scenarios using PRECIS* (p. 35). Exeter, UK: Met Office Hadley Centre.

KNMI-Climate Explorer. Website - http://climexp.knmi.nl/. (Access, 10 Jan 2017).

Kreft, S., Eckstein D., & Melchior, I. (2016). Global climate risk index 2017. ISBN 978-3-943704-49-5, 32p. Purchase order number: 17-2-01e. Germanwatch e.V.

Lee, Y. H., Pierce, J. R., & Adams, P. J. (2013). Representation of nucleation mode microphysics in global aerosol microphysics models.*Geoscientific. Model Development Discussions, 6*(1), 893–924. https://doi.org/10.5194/gmd-6-1221-2013

Masson-Delmotte, V., Schulz, M., Abe-Ouchi, A., Beer, J., Ganopolski, A., González Rouco, J. F., et al. (2013). Information from paleoclimate archives. In T. F. Stocker, D. Qin, G.-K. Plattner, M. Tignor, S. K. Allen, J. Boschung, A. Nauels, Y. Xia, V. Bex, & P. M. Midgley (Eds.), *Climate change 2013: The physical science basis. Contribution of Working group I to the Fifth assessment report of the intergovernmental panel on climate change* (pp. 383–464). Cambridge, UK and New York: Cambridge University Press. https://doi.org/10.1017/CBO9781107415324.013

Meinshausen, M., Smith, S. J., Calvin, K., Daniel, J. S., Kainuma, M. L. T., Lamarque, J. F., et al. (2011). The RCP greenhouse gas concentrations and their extensions from 1765 to 2300. *Climatic change, 109*(1–2), 213–241.

Moss, R. H., Edmonds, J. A., Hibbard, K. A., Manning, M. R., Rose, S. K., Van Vuuren, D. P., et al. (2010). The next generation of scenarios for climate change research and assessment. *Nature, 463*(7282), 747–756.

Nakicenovic, N. & Swart, R. (2000). Special report on emissions scenarios. In N. Nakicenovic & R. Swart (Eds.), *Special Report on Emissions Scenarios* (p. 612. ISBN 0521804930). Cambridge, UK: Cambridge University Press v1. July 2000.

Nazarenko, L., Schmidt, G. A., Miller, R. L., Tausnev, N., Kelley, M., Ruedy, R., et al. (2012). Regional climate downscaling: What's the point? *Eos, 93*(5), 52–53.

Pielke Sr., R. A., & Wilby, R. L. (2012). Regional climate downscaling: What's the point? *Eos.* Trans. AGU, *93*(5), 52.

Piontek, F., Müller, C., Pugh, T. A., Clark, D. B., Deryng, D., Elliott, J., et al. (2014). Multisectoral climate impact hotspots in a warming world. *Proceedings of the National Academy of Sciences, 111*(9), 3233–3238.

Rahmstorf, S. (2007). A semi-empirical approach to projecting future sea-level rise. *Science, 315*, 368–370.

Randall, D. A., et al. (2007). Climate models and their evaluation. In *Climate change 2007: The physical science basis. Contribution of working group I to the Fourth assessment report of the IPCC.* Cambridge, UK: Cambridge University Press.

Riahi, K., Grübler, A., & Nakicenovic, N. (2007). Scenarios of long-term socio-economic and environmental development under climate stabilization. *Technological Forecasting and Social Change, 74*(7), 887–935.

Rogelj, J., Luderer, G., Pietzcker, R. C., Kriegler, E., Schaeffer, M., Krey, V., et al. (2015). Energy system transformations for limiting end-of-century warming to below 1.5 [deg] C. *Nature Climate Change, 5*(6), 519–527.

Romanou, M. S., Shindell, D. T., Sun, S., Tsigaridis, K., Unger, N., Voulgarakis, A., et al. (2015). Future climate change under RCP emission scenarios with GISS ModelE2. *Journal of Advances in Modeling Earth Systems, 7*(1), 244–267. https://doi.org/10.1002/2014MS000403

Schaeffer, M., Hare, W., Rahmstorf, S., & Vermeer, M. (2012). Long-term sea-level rise implied by 1.5°C and 2°C warming levels. *Nature Climate Change, 2*(12), 867–870.

Sherwood, S. C., & Huber, M. (2010). An adaptability limit to climate change due to heat stress. *Proceedings of the National Academy of Sciences, 107*(21), 9552–9555.

Smith, S. J., & Wigley, T. M. L. (2006). Multi-gas forcing stabilization with minicam. *The Energy Journal, 27*, 373–391.

Stanton, J. C., Shoemaker, K. T., Pearson, R. G., & Akçakaya, H. R. (2015). Warning times for species extinctions due to climate change. *Global Change Biology, 21*(3), 1066–1077.

Strauss, B. H., Kulp, S., & Levermann, A. (2015). *Mapping choices: Carbon, climate, and rising seas — Our global legacy*. Princeton, NJ: Climate Central.

Taylor, K. E., Stouffer, R. J., & Meehl, G. A. (2012). An overview of CMIP5 and the experiment design. *Bulletin of the American Meteorological Society, 93*(4), 485–498.

Thomas, M. A. & Lin, T. (2015). Addressing uncertainty in ensemble sea-level rise predictions 12th. International conference on applications of statistics and probability in civil engineering, ICASP12 Vancouver, Canada, July 12–15, 2015.
Van Vuuren, D., den Elzen, M., Lucas, P., Eickhout, B., Strengers, B., van Ruijven, B., et al. (2007). Stabilizing greenhouse gas concentrations at low levels: An assessment of reduction strategies and costs. *Climatic change*. https://doi.org/10.1007/s/10584-006-9172-9
Van Vuuren, D. P., Edmonds, J., Kainuma, M., Riahi, K., Thomson, A., Hibbard, K., et al. (2011). The representative concentration pathways: An overview. *Climatic Change, 109*(1–2), 5.
Wise, M., Calvin, K., Thomson, A., Clarke, L., Bond-Lamberty, B., Sands, R., et al. (2009). Implications of limiting CO2 concentrations for land use and energy. *Science, 324*(5931), 1183–1186.
World Bank. (2012). *4°C: Turn Down the Heat.* A Report for the World Bank by the Potsdam Institute for Climate Impact Research and Climate Analytics, November 2012, Washington, DC 20433.

Chapter 3
Assessments and How an Increase in Temperature may Have an Impact on Agriculture in Brazil and Mapping of the Current and Future Situation

Eduardo Delgado Assad, Rodrigo Rudge Ramos Ribeiro, and Alan Massaru Nakai

3.1 Introduction

In this Chapter, the possible impacts of climate change in the vulnerability of agricultural production will be analysed, caused by an increase in temperature, considering an increase of 3 °C in temperature until the end of the century. In order to analyse agricultural production it is important to remember that the study of the effects of the increase in temperature, should also be done together with other factors such as, water availability and the increase of CO_2 concentration.

The goal of this study is to present an analysis, compilation and review of information gathered by a literature on the possible impacts on the Brazilian agricultural sector vis-à-vis the increase in temperature. The literature review was done based on articles published in international and national scientific journals and gray literature[1] (national and ministries' reports, PBMC, SAE, FBMC, IPCC and UNFCC-TCN reports, among others).

The increase of greenhouse gas emissions due to anthropic actions growing in Brazil, bring negative consequences to natural ecosystems. For this reason, climate change and global warming are being widely studied for the different agricultural sectors. In order to plants grow and develop, climate aspects like average air temperature (day and night), rainfall and solar radiation have to supply for each crop's requirements.

Through mathematical models based on data registered on oceans, biosphere and atmosphere, the IPCC, which was set up in 1988, has estimated an increase between 1.4 and 5.8 °C in global temperature until the end of the twenty-first century (IPCC, 2001; IPCC, 2007). An increase between 2 and 6 °C has been predicted for Latin

[1] Materials and research produced by organizations outside of the traditional commercial or academic publishing and distribution channels.

E. D. Assad (✉) · R. R. R. Ribeiro · A. M. Nakai
Embrapa, São Paulo, Brazil

C. A. Nobre et al. (eds.), *Climate Change Risks in Brazil*,
https://doi.org/10.1007/978-3-319-92881-4_3

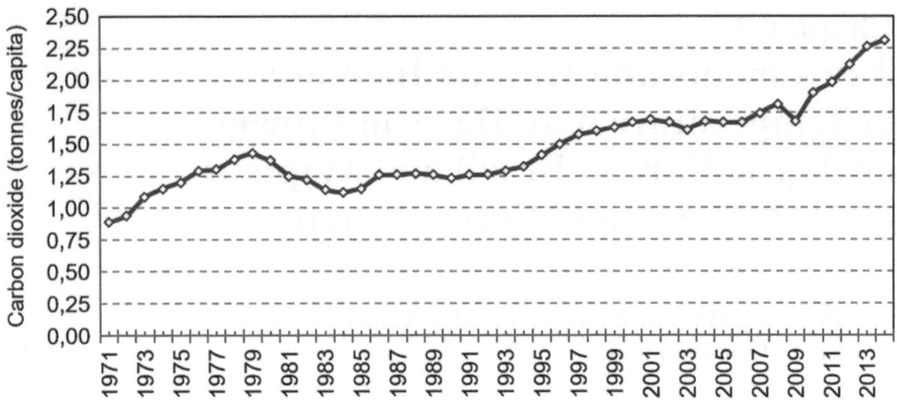

Fig. 3.1 Evolution of per capita CO_2 emissions in Brazil between 1971 and 2014 (Drafted by authors based on OECD, 2015 data)

America in terrestrial surface, as well as a rise in the frequency and intensity of extreme events like heat waves/cold, droughts, floods and above all, hurricanes and tropical storms.

According to the Organisation for Economic Co-operation and Development (OECD) database, in the most pessimistic scenario of CO_2 emissions increase, the planet's 3 °C increase limit would be surpassed in approximately 2055 and 4°C in around 2075. The OECD's data analysis of the evolution of per capita CO_2 emissions in Brazil between 1971 and 2014 shows that they have grown mainly in the last decade (Fig. 3.1). However, when compared to other countries like the United Kingdom and Argentina they show lower values.

Carbon emissions related to human activities have contributed significantly to increase of CO_2eq. concentration in the atmosphere and temperature. In case greenhouse gas emissions continue to grow at the current rates throughout the next few years, the planet's temperature may increase up to 4.8 °C this century, which could result in an up to 82-centimetre rise in sea level and cause important damage to most coastal regions.

The increase of CO2eq concentration has a direct effect in plants, possibly hindering their growth rate and crop yield, after all, concentration is a limiting factor to photosynthesis. If the increase in CO2eq concentration comes hand in hand with a rise in air temperature, there may be a reduction of crop cycles and a hike in plant tissue respiration, reducing or cancelling CO_2eq is positive effects. However, the effects of the increase of CO_2eq. concentration and temperature vary according to the crop analysed (Walter, Streck, Rosa, & Krüger, 2010).

The increase in temperature, together with changes in rainfall patterns will hinder agricultural production, reducing gains in yield and contributing to a rise in food prices, as supply gets ever more precarious (WEF, 2016).

All food security aspects are potentially affected by climate change, including access to food, its use and price stability. The increase in global temperature of approximately 4°C or higher at the end of the twenty-first century, together with an increase in demand, may represent great risks for food security at the regional and global level (IPCC, 2014b). Evidence suggests that agricultural production may be very much affected by climate change, but there is still little quantitative understanding of how these impacts would affect economic activities in countries like Brazil (Hertel, Burke, & Lobell, 2010). The effects of climate change on the world's food production will lead to a rise in product price and poverty in society. In some cases, climate change has a positive influence in poverty reduction, while in others it does not. Evidently, the impacts of climate change on food security is a complex issue, in the extent that it interacts with productive systems (Hertel, 2015).

Between 1980 and 2005, Brazil managed to reduce poverty in over 50%, with agriculture playing an important role. This was the result of an increase in public investment in research, extension work and education (Cervantes-Godoy & Dewbre, 2010).

The growth of Brazilian agricultural exports has been associated with a change in the market setting, where a change in traditional tropical products like orange juice and coffee, as well as soybean, sugar and meat has occurred.

A historical analysis of agriculture in Brazil shows that it is undergoing an expansion phase, in relation to the amounts produced, as well as planted area (Fig. 3.2). In fact, technological advances are evident when a rise in yield is seen (4.8% year) in relation to the increase in area (1.7% year.). Brazilian agriculture is moving from the south subtropical regions to the tropical savannah areas in the Central-West (Cerrado biome), where production is mainly done through dryland farming (Assad et al., 2015).

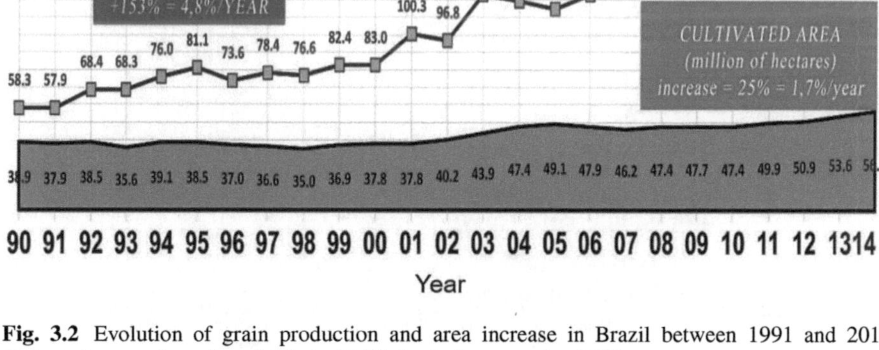

Fig. 3.2 Evolution of grain production and area increase in Brazil between 1991 and 2013 (Adapted source from Assad, Marin, Valdivia, & Rosenzweig, 2015)

The impact of climate change in brazilian agrobusiness depends on the effects on yield, as well as markets and prices. Climate change impacts are bigger on lower income farmers (FAO, 2016). In addition, climate change is already leading to impacts on food security and nutrition in more vulnerable communities (FAO, 2016).

The impact of climate factors, like temperature increase in crops may alter the different growth and development stages of each plant, leading to consequences that could aggravate yield related effects (Fig. 3.3). Besides to each plant's characteristics, the soil type, agricultural management and species' genotype are important features in relation to the increase in temperature and yield (Carter, 2013). In general, if a crop's requirements are supplied for, good yield levels are achieved, but if that does not happen, yield losses may be expected.

Climate changes will certainly interfere in agriculture, temperature alterations, rain, wind and sea level, possibly resulting in different impacts depending on for example, soil, type of plant, intensity of change (Fig. 3.4).

Climate variability is an important factor in agriculture and well known by farmers in relation to rain and drought (natural climate variations) (Rocha, 2008). Impacts caused by temperature increase in the agricultural sector, vis-à-vis a rise of 3 °C or over reinforce this concern.

The increase in temperature brought about by climate change leads to impacts in the ecosystem services, such as pollination done mainly by insects like bees. The alteration of the symbiosis plant/polliniser is sensitive to high temperatures and in tropical places like Brazil, these pollinisers are already close to their ideal temperature tolerance level (FAO, 2016).

In 2013, reports were presented by the IPCC. According to the document, most of the global warming seen since the twentieth century is due to the increase in greenhouse gases, resulting from human activities. The IPCC also points out those

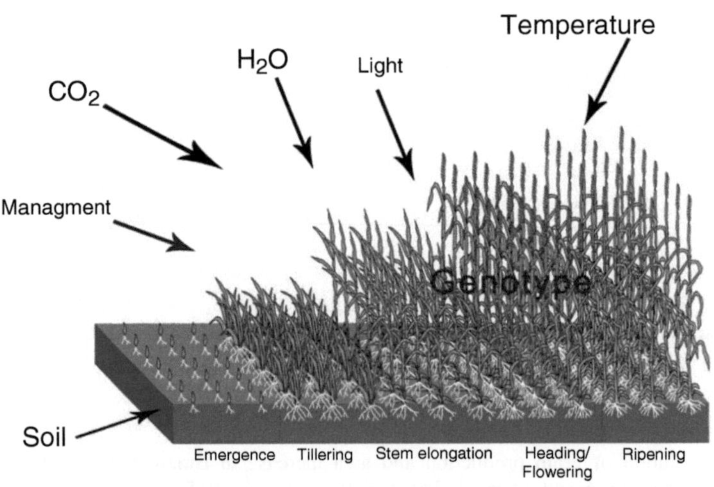

Fig. 3.3 Development and growth stages of a plant (Adapted source: Carter, 2013)

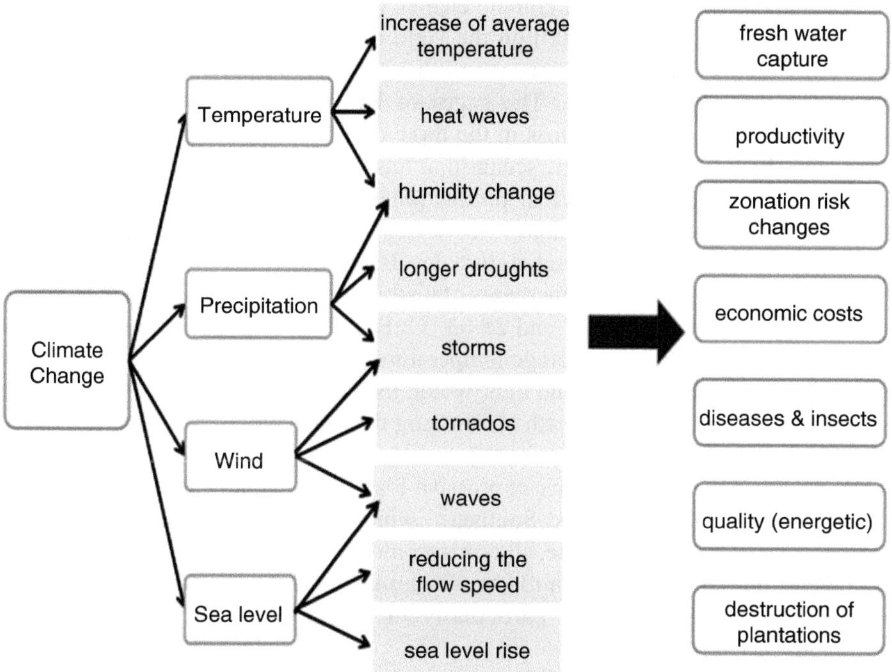

Fig. 3.4 Climate change characteristics and possible impacts on agriculture

extreme events (e.g. hurricanes, heat waves, droughts, flood and heavy rain) will become more frequent. According to IPCC, human action plays a clear role in the climate change process.

An increase of 3 °C or more in temperature is enough to make farming unviable in several areas and it is not yet possible to fully assess the impact this will have on Brazilian agribusiness. Studies based on crop simulation for rice, beans, maize, soybean, cotton and coffee (Assad, Pinto, Zullo Junior, Marin, & Pellegrino, 2008; Assad, Martins, de Beltrão, & Pinto, 2013; Zullo, Pinto, Assad, & de Avila, 2011), point to great losses in planted areas, when taking this scenario in consideration.

Agriculture is one of the sectors most affected by extreme natural events. With climate change, risks to food and nutritional security are multiplied by the expected increase in the frequency and intensity of extreme events, as well as weather related disasters (FAO, 2016). The rise of the average temperature also implies changes in rainfall and wind, among other factors.

Projections of the effects of climate change in agriculture in Brazil for 2050, indicate that this sector would experience expressive losses. The Brazilian regions most vulnerable to climate change would be the Amazon and Northeast.In the Amazon, temperature rise may reach 8 °C by 2100 where some models indicate the possibility of changes to the Amazon forest (so called savannisation). In the northeast region rise may 4°C by 2100.

An analysis of the effect of climate change (based on IPCC's third report, from 2040 to 2070 and 2070 to 2100) on the yield of maize, sugarcane and cassava in states of Brazil's Northeast region, shows that production levels would be higher if there was no climate change. The states of Rio Gande do Norte, Paraíba and Pernambuco may show yield loss in the three crops in the medium, as well as the long term. Even in an optimistic scenario, it has been observed that climate change would have a negative impact in production in municipalities in the south and central-south of the state of Bahia.

In relation to estimated temperature levels, by 2100 the Brazilian Northeast region may show temperature increases of up to 3.4 °C. The average future temperature will reach between 28.37 and 28.63 °C. Based on these estimates, crops like cassava, which is plated in average temperatures between 20 °C and 27 °C, would be susceptible to the weather and thus, would led to economic impacts in producing states of the Northeast. It is worth mentioning that on average, rainfall levels in the Northeast region may drop around 10%.

Agriculture would experience expressive losses in all states, with the exception of the coldest in the South and Southeast, which would see milder temperatures. With the exception of sugarcane, all crops would see a reduction in low risk production areas, particularly soybean (34 to 30%), maize (15%) and coffee (17 to 18%) (Table 3.1). Yield would fall particularly in livelihood crops in the Northeast (Margulis & Dubeux, 2010).

The reduction of agriculture's low risk area will lead to strong production losses. Indeed, temperature increase implies that the crops analysed in this study will suffer negative impacts (with the exception of sugarcane and cassava), showing a reduction in production, which could be dramatic in some regions (Pinto, Oliveira, & Fernande, 2008).

Considering the conditions expected for the next few years and depending on the species, crops' physiological responses suggest an accentuated growth dynamics, with small changes in development, like flowering and fruiting (Da Mattaa, Grandisb, Arenque, & Buckeridge, 2010). In addition, in relation to variations in food quality in a hotter environment, with high CO_2eq concentration for instance, a

Table 3.1 Agricultural losses resulting in climate change in Brazil (in 2008 Brazilian reais) in Northeast (NE), Midwest (MW), Southeast (SE), South(S) regions

Type of crop	*Variation% of low risk area (2050)*	*Impact on average yield per region*	*Annual economic loss (R$) 1 U$ equal 3.3 R$*
Rice	*−12%*	*−12% (MW and + 44% (S)*	*R$ 530 million/year*
Cotton	*−14%*	-	*R$ 408 million/year*
Coffee	*−17%*	-	*R$ 1597 million/year*
Beans	*−10%*	*−8% (MW) and + 37% (S)*	*R$ 363 million/year*
Soybean	*−34%*	*−0.7 (MW) and + 21% (S)*	*R$ 6308 million/year*
Maize	*−15%*	*−27% (NE) and − 10% (S)*	*R$ 1511 million/year*
Sugarcane	*139%*	*+66% (S) and + 34% (SE)*	-

Adapted source: Margulis & Dubeux, 2010

drop in the concentration of protein and mineral nutrients is expected, as well as alteration to the lipid composition (Da Mattaa et al., 2010).

A climate modelling study (model HadCM3 using A2 emission scenarios) for temperature change projections in Brazilian agriculture for 2040, indicates that there are some positive impacts resulting from such changes, because of the increase of the minimum temperature. However, negative impacts come about from the water stress resulting from evapotranspiration (Silva, Heinemann, & Stone, 2014).

The methodology used for modelling temperature changes may have an impact in agricultural vulnerability studies in Brazil. Studies show that the temperature estimated by the PRECIS-Br model, without correcting the Kalman filter will have a smaller impact on determining the evapotranspiration of crops grown from January to March, than the estimate made using the filter (Porto de Carvalho, Assad, & Pinto, 2011).

3.2 Methodology

This document was drafted based on the Fifth Assessment Report (AR5) of the Intergovernmental Panel on Climate Change (IPCC) for scenario RCP 8.5, which is the more pessimistic and takes into consideration over 3 °C in temperature warming (Fig. 3.5). In addition, other reports and bibliographical data were used, and agricultural information was consulted on the Instituto Brasileiro de Geografia e Estatística (IBGE) database. In the 2007 IPCC report (AR4), the most pessimistic scenario A2

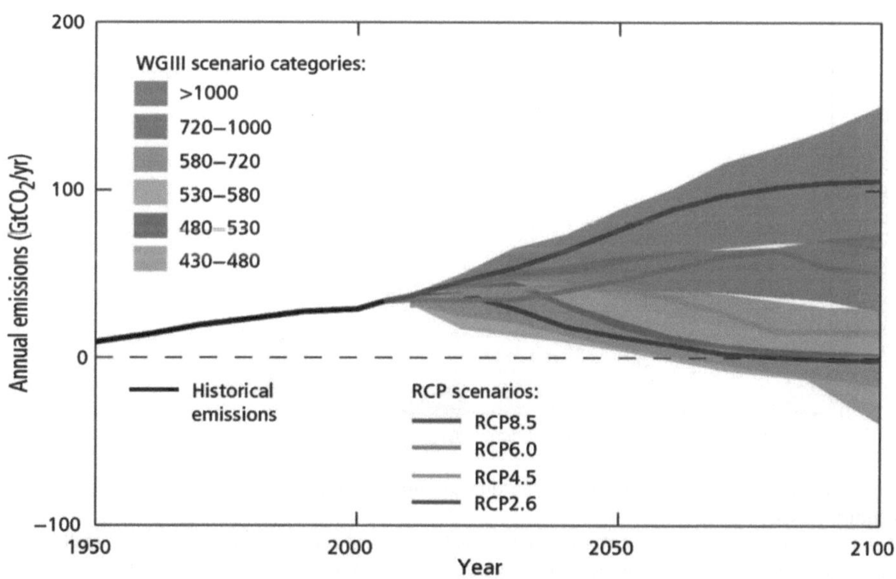

Fig. 3.5 IPCC CO_2 anthropogenic emissions scenario (IPCC, 2014a)

was considered, with an increase in temperature of over 3 °C (between 4 and 8 °C in the Amazon, 4°C in the Northeast, between 4 and 6 °C in the Central-West and 3 and 4 °C in the South) (Avila, 2007).

Therefore, temperature increase projections made by the IPCC Fifth Assessment Report (AR5) published in 2014, proposed several scenarios that may occur until 2100 .

3.3 Assessment of Possible Impacts Due to Temperature Increase: Projections for 2020, 2050 and 2080

In the Executive Summary Report of IPCC AR5 WGII, it Is Worth Mentioning that 'Relatively Few Studies Have Considered Impacts on Cropping Systems for Scenarios with Global Average Temperatures 4°C Higher or More' (IPCC, 2014b, p. 16).

As climate change advances through time, the predicted effect on crop yield is ever more negative, with the magnitude of such effect being highly uncertain (Porter, Montesino, & Semenov, 2015). This progression of changes and their uncertainty vis-à-vis cropping are shown on Fig. 3.6. After 2050, the risk of more severe impacts in crop yield increases and depends on the level of warming (IPCC, 2014b).

A great variety of projections is seen showing an increase in temperature of over 3 °C and yield for each crop in a certain region. However, different studies for the

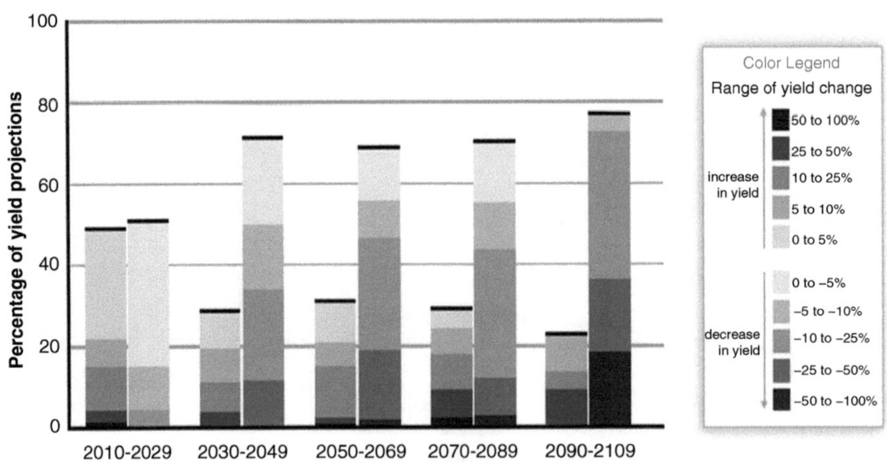

Fig. 3.6 Projection of crop yield change due to climate alterations throughout the twenty-first century (IPCC, 2014b)
The figure consider different emission scenarios for tropical and temperate regions, as well as for combined adaptation and non-adaptation cases for five time intervals on the axis. Crop yield changes are related to levels at the end of the twenty-first century. The data for each period adds to 100%.

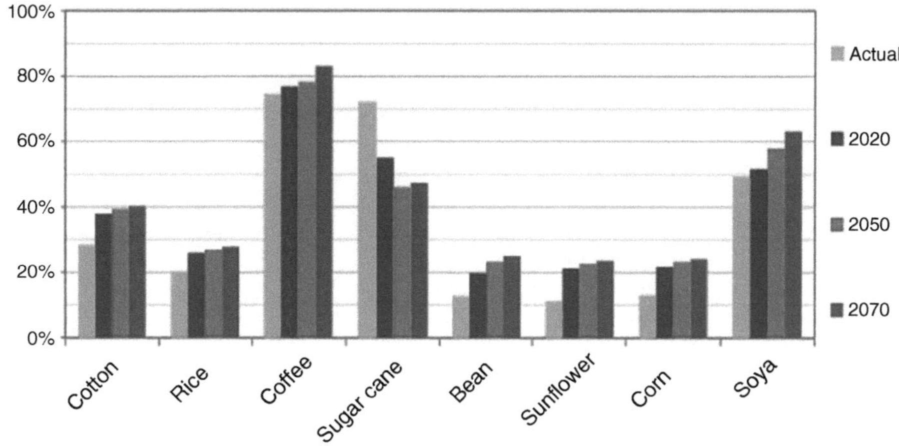

Fig. 3.7 Different scenarios for yield loss risk for different crops in Brazil (Adapted source: Pinto et al., 2008)

same region and crop result in differing impacts (production increase and reduction). Therefore, it is risky to make an analysis using aggregate data, where the range of possible results is not clearly visible. Yield loss due to high temperature, which induces the abortion of flowers is an important matter to consider, particularly for coffee and beans. Temperatures equal to or over 30–34 °C at the moment of flowering may inhibit pollen production and grain formation, resulting in unstable yield from one year to the next for maize and soybean (Porter et al., 2015).

An important aspect in the study of agricultural adaptation is identifying each crop's temperature limit. Results seen from case studies of temperature increase during flowering in other countries, point out that the increase in local, as well global temperature has a strong effect in adaptation capacity. Most studies use local temperature as an independent variable (Porter et al., 2015).

A yield loss risk analysis[2] with probability higher than 20% was done for different crops (cotton, rice, coffee, sugarcane, beans, sunflower, maize and soybean) for the current scenario (1976–2005), 2020, 2015 and 2070 using an average warming of 4°C or over for all municipalities in Brazil. The results show the impact of climate change on sugarcane, which reduces the risk of yield loss: 72% of Brazilian municipalities have probability higher than 20% of yield loss in the current scenario, dropping to 55% in 2020, 46% in 2050 and 48% in 2070 (Fig. 3.7).

The increase in temperature may have different impacts due to the development stages of crops or tillage. It is necessary to identify it, so a better diagnosis may be reached, in relation to appropriate consequences and recommendations. The following Figs. (3.8 to 3.13) simulate (ETA regional model; 40 km resolution) the increase in temperature in Brazil for different periods.

[2] Study done using the PRECIS model for scenario A2.

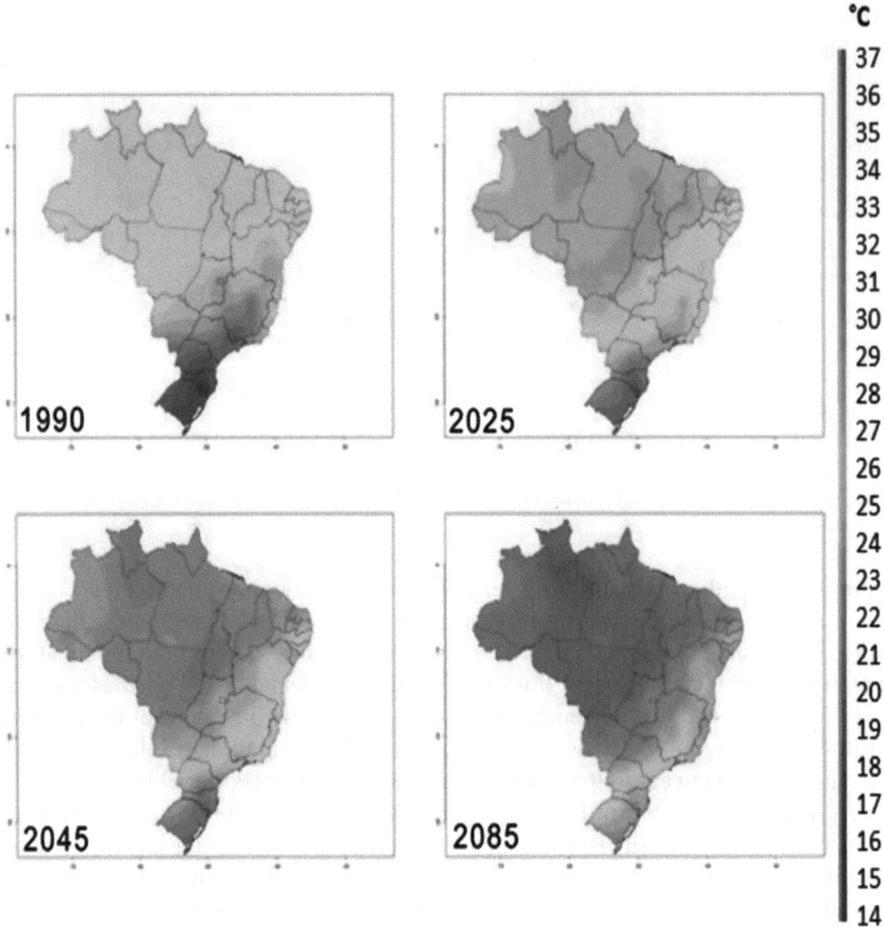

Fig. 3.8 Temperature results in 1990 (observed) and estimate for 2025, 2045 and 2085 (Delta ETA; RCP 8.5) in Brazil's regions (Assad et al., 2016)

Temperature variation is more intense in scenario RCP 8.5, which is within the estimate margin made by IPCC. Indeed, between 2071 and 2099, the hotspot[3] becomes the Cerrado and part of the Amazon (Fig. 3.10).

In this map sequence, average maximum and minimum temperature variations were plotted for scenario RCP 8.5 for the periods 2011–2040, 2041–2070 and 2071–2090, compared with the 1975–2005 baseline.

The following maps show temperature increase variations (ETA regional model) between two periods, in other words, temperature increase in a thirty-year period (Fig. 3.12).

[3] Areas with high temperature concentration.

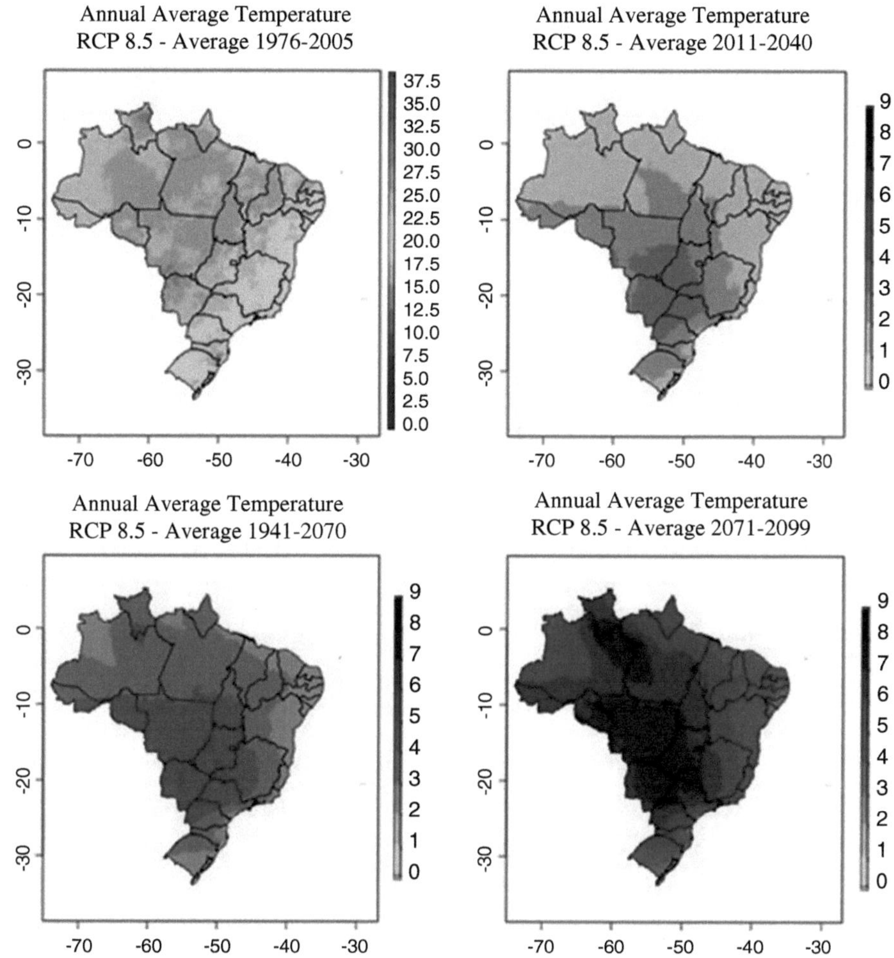

Fig. 3.9 Estimate of annual average temperature increase for the RCP 8.5 scenario in 1976–2005 (observed) and for the periods 2011–2040, 2041–2070 and 2071–2099 (Delta ETA; RCP 8.5) in Brazil's regions (Assad et al., 2016)

The consequences for agriculture of temperatures frequently over 34 °C (heat waves) are things like the abortion of flowers in coffee, orange and beans, as well as the death of chickens, abortion in pigs and reduction in milk production. It is worth pointing out that at current emission estimates, following the scenario RCP 8.5, puts part of Brazil's livestock and agricultural production at risk.

The average limit temperature of the crops analysed is of 37 °C, with coffee's maximum temperature limit standing at 34 °C, while rice and maize 45 °C (Table 3.2).

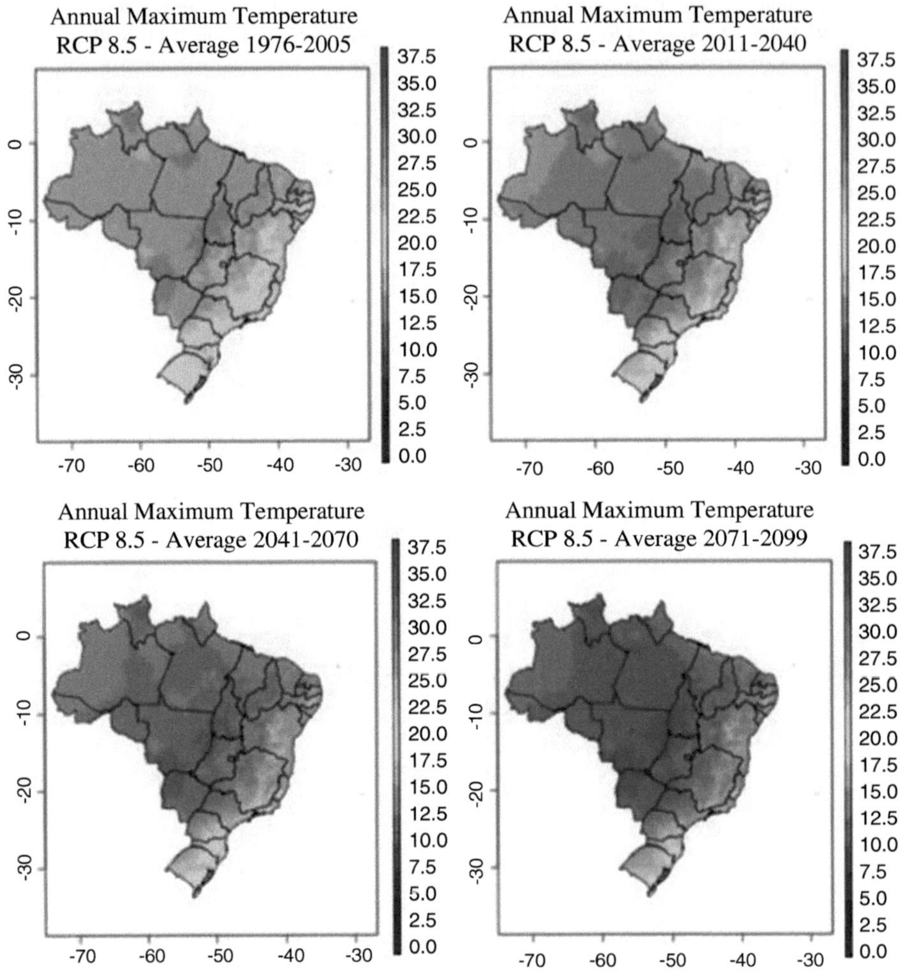

Fig. 3.10 Estimate of annual average maximum temperature increase for the RCP 8.5 scenario in 19,762,005 (observed) and for the periods 2011–2040, 2041–2070 and 2071–2099 (Delta ETA; RCP 8.5) in Brazil's regions (Assad et al., 2016)

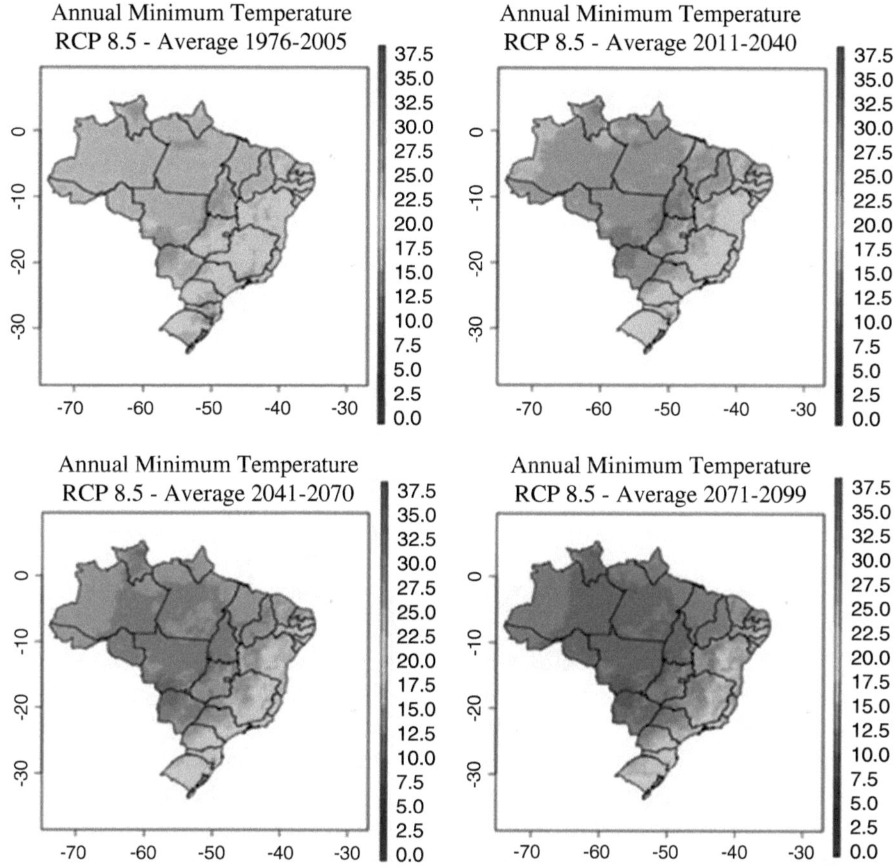

Fig. 3.11 Estimate of annual average minimum temperature increase for the RCP 8.5 scenario in 1976–2005 (observed) and for the periods 2011–2040, 2041–2070 and 2071–2099 (Delta ETA; RCP 8.5) in Brazil's regions (Assad et al., 2016)

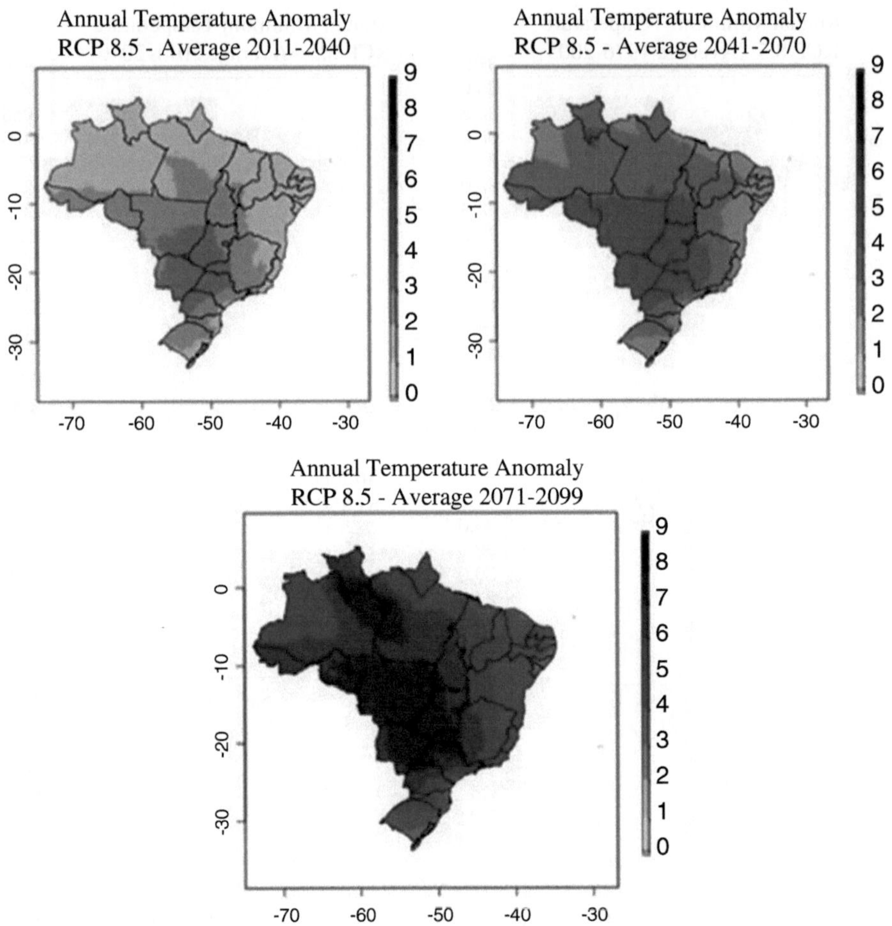

Fig. 3.12 Estimate of annual average temperature increase for the RCP 8.5 scenario for the periods 2011–2040, 2041–2070 and 2071–2099 (Delta ETA; RCP 8.5) in Brazil's regions (Assad et al., 2016)

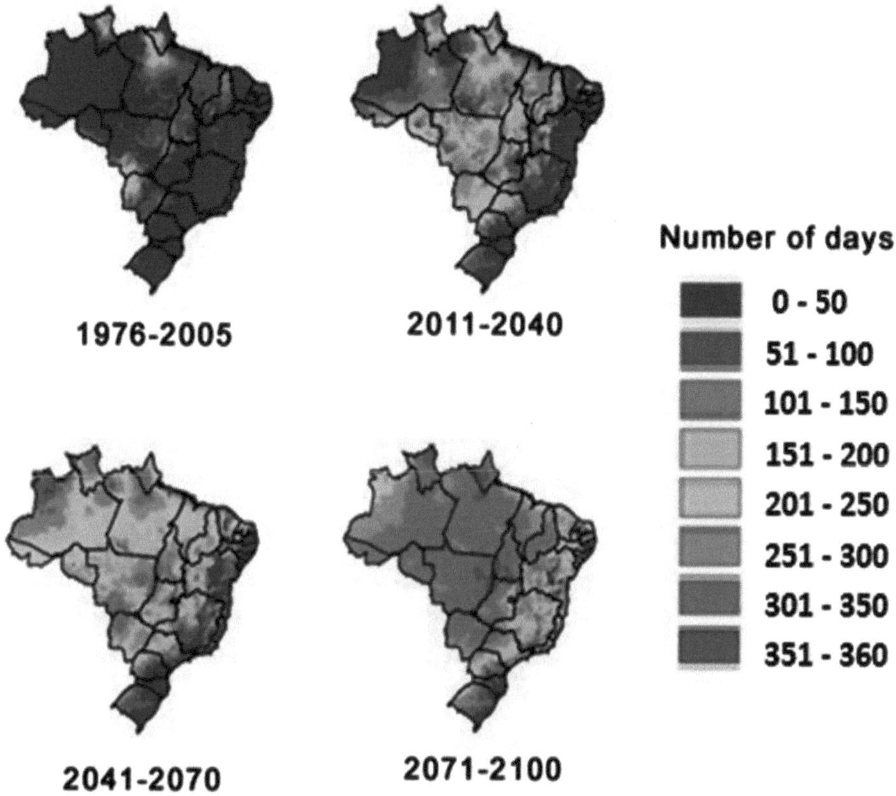

Fig. 3.13 Spatial distribution of how often the annual average occurs, with days with maximum temperature over 34 °C in the RCP 8.5 scenario (ETA-HADGEM2), for the current and future situation (Assad et al., 2016)

Table 3.2 Maximum temperature limits (°C) for crops

Sugarcane	35 °C	Beans	35 °C
Cotton	40 °C	Sunflower	40 °C
Rice	45 °C	Maize	45 °C
Potato	35 °C	Soybean	35 °C
Coffee	34 °C	Wheat	30 °C

Source: Monteiro (2009)

The Fig. 3.14 shows Brazil's regions with annual average temperature above 35 °C in the RCP 8.5 scenario for 2085. This temperature limit reaches several crops' threshold.

Taking into account a scenario with a temperature increase of 4°C in the next few years, the abortion of flowers in coffee, beans, orange may occur, as well as the death of chickens, abortion in pigs and reduction in milk production. A reduction in over 90% in the production of second-crop corn is predicted and 80% in soybean in the low risk area (based on 1990) (Assad et al., 2016). The Table 3.3 shows the

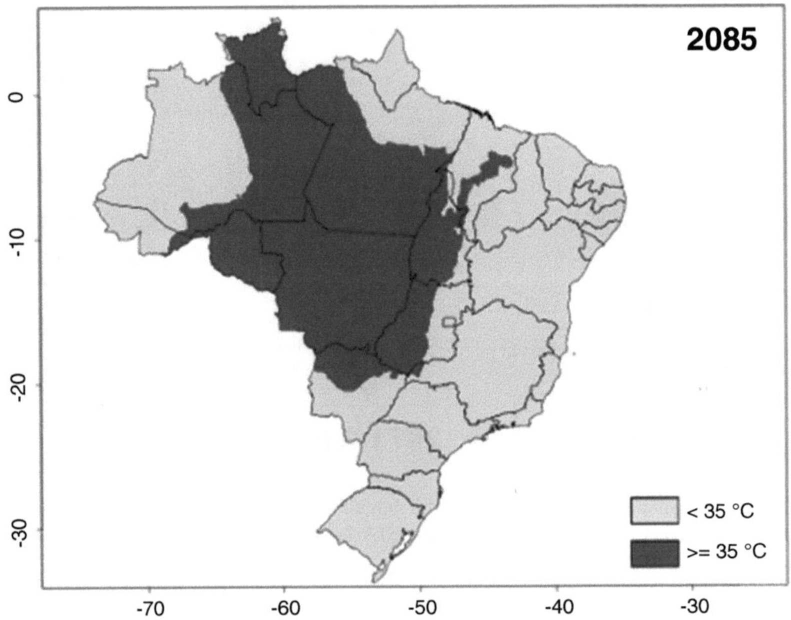

Fig. 3.14 Brazil map showing temperatures above 35 °C for 2085 (source: EMBRAPA CNPTIA, 2015)

Table 3.3 Impacts of temperature increase in Brazilian agriculture

Scenario RCP 8.5						
Crop	2025 (ha)	Δ (%)	2055 (ha)	Δ (%)	2085 (ha)	Δ(%)
Rice	2,238,483	−7.2	2,231,870	−7,5	2,077,094	0
Second-crop corn	1,751,641	−76.5	1,128,835	−84,9	204,339	−97.3
First-crop corn	6,661,951	12.3	6,646,863	−12,5	5,908,882	−22.2
First-crop beans	1,124,132	42.6	1,064,133	−45,6	838,874	−57.1
Second-crop beans	423,463	58.5	396,056	−61,2	286,938	−71.9
Soybean	8,910,284	64.4	8,556,636	−65,7	4,693,604	−81.2
Wheat	1,501,642	21.5	1,596,339	−16,5	1,457,725	−23.8

Source: Assad et al. (2016)

impact of temperature increase in low risk areas, according to the Eta regional model run with the CIMP5 HadGEM2 ES global model.

3.4 Assessment of Possible Impacts Due to Temperature Increase: Projections for Different Crops

3.4.1 Sugarcane

Brazil is one of the biggest sugarcane producers in the world, with a harvest area of approximately 10 million hectares in 2016 (FIESP, 2017) a 200% increase in relation to 1990 (IBGE).

A climate risk study[4] for sugarcane in Brazil's Northeast region shows that there is a relevant difference between the warming scenario and current climate change, in terms of predicted temperature variation (a scenario with a temperature increase between 3 and 5 °C) in cropped areas. For current weather conditions in Brazil's Northeast region, drought risk analysis and its impact on agriculture indicate that low and medium climate risk areas are inferior when compared to high climate risk for the growing of sugarcane. Analysing the scenario with a temperature increase of 3 °C, high climate risk areas are more significant, cancelling low climate risk areas in relation to cropping in the period of April and May. Despite this, medium risk areas ensure a small variation between 0.6 and 0.1%. On the other hand, August shows a growth in medium risk areas in relation to the previous scenario from 15.4 to 17.6%. Therefore, sudden increases in temperature may also lead to the reduction of medium and low climate risk areas. In fact, the scenario with 5 °C temperature increase provides evidence of this reduction (Silva, Oliveira, Santos, & Silva, 2013).

The production loss risk analysis for sugarcane in Brazil's regions, with probability higher than 20% is shown on Fig. 3.15. The results indicate that production loss risks drop when there is a shift to the South region.

Contrary to what will happen to the other crops assessed, a 4°C temperature increase may benefit sugarcane. With the increase in temperature, the low climate risk area for this crop should double. Areas located in higher latitudes, which today present restrictions due to the high risk of frost lose this feature, particularly the state of Rio Grande do Sul, turning into potentially productive regions in 10 to 20 years. The Central-West region, which today presents a high productive potential, will remain a low risk area. However, it will depend more on supplementary irrigation of around 50 mm/month in the driest period, in order to ensure germination.

In the A2 climate change scenario, the sugarcane appropriate area could double by 2020 (Assad et al., 2015). In another study on the impact in production of the

[4]This study used the SARRA model (*Systeme d'Analyse Regionale dês Risques Agroclimatiques*) to assess drought risks and their impact on agriculture in the studied area.

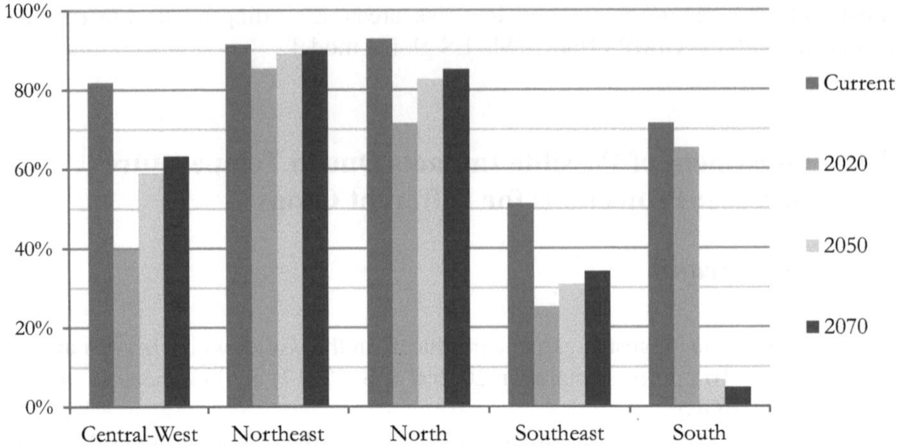

Fig. 3.15 Different production loss risk scenarios for sugarcane in Brazil's regions (Adapted source: Pinto, Assad, et al., 2008)

4°C increase in the state of Goiás, a trend of yield increase (up to 4%) was identified. Nevertheless, in some environments with higher water restrictions, production is expected to remain stable or even drop (Evangelista, Marin, & Junior, 2009).

3.4.2 Cotton

The production loss risk analysis for cotton in Brazil's regions, with probability higher than 20% is shown on Fig. 3.16. The results point out that the Northeast is the most affected region, going from 23% risk in the current scenario to 62% in 2070.

The effect of the increase in temperature in the A2 scenario using the PRECIS model, in relation to the climate risk for cotton production in Brazil shows that the Northeast is the region with the biggest reduction in low risk areas for this crop, when compared to other regions in the country. The Southeast and Central-West regions show a small reduction in low climate risk areas for cotton until 2040, while the North did not present any reduction in these areas. According to Assad et al. (2013), the rise in temperature has a negative impact on cotton production in Brazil, because the low risk area for this crop may diminish.

Cotton should suffer a reduction in low risk areas, particularly in the northeastern Agreste [5] and Cerrado regions, made up of the south of the Maranhão state, south of the Piauí state and west of the Bahia state (FGV-GVces, 2013).

[5] The **agreste** is a narrow zone of **Northeast Brazil** in the states of Rio Grande do Norte Paraíba, Pernambuco, Alagoas, Sergipe and Bahia between the coastal forest zona da mata and the semiarid sertão.

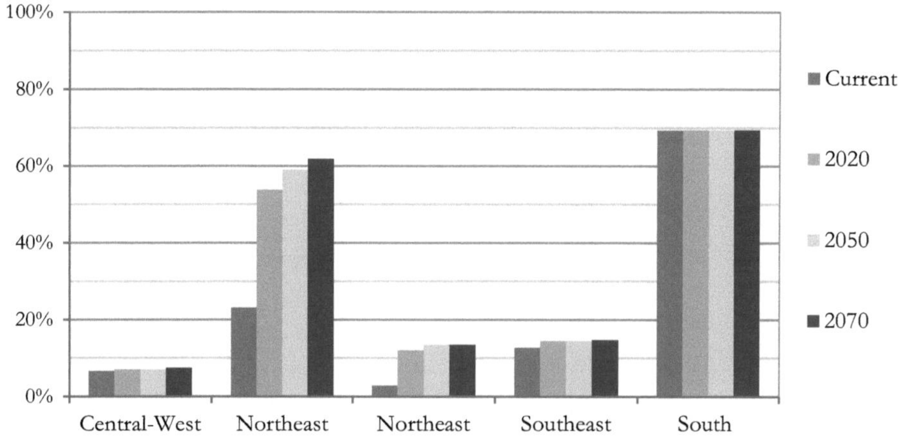

Fig. 3.16 Different production loss risk scenarios for cotton in Brazil's regions (Adapted source: Pinto, Assad, et al., 2008)

3.4.3 Rice

The production loss risk analysis for rice in Brazil's regions, with probability higher than 20% is shown on Fig. 3.17. The results point out that the Northeast is the most affected region, going from 47% risk in the current scenario to 79% in 2070. A risk reduction is also seen in the South region, going from 22% in the current scenario to 10% in 2070.

If there is a 5.8 °C increase in global temperature, the low risk areas for rice would be reduced to 2.4 million ha (Assad et al., 2016).

A 4°C temperature increase should lead to the reduction of low risk areas for rice, particularly in the northeastern Agreste and Cerrado regions, made up of the south of the Maranhão state, south of the Piauí state and west of the Bahia state. Even with the damage caused by climate, it should be possible to keep production stable, at the levels seen today, as well as concentrated in the Central-West, South and Southeast regions.

High temperatures during the day and night may reduce rice's productive potential drastically, as it shortens the crop's cycle, increasing respiration and spikelet sterility. This trend may be mitigated by choosing genotypes more resistant to high air temperatures during flowering and seeding.

Production loss studies for irrigated rice in the state of Rio Grande do Sul (Infocrop model), when a 4 °C increase in air temperature was included, pointed to a loss in yield of 7 to 8.5% (Walter et al., 2010). Limitations imposed on rice yield by CO_2 and high temperature may be attenuated at least in part, by altering seeding time and selecting genotypes that present higher fertility in high temperatures (Krishnan, Swain, ChandraBhaskar, Nayak, & Dash, 2007).

For rice, estimates show that on average, productivity could increase. One of the main reasons for this positive aspect seems to be related to the fact that rice is a crop

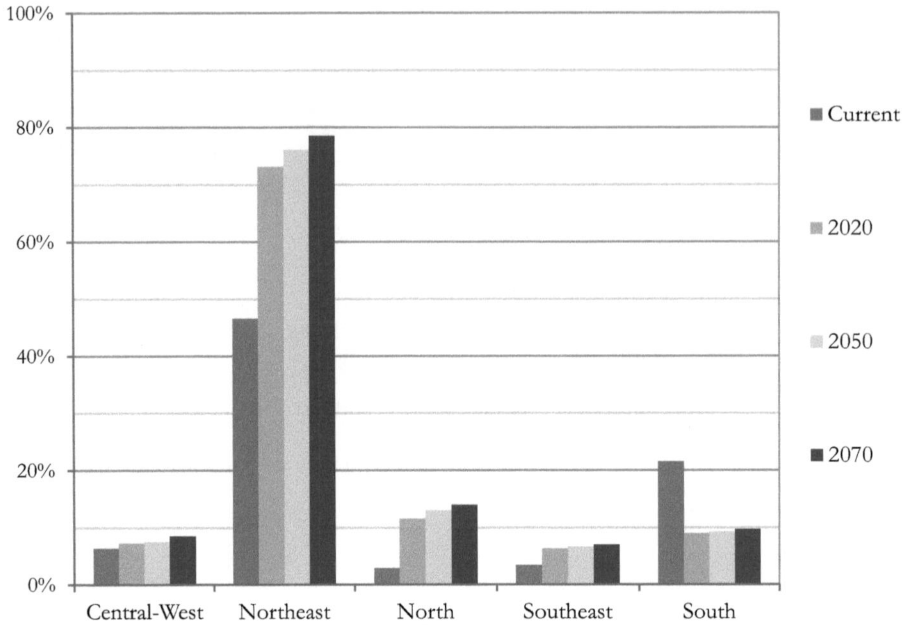

Fig. 3.17 Different production loss risk scenarios for rice production in Brazil's regions (Adapted source: Pinto, Assad, et al., 2008)

from a humid/irrigated region. With warmer conditions, the negative effect in the shorter grain filling period would be compensated by higher biomass accumulation rates, because of more favourable photosynthesis conditions (Fernandes, Soliman, Confalonieri, Donatelli, & Tubiello, 2012).

The increase in rice productivity depends on the added biomass, which is generally affected by the incidence of diseases. Climate change may influence pathogen multiplication rates, as well as susceptibility to the host (Prabhu, da Silva, & de Filippi, 2008). The most important rice diseases in Brazil are: brusone, scald, stained grain and rice sheath blight, the latter restricted to irrigated rice. The climate impacts of the diseases are:

- **Brusone:** for Brazil's Central-West region, brusone is expected to be less severe in 2080 due to the increase in minimum temperature, varying between 22 and 24°C. In addition, maximum temperature and rainfall should increase, favouring a drop in brusone severity and incidence. In the state of Rio Grande do Sul, brusone will not reduce productivity significantly. However, the increase in maximum and minimum temperatures may be a limiting factor for the crop's yield (Prabhu et al., 2008).
- **Scald:** despite there being no information on the effect of temperature rise on the disease's development, the fungus may possibly adapt to high temperatures (once the disease is endemic in the states of Pará and Amazonas). Changes in

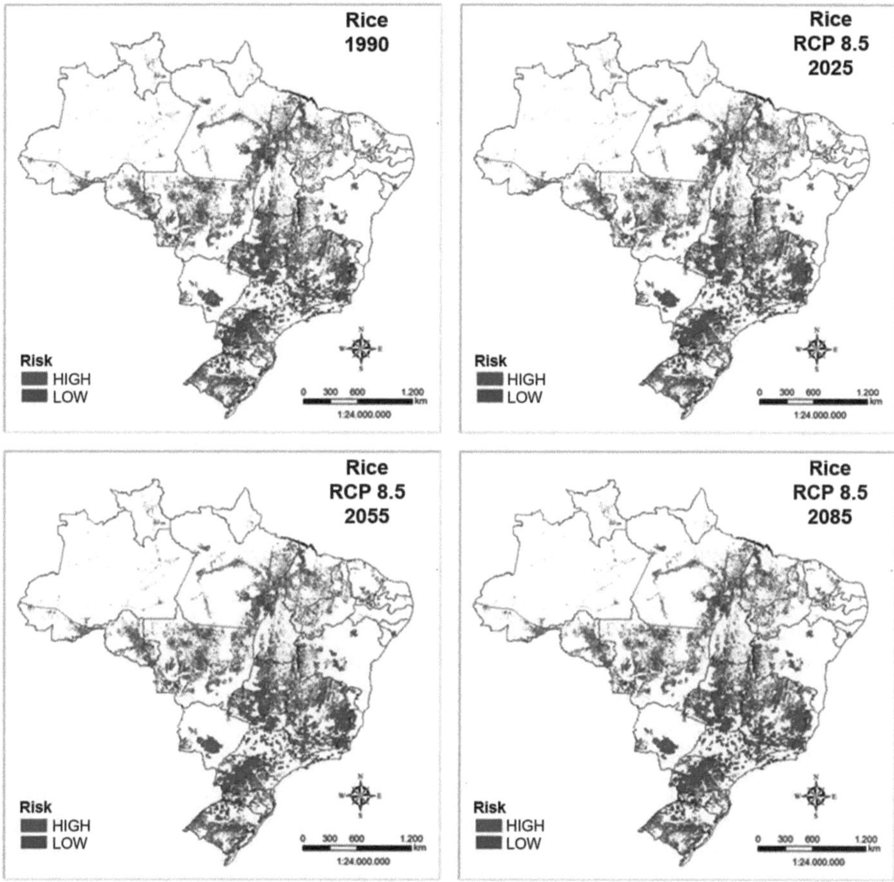

Fig. 3.18 Impact of the reduction of low risk areas for rice in comparison to 1990, in the RCP 8.5 scenario (Assad et al., 2016)

rainfall may increase the disease's severity and damage yield (Prabhu et al., 2008).

- **Stained grain:** the disease may occur more often and at higher intensity in Brazil's South, where most of the Brazilian rice production is found (Prabhu et al., 2008).
- **Sheath Blight:** with the increase in maximum and minimum temperatures in Brazil's South region, sheath blight may become a very important disease in Brazil, together with brusone (Prabhu et al., 2008).

The Fig. 3.18 shows the impact in the reduction of low risk areas, compared to 1990. In the case of the RCP 8.5 scenario, loss of low risk area reaches over 13%. Losses become more intensified, limiting the production of rice to irrigated areas only, with reasonable amount of rainfall (state of Goias, north of Mato Grosso and Paraná).

3.4.4 Potato

Potato is grown in Brazil mainly in temperate climate regions, usually in the country's South region. However, despite having a wide geographic reach in the country, it is considered to be particularly vulnerable to global warming because of the relatively small temperature range, between 10 and 30 °C, it may be grown in. High temperatures outside of the ideal interval affects the plant's metabolism directly. Producing municipalities located in colder regions will not be affected significantly, although some changes will be necessary at planting seasons (Lopes, Silva, Cruz, Assad, & Pereira, 2011). A temperature increase leads to a rise in plant transpiration, thus, affecting its demand for water and reducing plant yield. The increase in temperature will be made worse by rainfall distribution and yield reduction in tropical regions may reach 20–30%. Contrary to this situation, temperature increase in high altitude regions may have favourable effects (Info – Resources Focus, 2008).

3.4.5 Cassava

According to information made available by EMBRAPA (2015), the ideal temperature range for the growing of cassava is between 20 and 27 °C (annual average). For the 2070–2100 period, no state in the Northeast region is expected to register temperatures below 27 °C on average.

A 4°C rise in temperature may benefit the crop in the next decades, as the area in the country suitable for its growth will increase, also bringing up production value. However, the positive figures hide the strong setback the crop will experience in the northeastern Semiarid and Agreste, where most of its consumers are and where its production is strongly related to food security. In this region, the area suitable for its growth shall decrease drastically, contrary to what will be seen in the rest of the country. In the following decades, the situation will improve in the country's southern areas, because of reduced risk of frost, as well as in the Amazon, due to a drop in water surplus.

Therefore, states like Bahia will show yield gains lower than the average recorded for the whole of the Northeast region. Results point to a relative increase in cassava productivity levels in the states of Maranhão, Ceará, Alagoas, Sergipe and Piauí (Araújo, Silva, Gomes, Féres, & Braga, 2014).

3.4.6 Coffee

There are studies that consider a 15% increase in rainfall and a temperature rise between 3 and 5.8 °C in the state of São Paulo, for the climate risk zoning of coffee. This is based on the monthly and annual average between 1961 and 1990, and indicates that São Paulo areas, classified as per how suited they are for growing coffee vary between 78.7 and 30.3%, taking into account a temperature rise between 3 and 3.3 °C (Junior, Pinto, & Assad, 2006).

The Fig. 3.19 shows the impact of the temperature increase in relation to the risk of coffee production loss in all of Brazil's regions.

The particularly specie of *coffee arabica* is the crop that will most clearly suffer with the 4 °C increase in temperature. If today the main producing states are Minas Gerais, Espírito Santo and São Paulo, in the future, the grain may migrate to Paraná, Santa Catarina and Rio Grande do Sul. Nevertheless, even in the latter two, which will present a significant increase in low risk areas for the crop by 2050, there will be a reduction in area as temperatures continue to rise in 2070. The crop may be hit by lack of water or high temperatures in regions where it is traditionally grown (Fgv-GVces, 2013).

The impact assessment of climate change in coffee's agro-climatic zoning in Brazil, points to a reduction in low risk areas in relation to its cropping in some states, Table 3.4 (95% reduction in Goiás, Minas Gerais and São Paulo), around 75% in Paraná if the temperature reaches 5.8 °C (Assad et al., 2004).

The state of Minas Gerais is a territory particularly exposed to intense rainfall and heat waves. As it is also Brazil's biggest coffee producer, an increasing number

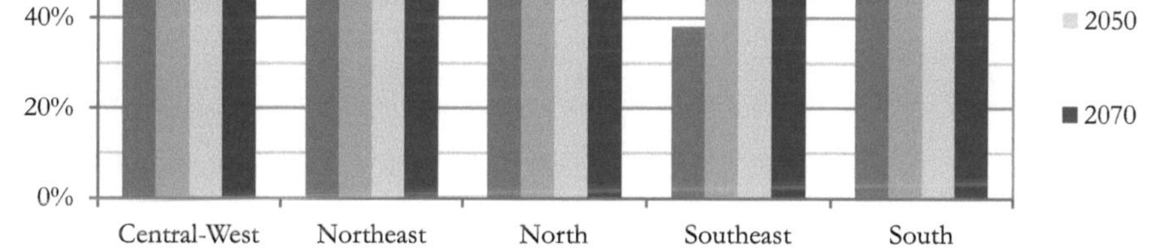

Fig. 3.19 Different production loss risk scenarios for coffee production in Brazil's regions (Adapted source: Pinto, Assad, et al., 2008)

Table 3.4 Percentage of area with coffee arabica in some states and possible changes with the increase in temperature

	Goiás (%)	Minas Gerais (%)	Paraná (%)	São Paulo (%)
Current	38.4	75.9	70.4	76.1
+ 3 °C	0.1	23.7	66.7	30.4
+ 5.8 °C	0.0	2.6	25.2	3.4

Source: Assad, Pinto, Junior, and Ávila (2004).

of municipalities have declared an abnormal situation due to dry weather/droughts in the 2004–2012 period.

Considering the incidence of nematodes (*Meloidogyne* incognita species) and the coffee leaf miner (*Leucoptera coffeella*), projections for 2020, 2050 and 2080 indicate that there could be an increase in the infestation of the nematode, as well as the pest, due to the higher number of generations per month, when compared to normal climate standards for 1961–1990 (Ghini, Hamada, Pedro Júnior, Marengo, & Gonçalves, 2008).

3.4.7 Beans

For the states of Goiás and Mato Grosso, with two water capacity estimates for the soil (30 and 50 mm) and three air temperature levels (without warming and with warming of 3 and 5.8 °C), it may be seen that in soil with low water capacity available (30 mm), negative impacts resulting from global warming are bigger. In soil with 50 mm water storage and an extra 5.8 °C in air temperature there might be a considerable reduction in areas suited to the growing of beans, particularly in the state of Goiás. However, in the state of Mato Grosso, negative impacts are smaller because rainfall quantity and distribution are better that in Goiás (Silva et al., 2014).

The production loss risk analysis, with probability higher than 20% for beans in Brazil's regions (Fig. 3.20), shows that in the North and Northeast, the risk of production loss increases, while in other regions it remains practically stable.

With a 4 °C increase, there will be a reduction in the low risk areas for the crop. As with other grains assessed by this study, the most significant loss will take place in the Northeast, particularly in the Agreste and south of the states of Maranhão and Piauí, as well as the west of Bahia. For the RCP 8.5 scenario, the situation for beans

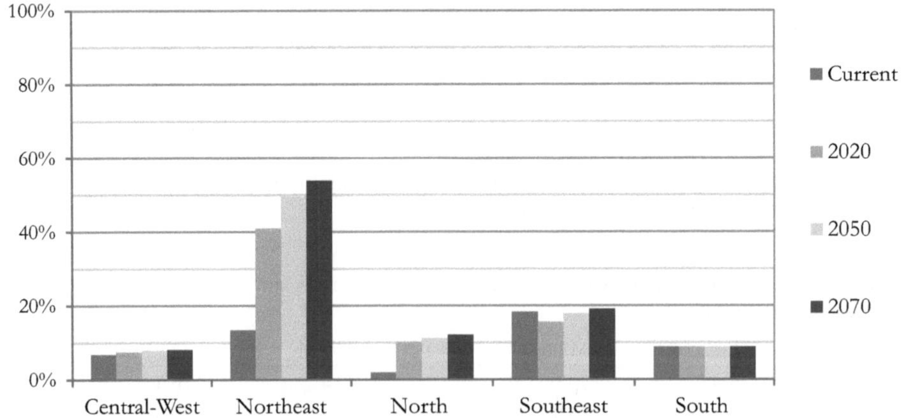

Fig. 3.20 Different production loss risk scenarios for beans production in Brazil's regions (Adapted source: Pinto, Assad, et al., 2008)

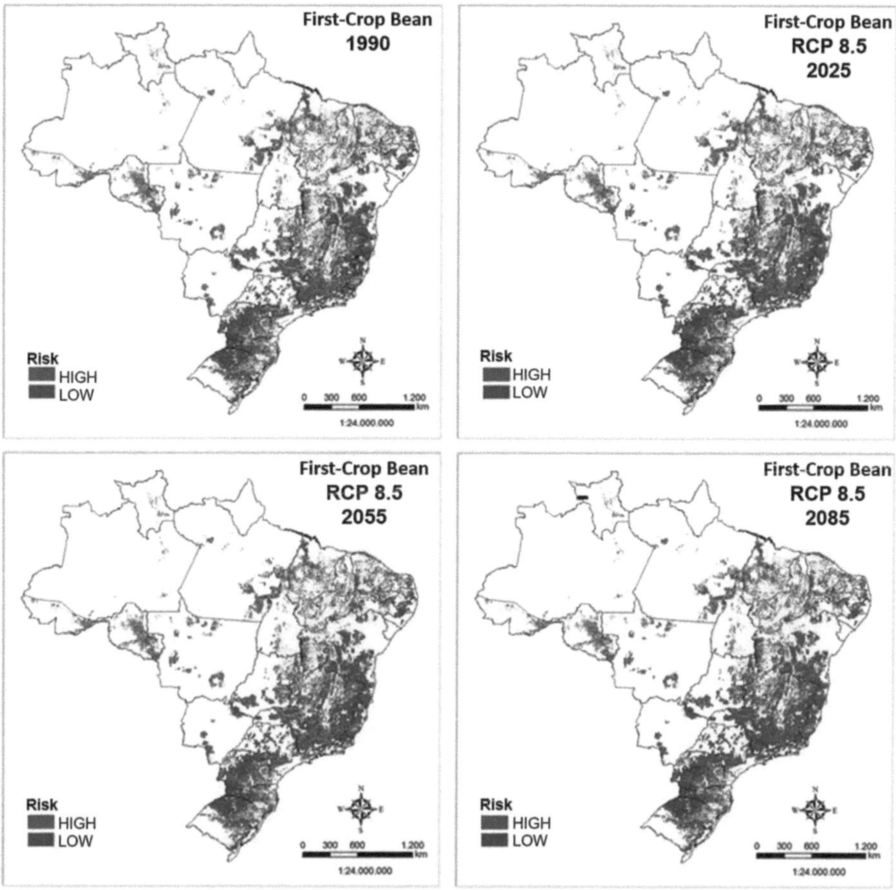

Fig. 3.21 Impact of the reduction of low risk areas for beans in comparison to 1990, in the RCP 8.5 scenario (Assad et al., 2016)

gets more critical, perhaps losing 57% of its low risk areas (Fig. 3.21). In national terms, the trend is to intensify production in the states of Paraná, Santa Catarina and Rio Grande do Sul, where there is a reduced water deficit and temperatures are lower. Part of the south of the state of Minas Gerais may be kept as a producing area.

3.4.8 *Sunflower*

Sunflower has growth potential and may supply for the demand of biofuels. In addition, it could be an important crop for family farming. The production loss risk analysis with probability higher than 20%, shows that the Northeast region goes from a current risk of 24% to 63% in 2070 (Fig. 3.22).

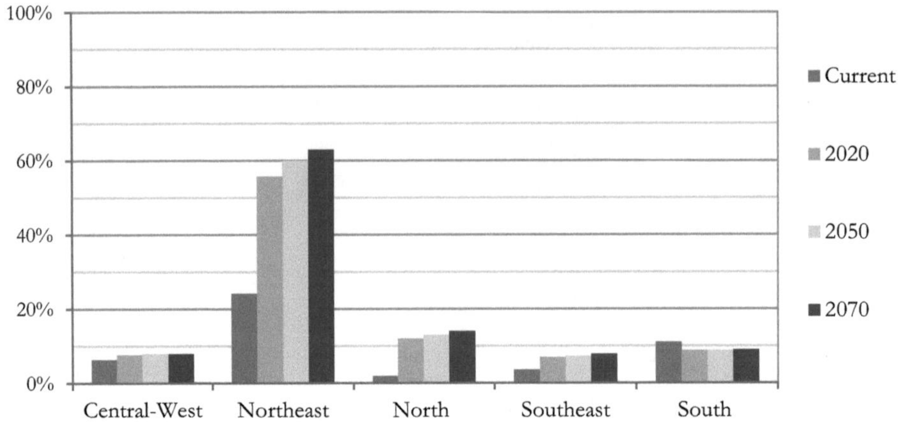

Fig. 3.22 Different production loss risk scenarios for sunflower production in Brazil's regions (Adapted source: Pinto, Assad, et al., 2008)

For 2020, a 14% fall is expected in low climate risk areas, which may increase to 16% by 2050, reaching 18% in 2070. The most affected regions will be the northeastern Agreste and Cerrado (Fgv-GVces, 2013).

3.4.9 Maize

The increase in temperature will lead to a reduction in low risk area for the growing of maize in Brazil. There are simulation studies that consider a 15% increase in rainfall and a rise in temperature between 3 and 5.8 °C in the state of São Paulo, for the climate risk zoning of maize. This is based on the monthly and annual average between 1961 and 1990, and indicates that the low risk area for clayey soil, taking into consideration an average temperature rise of up to 3 °C, will lead to a maximum reduction of 5% in the planting area. The 5.9 °C increase in temperature will reduce low risk areas, thus making the production of maize fall dramatically, regardless of the type of soil (Zullo Junior et al., 2006).

In a scenario with a 4°C increase, in 2050 a reduction in the incidence and severity of foliar diseases by fungus and bacteria should occur, while the incidence of stunting will grow. In the same conditions, new diseases may establish themselves by 2080. It is always important to point out that there will be favourable conditions to ear of corn disease and the production of burnt grains by fungus, hindering their quality (Pinto, Oliveira, & Fernande, 2008).

If the global temperature increases close to 5.8 °C, the low risk area for the planting of maize would be reduced to 5.9 million hectares (Assad et al., 2016). The production loss risk analysis with probability higher than 20% for maize in Brazil's regions, indicates that the Northeast would go from a current risk of 31% to 64% by 2070 (Fig. 3.23).

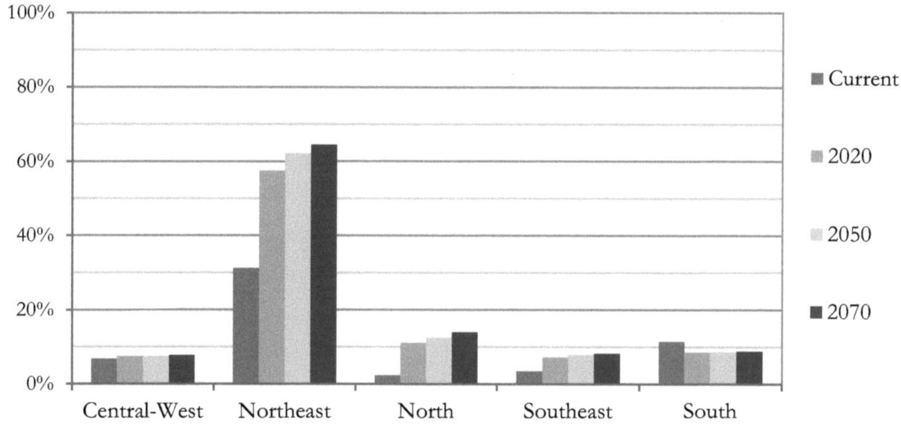

Fig. 3.23 Different production loss risk scenarios for maize production in Brazil's regions (Adapted source: Pinto, Assad, et al., 2008)

Maize will be one of the crops to suffer the most in the country in terms of production with the 4 °C increase. With this temperature rise, the expectation is for the degree/day quantity to be reached quickly, shortening the crop's cycle. The northeastern Agreste Brazilian region, today is responsible for most of the crop's production in the area, will incur a strong reduction in low risk areas for maize, as will the south of the states of Maranhão and Piauí, the west of Bahia and the country's Central-West. One of the main foodstuff for birds and bovine, maize will arrive in 2020 with a 12% reduction in its low risk area, figure that will rise to 15% in 2050 and 17% in 2070 (Fgv-GVces, 2013).

According to scenario A2 for the state of Minas Gerais, a 4°C temperature increase associated to a rise in CO2eq concentration will lead to up to 30% reduction in the production of maize in 2050 and 2080 in Brazil, when compared to the year 2000. A temperature increase close to 4 °C is expected in the western part of the Minas Gerais (Costa et al., 2009).

In a scenario with a 4 °C increase or more in temperature, a reduction in grain yield is seen in Santa Maria (state of Rio Grande do Sul) (Streck & Alberto, 2006).

3.4.10 *Second-Crop Corn*

Already a high climate risk crop. In global warming scenarios, risk increases substantially because of the increase in temperature and lack of water. In the South region, reduction in frost will provide favourable conditions to keep production levels. For the RCP 8.5 scenario (Fig. 3.24), there is a high temperature increase, leading to lack of water when second-crop corn needs it. Production restrictions are limited to almost all of the national territory.

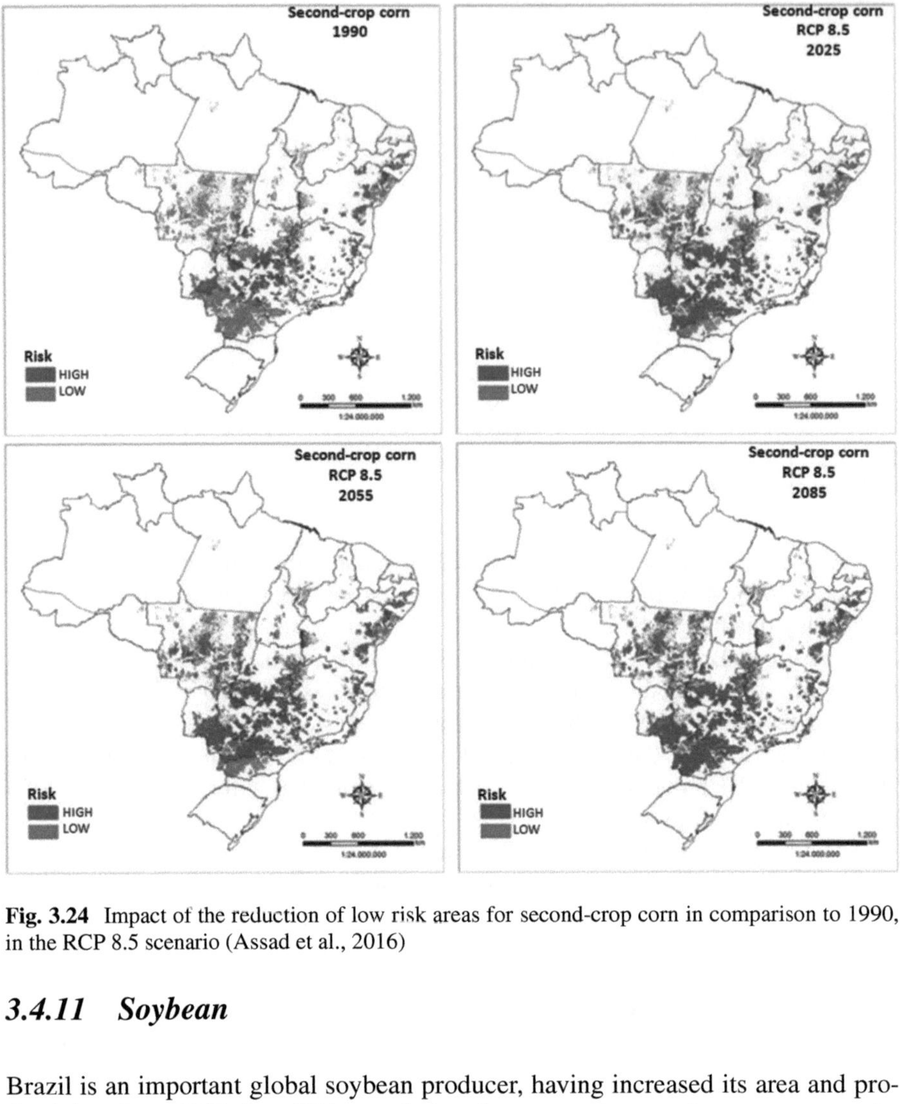

Fig. 3.24 Impact of the reduction of low risk areas for second-crop corn in comparison to 1990, in the RCP 8.5 scenario (Assad et al., 2016)

3.4.11 Soybean

Brazil is an important global soybean producer, having increased its area and productivity since 1961, in addition to being the country's most important crop. An analysis done with case studies between 1980 and 2011 showed that the main cause for yield difference in soybean was triggered by the lack of water (water deficit), particularly in Brazil's South region (Sentelhas et al., 2015).

Soybean adapts better to regions where temperatures vary between 20 and 30 °C, with its ideal development temperature standing at 30 °C. Regions with temperatures lower than 10 °C are not appropriate for growing soybean and those with temperatures over 40 °C, lead to impacts in flowering and reducing the pods' retention capacity (Farias, Nepomuceno, & Neumaier, 2007).

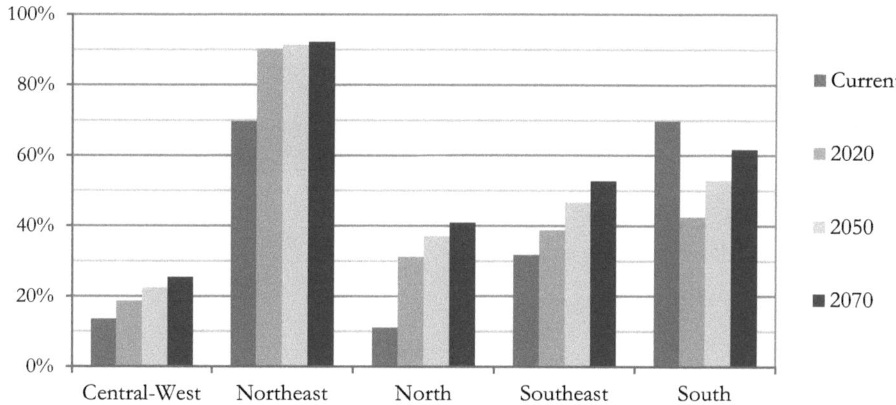

Fig. 3.25 Different production loss risk scenarios for soybean in Brazil's regions (Adapted source: Pinto, Assad, et al., 2008)

The production loss risk analysis, with probability higher than 20% for soybean in Brazil's regions, indicates that the Northeast would go from a current risk of 11% to 41% by 2070. The production loss risk is reduced in the South region, going from 70% in the current scenario to 62% in 2070 (Fig. 3.25).

Soybean will suffer with 4°C warming, if planting conditions remain as they are and no genetic modification is made. At least this should not be the case, as new high temperature resistant crops are already being studied. In 2070, the country's low risk area may drop to 60% in relation to the one currently available, due to the increase in water deficiency and possibly more intense Indian summers. The South region and the northeastern Cerrado will be the most affected areas (Fgv-GVces, 2013).

In the case of an increase of 5.8 °C in the global temperature, only 4.7 million hectares would be considered low risk for growing soybean, in other words, there would be an 81.2% reduction in such areas (Assad et al., 2016).

In relation to the increasing severity of soybean rust in Brazil's South region, in reference to normal climate standard records (1961–1990), projections indicate a rise in areas with 40% severity due to growing rainfall associated to 4°C warming (Chevarria, Del Ponte, Jahnke, & Hamada, 2012). Temperatures over 38 °C during the growing stages are harmful to the plants (Centurion & Ghini, 2011).

For soybean, yield may fall due to climate change by 2020 and even more so in 2050, although with different magnitudes throughout the region. Yield loss may be big in Brazil (over 30% in relation to the 1960–1990 baseline).

There are studies on the impact of climate change on the incidence of the soybean rust disease (caused by the fungus *Phakopsora pachyrhizi Sidow*), main soybean disease in Brazil's South region. Indeed, using the climate standard database for the period 1961–1990, as well as projections for 2020, 2050 and 2080, in a 3.4 °C warming scenario (A2 with temperature variation between 2.0 and 5.4 °C), the studies indicate that in the period with the highest occurrence of soybean rust

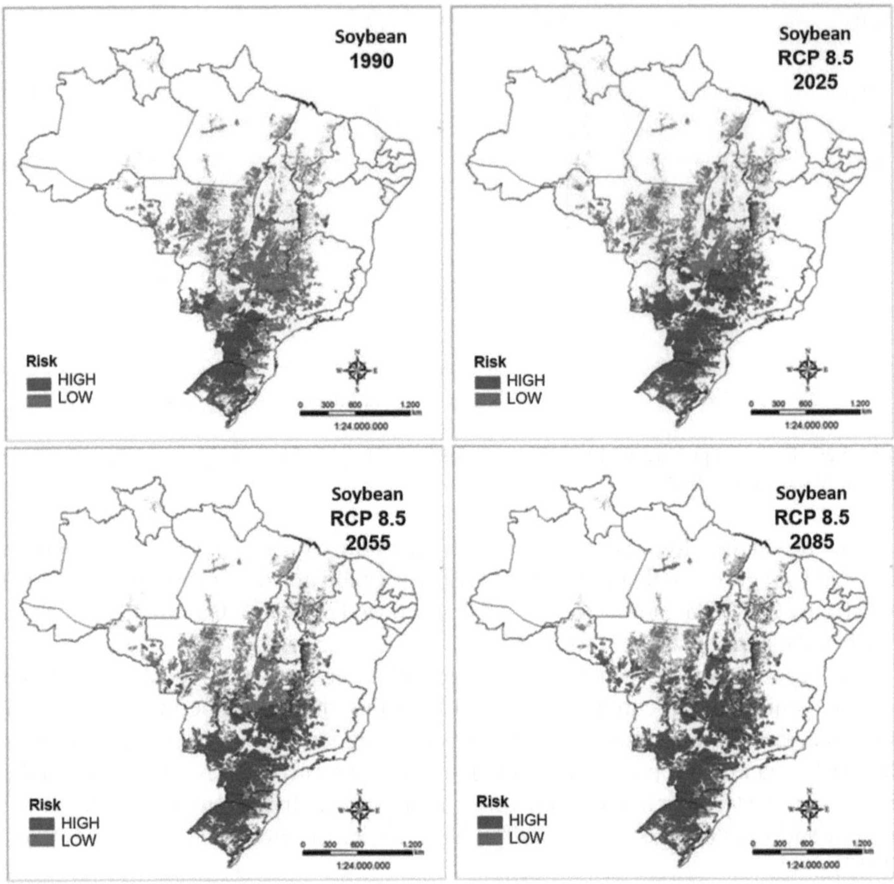

Fig. 3.26 Impact of the reduction of low risk areas for soybean in comparison to 1990, in the RCP 8.5 scenario (Assad et al., 2016)

(December to February) in the current scenario, the highest percentage of area in the region (from 70.19 to 81.74%), presented average severity between 40 and 60%. In future scenarios however, for the same period, models predict a small growth in the same area class with a variation of 71.11 to 84.01% in 2020, 74.33 to 83.55% in 2050 and 76.52 to 89.02% in 2080 (Chevarria et al., 2012).

For the RCP 8.5 scenario, the situation for soybean gets even more critical, perhaps losing 81% of its low risk areas (Fig. 3.26). In national terms, the tendency is to move to the north of the state of Mato Grosso, which is already happening. However, crops with high tolerance to drought and water deficit will be released into the market and might minimise the effects of global warming, as well as changes made to production systems, more balanced in relation to keeping the water in the soil and carbon sequestration.

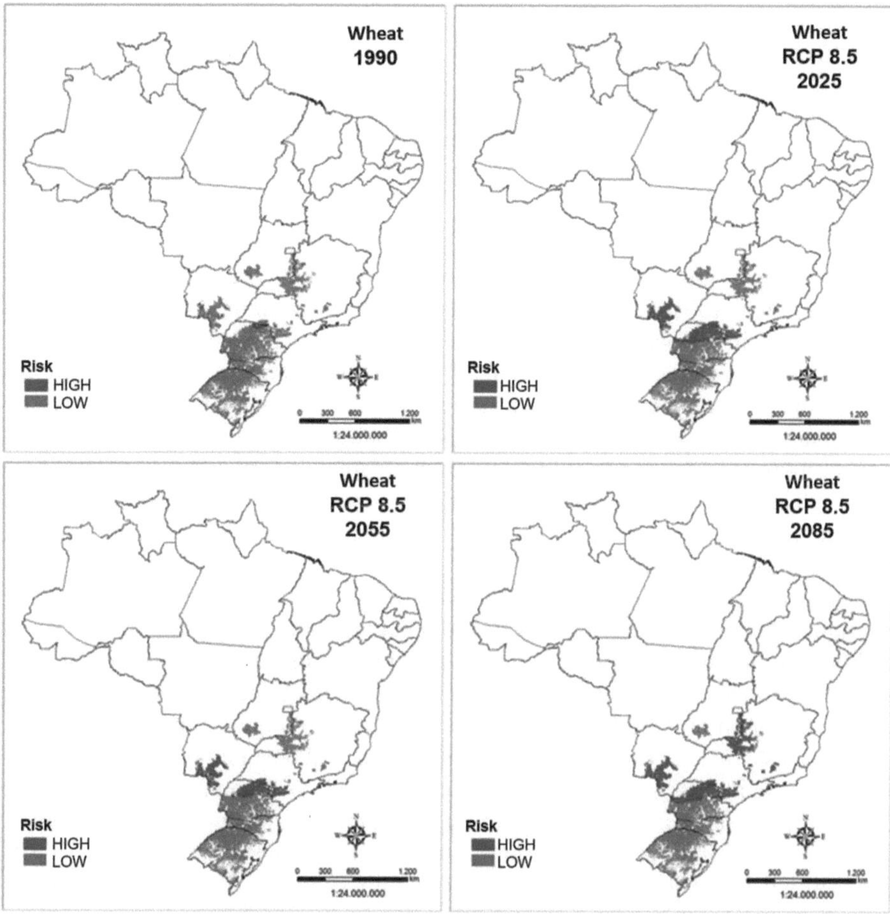

Fig. 3.27 Impact of the reduction of low risk areas for wheat in comparison to 1990, in the RCP 8.5 scenario (Assad et al., 2016)

3.4.12 *Wheat*

In relation to wheat, yield may be significantly affected by climate change, regardless of the emission scenario or general circulation model. Percentage yield falls are predicted to be more profound in parts of Brazil's Northeast (Fernandes et al., 2012).

Wheat is also very affected, perhaps losing up to 23.8% in scenario RCP 8.5 (Fig. 3.27). This may be explained based on warmer winters and the increase in night temperatures. Solutions are tolerant varieties, changing the crop's management cycle, introducing integrated systems.

In the city of Santa Maria, in the South of Brazil in the state of Rio Grande do Sul, increases of 5 and 6 °C have reduced wheat yield significantly, as well as the favourable effects of the rise in CO_2 with an up to 2 °C temperature rise and deleterious effects, resulting from an over 5 °C spike in temperature (Streck & Alberto, 2006).

3.5 Conclusions

Climate change impacts lead to the need to plan in order to mitigate effects of possible alterations to Brazil's climate risk zoning. Negative impacts resulting from a 3 °C increase in temperature outweigh the positive effects in agriculture.

Most positive impacts associated to a 4°C increase in temperature are linked to a minimum rise in temperature, which favours some crops, while negative effects are particularly connected to the rise in evapotranspiration rates. Implications with a 3 °C or more increase in agricultural production are still very difficult to determine due to changes related to extreme events, pests and diseases for instance. Even if adaptation measures are taken, they will imply in high economic costs and altering the climate risk zoning.

The potential risks imposed by climate change to Brazilian agriculture justify investments in farming research, particularly in the genetic modification of crops. In addition, studies to quantify the nature of extreme events, for periods between 2050 and 2011 have to be conducted. Technological advances in the management of crops and the adoption of good farming practices may also minimise impacts expected.

Lastly, as a final recommendation, the importance of developing crops tolerant to higher temperatures and water deficit is emphasised. With temperature increasing over 4 °C, impacts suffered will be strong at magnitudes not yet known by science. The main conclusion is that in terms of food supply, temperature rises above 4°C place Brazil in a very vulnerable situation, compromising its current role of the future's main player in relation to providing food to the world.

References

Araújo, P. H. C., Silva, F. F., Gomes, M. F. M., Féres, J. G., & Braga, M. J. (2014). Uma análise do impacto das mudanças climáticas na produtividade agrícola da região Nordeste do Brasil. *Rev. Econ. NE, Fortaleza, 45*(3), 46–57.

Assad, E. D., Marin, F. R., Valdivia, R. O., & Rosenzweig, C. (2015). AgMIP regional activities in a global framework: The Brazil experience. In a Global framework: The Brazil experience, AgMIP regional activities. In C. Rosenzweig & D. Hillel (Eds.), *Handbook of climate change and agroecosystems: The agricultural model intercomparison and improvement project (AgMIP) Integrated crop and economic assessments, Part 2, ICP Series on Climate Change Impacts, Adaptation and Mitigation* (Vol. 3, pp. 355–374). London: Imperial College Press. https://doi.org/10.1142/9781783265640_0023

Assad, E. D., Martins, S. M., de M Beltrão, N. E., & Pinto, H. S. (2013). Impacts of climate change on the agricultural zoning of climate risk for cotton cultivation in Brazil. *Pesquisa Agropecuária Brasileira, 48*(1), 1–8. https://doi.org/10.1590/S0100-204X2013000100001

Assad, E. D., Oliveira, A. F., Nakai, A. M., Pavão, E., J Monteiro, J. E., & Pellegrino, G. Q. (2016). Impactos e vulnerabilidades da agricultura brasileira as mudanças climáticas. In *Modelagem climática e vulnerabilidades setoriais à mudança clima no Brasil* (Vol. 4, pp. 127–188). Brasil: MCTIC.

Assad, E. D., Pinto, H. S., Junior, J. Z., & Ávila, A. M. H. (2004). Impacto das mudanças climáticas no zoneamento agroclimático do café no Brasil. *Pesquisa Agropecuária Brasileira, 39*(11), 1057–1064.

Assad, E. D., Pinto, H. S., Zullo Junior, J., Marin, F. R., & Pellegrino, G. Q. (2008). Mudanças climáticas e a produção de grãos no Brasil: avaliação dos possíveis impactos. *Plenarium, 5*(5), 96–117.

Avila, A. M. H. (2007). Uma Síntese do Quarto Relatório do IPCC. Revista Multiciência, Campinas, Edition n° 8, Mudanças Climáticas.

Carter, T. R. (2013). Agricultural impacts: Multi-model yield projections. *Nature Climate Change, 3*, 784–786.

Centurion, M. A. P. C., & Ghini, R. (2011). Impacto potencial das mudanças climáticas sobre as doenças e desenvolvimento da soja no Brasil. In R. Ghini, E. Hamada, & W. Bettiol (Eds.), *Impactos das mudanças climáticas sobre doenças de importantes culturas no Brasil.* Jaguariúna, Brazil: Embrapa Meio Ambiente.

Cervantes-Godoy, D., & Dewbre, J. (2010).Economic importance of agriculture for poverty reduction. OECD food, agriculture and fisheries working papers, n° 23, OECD Publishing. doi: https://doi.org/10.1787/5kmmv9s20944-en.

Chevarria, V. V., Del Ponte, E. M., Jahnke, S. M., & Hamada, E. (2012). Mapeamento da severidade da ferrugem asiática da soja na região sul do Brasil em cenários de clima atual e futuro. In *Workshop sobre Mudanças Climáticas e Problemas Fitossanitários, Jaguariúna. Anais.* Jaguariúna, Brazil: Embrapa Meio Ambiente http://www.alice.cnptia.embrapa.br/handle/doc/951451

Costa, L. C., Justino, F., Oliveira, L. J. C., Sediyama, G. C., Ferreira, W. P. M., & Lemos, C. F. (2009). Potential forcing of CO2, technology and climate changes in maize (Zea mays) and bean (Phaseolus vulgaris) yield in Southeast Brazil. *Environmental Research Letters, 4* 014013, 10.

Da Mattaa, F. M., Grandisb, A., Arenque, B. C., & Buckeridge, M. S. (2010). Impacts of climate changes on crop physiology and food quality. *Food Research International, 43*(7), 1814–1823 Climate Change and Food Science. http://www.sciencedirect.com/science/article/pii/S0963996909003421

Embrapa CNPTIA. (2015). Laboratório de Modelagem Ambiental. http://ainfo.cnptia.embrapa.br

Evangelista, B. A., Marin, F. R., & Junior, J. Z. (2009). Impacto das mudanças climáticas sobre a produção de cana-de-açúcar no Estado de Goiás. Presented at thc 16th Brazilian congress of Agrometeorology - 22–25 September 2009, Belo Horizonte, Minas Gerais.

FAO. 2016. Climate change and food security: Risks and responses., 88 pgs. www.fao.org/3/a-i5188e.pdf

Farias, J. R. B., Nepomuceno, A. L., & Neumaier, N. (2007). Ecofisiologia da soja. Circular Técnica, Embrapa 48, pp. 9. ISSN: 1516-7860. https://www.embrapa.br/soja/busca-de-publicacoes/-/publicacao/470308/ecofisiologia-da-soja

Fernandes, E. C. M., Soliman, A., Confalonieri, R., Donatelli, M., & Tubiello, F. (2012). *Climate change and agriculture in Latin America, 2020–2050: Projected impacts and response to adaptation strategies.* Washington, DC: World Bank https://openknowledge.worldbank.org/handle/10986/12582

Fgv-GVces. (2013). Diagnóstico preliminar das principais informações sobre projeções climáticas e socioeconômicas, impactos e vulnerabilidades disponíveis em trabalhos e projetos dos atores mapeados. FGV, Centro de Estudos de Sustentabilidade.

FIESP. (2017). Outlook Fiesp, projeções para o agronegócio Brasileiro 2027, Federação das industrias do Estado de São Paulo- SP-FIESP-2017, 86.p, ISBN:978-85-7201-029-0.

Ghini, R., Hamada, E., Pedro Júnior, M. J., Marengo, J. A., & Gonçalves, R. R. V. (2008). Risk analysis of climate change on coffee nematodes and leafminer in Brazil. *Pesquisa Agropecuária Brasileira, 43*(2), 187–194 Embrapa Meio Ambiente.

Hertel, W. T., Burke, M. B., & Lobell, D. B. (2010). The poverty implications of climate-induced crop yield changes by 2030. GTAP working paper no. 59.

Hertel, W. T., Verma, M., Ivanic, M., Magalhães, E., Ludena, C., & Rios, A. R. (2015). GTAP-POV: A framework for assessing the National Poverty Impacts of global economic and environmental change. GTAP technical paper no. 31.

Intergovernmental Panel Of Climate Change (IPCC). (2007). The physical science basis: summary for policymakers. Geneva: IPCC, 2007. 18p. Available at: <http://www.ipcc.ch/publications_and_data/ar4/wg1/en/spmsspm-projections-of.html>. Accessed on: 03 Sep 2015.

IPCC. (2001). In J. T. Houghton, Y. Ding, D. J. Griggs, M. Noguer, P. J. van der Linden, X. Dai, K. Maskell, & C. A. Johnson (Eds.), *Climate change 2001: The scientific basis. Contribution of working group I to the third assessment report of the intergovernmental panel on climate change* (p. 881). Cambridge, UK and New York: Cambridge University Press.

IPCC. (2014a). Climate change 2014: Synthesis report. *Contribution of working groups I, II and III to the fifth assessment report of the intergovernmental panel on climate change* [Core Writing Team, R.K. Pachauri and L.A. Meyer (eds.)]. IPCC, Geneva, Switzerland, 151.

IPCC. (2014b). Summary for policymakers. In C. B. Field, V. R. Barros, D. J. Dokken, K. J. Mach, M. D. Mastrandrea, T. E. Bilir, M. Chatterjee, K. L. Ebi, Y. O. Estrada, R. C. Genova, B. Girma, E. S. Kissel, A. N. Levy, S. MacCracken, P. R. Mastrandrea, & L. L. White (Eds.), *Climate change 2014: Impacts, adaptation, and vulnerability. Part A: Global and sectoral aspects. Contribution of working group II to the fifth assessment report of the intergovernmental panel on climate change* (pp. 1–32). Cambridge, UK and New York: Cambridge University Press.

Junior, J. Z., Pinto, H. S., & Assad, E. D. (2006). Impact assessment study of climate change on agricultural zoning. *Meteorol. Appl. (Supplement), 13*, 69–80. https://doi.org/10.1017/S135048270600257X

Krishnan, P., Swain, B., ChandraBhaskar, B., Nayak, S. K., & Dash, R. N. (2007). Impact of elevated CO_2 and temperature on Rice yield and methods of adaptation as evaluated by crop simulation studies. *Agriculture, Ecosystems and Environment, 122*, 233–242. https://doi.org/10.1016/j.agee.2007.01.019

Lopes, C. A., Silva, G. O., Cruz, E. M., Assad, E., & Pereira, A. S. (2011). Uma análise do efeito do aquecimento global na produção de batata no Brasil. *Horticultura Brasileira, 29*, 7–15.

Margulis, S., & Dubeux, C. B. S. (2010). *Economia da Mudança do Clima no Brasil: Custos e Oportunidades /editado; coordenador geral Jacques Marcovitch. São Paulo: IBEP Gráfica* (p. 82). https://www.scribd.com/fullscreen/34595160?access_key=key-o7i2sr431843t6irzyp

Monteiro, J. E. B. A. (2009). *Organizador. Agrometeorologia dos cultivos: O fator meteorológico na produção agrícola* (p. 530). Brasília D.F., Brazil: INMET ISSN 978-85-52817-00-7.

OECD. (2015). Air and GHG emissions (indicator). doi: https://doi.org/10.1787/93d10cf7-en. Accessed on: 04 Dec 2015.

Pinto, H. S., Assad, E. D., Junior, J. Z., Evangelista, S. R. M., Otavian, A. F., Ávlia, A. M. H., et al. (2008). *Aquecimento global e a nova geografia da produção agrícola no Brasil*. São Paulo, Brazil: Embrapa.

Pinto, N. F. J. A., Oliveira, E., & Fernande, F. T. (2008). Impacto potencial das mudanças climáticas sobre as doenças de milho no Brasil. In R. Ghini & E. Hamada (Eds.), *Mudanças climáticas: impacto sobre doenças de plantas no Brasil* (pp. 141–158). Brasília, DF, Brazil and Jaguariúna, Brazil: Embrapa Informação Tecnológica and Embrapa Meio Ambiente. isbn:978-85-7383-427-7.

Porter, J. R., Montesino, M., & Semenov, M. (2015). The risk of climate change for crop production. In D. King, D. Schrag, Z. Dadi, Q. Ye, & A. Ghosh (Eds.), *Climate change: A risk assessment*. London: Foreign & Commonwealth Office.

Porto de Carvalho, J. R., Assad, E. D., & Pinto, H. S. (2011). Kalman filter and correction of the temperatures estimated by PRECIS model. *Atmospheric Research, 102*, 218–226.

Prabhu, A. S., da Silva, S. C., & de Filippi, M. C. (2008). Impacto potencial das mudanças climáticas sobre as doenças de arroz no Brasil. In R. Ghini & E. Hamada (Eds.), *Mudanças climáticas: impacto sobre doenças de plantas no Brasil* (pp. 141–158). Brasília, DF, Brazil and Jaguariúna, Brazil: Embrapa Informação Tecnológica and Embrapa Meio Ambiente. isbn:978-85-7383-427-7.

Rocha, M. T. (2008). Muda o clima muda a agricultura. Revista Agrianual, Anuários Informa Economics FNP, pp. 11–13.

Sentelhas, P. C., Battisti, R., Câmara, J. R. B., Farias, A. C., Hampf, A. C., & Nendel, C. (2015). Climate change and agriculture research paper. The soybean yield gap in Brazil – Magnitude, causes and possible solutions for sustainable production. *Journal of Agricultural Science*, 1–18. Cambridge University Press 2015. https://doi.org/10.1017/S0021859615000313

Silva, C. S., Heinemann, A. B., & Stone, L. F. (2014). Risco Climático para o Feijoeiro Conforme Aumento da Temperatura do Ar nos Estados de Goiás e Mato Grosso. Comunicado Técnico Embrapa 221: SSN 1678-961X, http://www.infoteca.cnptia.embrapa.br/infoteca/bitstream/doc/1000526/1/Comtec221.pdf

Silva, V. P. R., Oliveira, S. D., Santos, C. A. C., & Silva, M. T. (2013). Risco climático da cana-de-açúcar cultivada na região Nordeste do Brasil. *Revista Brasileira de Engenharia Agrícola e Ambiental, 17*(2), 180–189.

Streck, N. A., & Alberto, C. M. (2006). Estudo numérico do impacto da mudança climática sobre o rendimento de trigo, soja e milho. *Pesquisa Agropecuária Brasileira, 41*(9), 1351–1359.

Walter, L. C., Streck, N. A., Rosa, T. H., & Krüger, C. A. M. B. (2010). Mudança climática e seus efeitos na cultura do arroz. Ciência Rural, Santa Maria, v.40, n° 11, p. 2411–2418. ISSN 0103-8478.

WEF. (2016). The global risks report 2016, 11th Edition. Published by the World Economic lForum within the framework of the global competitiveness and risks team. REF: 080116.

Zullo Jr., J., Pinto, H. S., Assad, E. D., & de Avila, A. M. H. (2011). Potential for growing Arabica coffee in the extreme south of Brazil in a warmer world. *Climatic Change, 1*, 1573–1480.

Chapter 4
A Review of the Health Sector Impacts of 4 °C or more Temperature Rise

Sandra de Souza Hacon, Beatriz Fátima Alves de Oliveira, and Ismael Silveira

4.1 Introduction

Global temperature rise is projected to exceed 2 °C over the next few decades. As a result, climate change related impacts would be expected over the world. Such impacts are likely to have deep implications for the population, affecting several economic sectors, water resources, biodiversity, agriculture, energy, transport, and human health (Giang, Dung, Bao Giang, Vinhc, & Rocklöv, 2014). Between 2030 and 2050, climate change is expected to account for approximately 250,000 additional deaths per year due to malnutrition, vector diseases, diarrhoea and cardiovascular diseases (WHO, 2015a).

Climate change can impact on human health in different ways, including direct and indirect impacts, as well as those that are socially generated in response to climate alterations. Direct impacts are those primarily associated with extreme events such as heat waves, droughts, and more intense rainfall regimes. Indirect impacts, on the other hand, derive from changes in a range of physical, ecological and social mechanisms and biogeochemical cycles, which may alter the distribution of vector-borne and water-borne diseases, as well as increase air pollutant emissions. Socially generated impacts are those associated with economic changes, which may result in occupational impacts, malnutrition, psychosocial problems and forced migration (Bennet et al., 2014; IPCC, 2014). Despite its clear and detailed classification, climate change impacts on human health are complex and can be maximised or

S. de Souza Hacon (✉) · B. F. A. de Oliveira
National Public Health School at the Oswaldo Cruz Foundation (FIOCRUZ), Rio de Janeiro, Brazil

I. Silveira
National Public Health School at the Oswaldo Cruz Foundation (FIOCRUZ), Rio de Janeiro, Brazil

Institute of Social Medicine at the University of the State of Rio de Janeiro, Rio de Janeiro, Brazil

C. A. Nobre et al. (eds.), *Climate Change Risks in Brazil*,
https://doi.org/10.1007/978-3-319-92881-4_4

minimised as a result of individual and collective determinants inherent to a given form of social organisation and its interrelations.

Over the past few decades, studies on the impacts of temperature rise on human health have focused on retrospective analyses, such as the impacts of heat waves on mortality rates in Europe in 2003, and in Russia in 2010 (Matsueda, 2011; Robine, Cheung, Le Roy, et al., 2008), and on studies aiming at assessing exposure-response ratios, and establishing a relative risk between temperature and several health outcomes, in particular overall mortality rates (Basagaña et al., 2011; D'Ippoliti et al., 2010; Gasparrini et al., 2015; Kaiser et al., 2007; Kinsgley, Eliot, Gold, Vanderslice, & Wellenius, 2015; Le Tertre et al., 2006). Moreover, many of these studies have tried to establish temperature thresholds where morbi-mortality outcomes start being observed, whether in extreme heat or extreme cold. Both relative risk values and temperature thresholds have been used to quantify climate change impacts through estimates of excess deaths or hospitalisations in different climate change scenarios (Hajat & Kosatky, 2014; Huynen & Martens, 2015; Kinsgley et al., 2015; Wu et al., 2014).

Potential health impacts associated with climate change have been, in general, estimated for a warming risk scenario up to 2 °C (IPCC, 2013). However, this scenario might be too optimistic. Depending on the emission policies adopted, climate models estimate global average temperature rises between 0.3 °C and 4.8 °C for the period 2081–2100 relative to the baseline period 1986–2005 (IPCC, 2013). In addition, given the unpredictability of environmental, social, demographic and geopolitical changes, some parts of the world (including some areas in Brazil) may experience a temperature rise from 4 °C to 7 °C. In such scenario, a significant part of the world might become uninhabitable, which would affect millions of people (IPCC, 2014). If Greenhouse Gases (GHG) emissions continue at current levels, the earth seems to be moving towards a global average warming above 3 °C, with a 20% chance of reaching 4 °C (World Bank, 2012).

With regards to the health sector, and public health in particular, the consequences of temperature rise (considering a 4 °C warming as the worst case scenario) are still hard to predict. This is not an easy task, as the relationship between climate and health is neither linear nor symmetrical, and it is influenced by a number of other factors. In most cases, the analyses require a major modelling effort, together with approaches that faithfully represent the heterogeneity of and links between several components involved in the health-disease process. Thus, the effects of climate change on the human population may be exacerbated or intensified depending on the relationship between social, economic, biological, ecological and physical systems.

For a warming above 4 °C, climate models show a trend of increasing frequency and/or intensity of extreme events (IPCC, 2013; IPCC, 2014), with consequent direct implications for human health, specially for the most vulnerable populations. Direct impacts on health could be related, in particular, to heat waves, prolonged droughts, extreme rainfall and coastal floods. The latter are considered high risk at a 1 °C warming.

In Brazil, several extreme events have been recorded in the past 15 years, such as El Niño in 1997–1998; droughts in the Brazilian Amazon in 2005 and 2010, in the semiarid areas of the Northeast from 2012 to 2015, and in the Southeast in 2014; and floods in the Amazon in 2009, 2013 and 2014, in the South in 2008, 2014 and 2015, and in the Northeast in 2010 and 2011. Between 2001 and 2015, it is estimated that approximately 40 million people may have been affected by some extreme event in Brazil, such as droughts, floods or extreme temperatures (EM-DAT, 2015). In addition to the loss of life, other serious public health problems may occur, including an increase in water-borne and vector-borne diseases, post-traumatic psychosocial problems, cardiovascular diseases, material losses, damages to local infrastructure, and social costs related to education and housing (Coelho & Massad, 2012; Gracie, Barcellos, Magalhães, Souza-Santos, & Barrocas, 2014; Gomes, Nobre, & Cruz, 2012; Guimarães et al., 2014; Oliveira et al., 2012). However, these impacts tend to be exacerbated not only by climate change, but also due to the health sector's inability to deal with emergency situations, extreme risks and social inequality (Barcellos et al., 2009).

Although the risk of extreme events is more evident, gradual temperature rise may also introduce long-term risks to human health. In general, the perception of such risks is not regarding as a real health problem, and therefore they might be ignored by the population and by health authorities. Among the impacts, changes in the distribution of vector-borne and water-borne diseases such as dengue, malaria and diarrhoea are the most evident. Changes in temperature, humidity and rainfall regime may also, for example, influence the transport of microorganisms, the emission of pollutants and the production of pollen, and consequently affect human health (Barcellos et al., 2009; Moreno, 2006).

Most studies that relate climate change and increase in pollutants have focused on the effects of ozone and particulate matter (Doherty et al., 2009; Jacob & Winner, 2009), and temperature is used as confounding in several studies about air pollution and its effects on human health (Gouveia & Fletcher, 2000; Gouveia, Freitas, Martins, & Marcilio, 2006; Ignotti et al., 2010). In Brazilian urban areas, air pollution effects may be maximised due to thermal inversion, while in the Brazilian Amazon such impacts derive from an increase in biomass burning during dry periods. In the Brazilian Amazon, the increased concentration of pollutants deriving from burning biomass and its effects on health have been reported among the most vulnerable groups, such as children and the elderly (Ignotti et al., 2010; Oliveira, Marinho, Costa Neto, & Kligerman, 2012; Smith, Aragão, Sabel, & Nakaya, 2014). As we move towards a 3 °C temperature rise, with a 4 °C warming in some regions in Brazil, intensified droughts tend to increase even further the risk of exposure to air pollutants from burning biomass.

When assessing climate-change-related health outcomes, very few studies discuss the human body's 'tolerance' or adaptability limits, i.e., the levels where the impacts of temperature rise on health become irreversible. Once these levels are reached, they might dramatically influence people's daily routine, safety, and quality of life and health (Bennet et al., 2014; IPCC, 2014; Patz, Frumkim, Holloway, Vimont, & Hainese, 2014). The challenge is exactly to estimate a threshold or a critical warming limit for Brazilian biomes where health impacts will become crucial. However, this is a highly limited and complex exercise due to people's individual biological susceptibility and vulnerability. Sherwood and Huber (2010) explored the human body's physiological tolerance in extreme external temperature conditions through wet bulb globe temperature (WBGT) estimates – a widely used indicator to assess heat stress, especially on workers' health. The authors estimated that a 7 °C global warming would create small areas where the dissipation of metabolic heat would become impossible, including some regions in Brazil, such as the Northeast and the Central-west.

This chapter aims at reviewing the literature on the health sector impacts of a warming up to 4 °C in Brazil. The following aspects have been considered for this review: 1) Publications reporting on the health impacts of temperature rise in Brazil, presenting the current situation and potential future scenarios; 2) Publications assessing whether climate change has already generated impacts (or not) on the health sector in Brazil over the past 50 years; 3) Publications on observed and projected climate change impacts on the health sector as a consequence of a 4 °C warming or more; and 4) Dissemination of useful information to manage and reduce climate risks associated with temperature rise, and their impacts on the health sector.

4.2 Material and Methods

4.2.1 Methodological Considerations

This integrated literature review has covered the main electronic databases, the Intergovernmental Panel on Climate Change (IPCC), the Brazilian Panel on Climate Change (PBMC), the World Bank and the World Health Organization (WHO). The review has included articles and reports with the following characteristics: (1) describing the current situation of climate-change-related health outcomes in Brazil; and (2) containing estimates and projections for future impacts on the health sector of a 4 °C warming or more in Brazil. The following health impacts have been included: (a) direct impacts: heat stress, mortality by extreme events such as heat waves, prolonged droughts and floods, and sea level rise; (b) indirect impacts: mortality and morbidity by all causes and specific causes (cardiovascular and respiratory); distribution of vector-borne diseases with emphasis on leishmaniasis, dengue and malaria; distribution of water-borne diseases with emphasis on diarrhoea; and c) socio-demographic and economic vulnerability. The factors considered for this review are described in Box 4.1.

Given the complexity of projecting future scenarios for the impacts of a temperature rise above 4 °C, and in the absence of studies considering that scenario for Brazil, we have decided to focus on studies and methodologies that could be replicated. These studies were considered 'key' to analyse the impacts on the health sector, as they provide methodologies and relative risk references that can be used to estimate the impacts on human health.

Box 4.1 Factors considered in this review *Descriptive studies referring to the current situation of health outcomes to characterise the baseline scenario*

- *Descriptive studies presenting qualitative risks for the health sector deriving from a temperature rise above 4 °C for Brazil*
- *Studies presenting projections of future impacts for a temperature rise above 4 °C for Brazil*
- *Estimated and projected impacts for Representative Concentration Pathway 8.5 (RCP 8.5) scenarios for Brazil*
- *Key studies that may be used as examples to estimate the effects of a temperature rise above 4 °C*

4.2.2 Data Sources

For this review, the following data sources have been used:

- Electronic databases: (a) PubMed – *National Library of Medicine, Bethesda, MD, USA*; (b) Bibliographic Database Scopus; (c) Web of Science; and (e) Scientific Electronic Library Online/SciELO, used to incorporate national publications. We have searched for descriptors in Portuguese, English and Spanish in the titles or abstracts of the studies. Several combinations of such descriptors and keywords have been used with the following terms: *'climate change'; 'temperature'; 'temperature wet-bulb'; 'Representative Concentration Pathways 8.5'; 'heat waves'; 'heat stress'; 'heat-mortality'; 'floods'; 'drought'; 'sea-level rise'; 'heat-morbidity'; 'cardiovascular disease'; 'transmissible disease'; 'respiratory disease'; 'malaria'; 'dengue'; 'diarrhoeal disease'; 'leptospirosis'; 'leishman*'; 'Brazil'*. The period covered for national literature was from 2000 to 2015, and for international literature, from 2007 to 2015;
- Intergovernmental Panel on Climate Change (IPCC): "*Central and South America: In: Climate Change 2014: Impacts, Adaptation, and Vulnerability. Part B: Regional Aspects; Contribution of Working Group II to the Fifth Assessment Report of the Intergovernmental Panel on Climate Change*";
- Brazilian Panel on Climate Change (PBMC): "First Report on Climate Change – Chapter 6: Human Health, Welfare and Safety";
- Third National Communication to the UNFCCC (TNC-UNFCCC): "Climate change vulnerability, risks and impacts on health in Brazil";
- World Bank Group: "*4° Turn Down the Heat: Confronting The New Climate Normal – Regional Results";*
- World Health Organization: "*Heatwaves and Health: Guidance on Warming-System Development – 2015*" and "*Climate and Health Country Profile – 2015*" report.

4.3 Analysis of Studies and Presentation of Results

The publications were reviewed in the following contexts: temperature variation (a temperature rise of 4 °C or more); setting temperature thresholds for certain health outcomes and exposure scenarios, considering IPCC's RCP 8.5 (Representative Concentration Pathways 8.5); type of health impact (direct or indirect); impacted population (children, adults, elderly); impacted regions or places.

The results have been compiled in tables presenting a summary of the main findings. For ease of understanding, the results are presented in the following order:

(a) Direct impacts: this section includes impacts related to heat stress, extreme events such as heat waves, and natural disasters with emphasis on prolonged droughtsand floods;
(b) Indirect impacts: this section includes impacts related to mortality by all causes for all age groups, as well as specific causes (cardiovascular and respiratory) for the elderly; distribution of vector-borne diseases with emphasis on leishmaniasis, dengue and malaria; distribution of water-borne and food-borne diseases, with emphasis on leptospirosis and diarrhoea.

For each health outcome, there is a brief description of the studies developed in Brazil, which aims at providing data to characterise the current scenario or a baseline period for health impacts. Subsequently, we present the main results of studies projecting impacts on the health sector in Brazil associated with a temperature rise above 4 °C, considering the RCP 8.5 scenario. In the absence of more refined projections for Brazil, we have decided to replicate some methodologies, as applicable. These methodologies are mentioned before their application.

4.4 Direct Impacts of Climate Change on the Health Sector as a Response to Exposure to a Warming of 4 °C or More

4.4.1 Heat Stress

High temperatures and air humidity, together with extenuating human physical labour and lack of acclimatisation, may lead to heat stress. This condition results from exposure to an environment or circumstance that blocks the body's natural ability to release internal heat and keep a constant body temperature, causing damages to health such as discomfort, fatigue, drowsiness, risk of accidents and cardiac decompensation (Carmargo & Furlan, 2011).

Many different measures have been used to assess heat stress, including the Heat Index (HI), Humidex, the Apparent Temperature, the Wet-Bulb Globe Temperature (WBGT), and more recently the Universal Thermal Climate Index (UTCI). In general, these methods combine the values of air temperature and relative humidity (WHO, 2015b), using empirical approximations of how a person would feel within a certain temperature and humidity range. Of the methods mentioned above, two stand out: WBGT, widely used in occupational climate conditions; and UTCI, developed by the International Society of Biometeorology (ISB) based on a human physiological model.

WBGT is an indicator that sets an absolute limit for metabolic heat transfer based on the laws of physics, instead of extrapolating empirical approximations (Sherwood, 2014). With this indicator, Sherwood and Huber (2010) identified human limits for adaptability to warming derived from climate change, setting 35 °C as the physical and thermodynamic limit for the human body to deal with excessive heat. Thus, WBGT values above 35 °C might cause mental fatigue, physical exhaustion, dehydration and cardiovascular compromise, leading to collapse and even death.

UTCI is a modern index based on the most recent human thermal regulation, biophysics and heat exchange models. This index provides a valid estimate of human physiological response to heat stress, ranging from extreme cold to extreme heat. UTCI values between 18 °C and 26 °C are considered thermal comfort; from 26 °C to 32 °C, moderate heat stress; and above 32 °C, strong heat stress (Bröde, Fiala, & Blazejczyk, 2011).

In Brazil, publications have explored both WBGT and UTCI (Nassis et al., 2015, Maia et al., 2015, Rossi, Krüger, & Bröde, 2012). Nassis, Brito, Dvorak, Chalabi, and Racinais (2015) used WBGT to assess the effects of heat stress in high performance sports during the 2014 World Cup. The results showed that 25% of the matches took place in circumstances considered as of high heat stress (between 28 °C and 30 °C). In at least 2 matches, the results pointed towards values between 30 °C and 32 °C, or above 32 °C, classified as very high heat stress, and unsuitable for intense physical activities. During these matches, there was a reduction in high intensity activities and in the distance run by the athletes. Concerning UTCI, in Curitiba, a capital in the South of Brazil, discomfort caused by heat was predicted for temperatures above 27 °C, indicating a need to adjust thermal comfort and discomfort ranges for this indicator, which proposes a threshold at 26 °C according to local climate conditions (Rossi, Krüger, & Bröde, 2012).

Using WBGT as reference, up to 2012 the maximum value at global level was 30 °C, reaching 31 °C in some regions, such as India and the Amazon. In Brazil, between 1999 and 2008, annual WBGT peaks in the afternoon ranged from 25 °C to 30 °C, and the highest temperatures were observed in the North region of the country (Sherwood & Huber, 2010) (Fig. 4.1). However, data from meteorological stations available online at NOAA's Global Summary since 1980 show that, among all Brazilian state capitals, maximum WBGT exceeded 31 °C for 46, 19 and 73 days in Rio Branco (AC), Teresina (PI) and Cuiabá (MT), respectively (Fig. 4.2).

At WBGT above 31 °C, for example, it is recommended that maximum exposure should not exceed 1 h for people in non-acclimatised environments and performing light work activities, with metabolic expenditure up to 233 W (ACGIH, 1996; ISO7243, 1982). In the cities mentioned, where high WBGT values have been reached, the situation may be aggravated in a global warming scenario above 4 °C, since according to regionalised Eta/HadGEM2-ES projections (Chou et al., 2014)

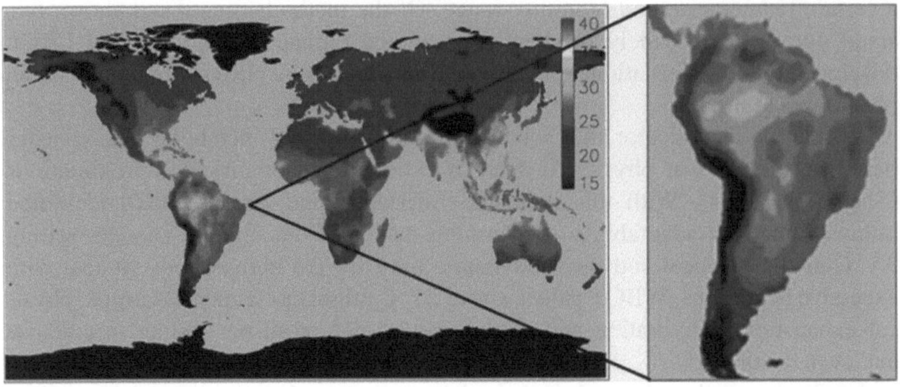

Fig. 4.1 Annual WBGT peaks in the afternoon (1999–2008) Published in Sherwood and Huber (2010) and made available in an improved version in 2012a (Data source: http://web.science.unsw.edu.au/~stevensherwood/wetbulb.html)

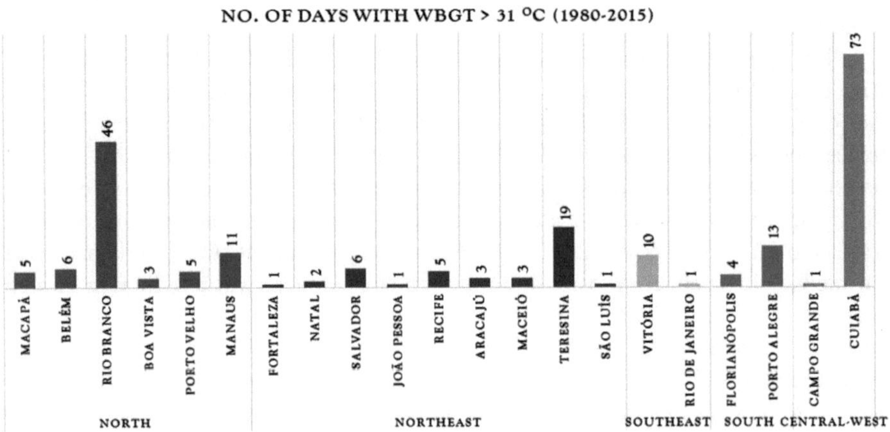

Fig. 4.2 Number of days when WBGT exceeded 31 °C since 1980 (Data source for the chart: NOAA's Global Summary. Data accessed via Hothaps-Soft (Otto et al., 2014))

and their municipalisation (Costa et al., 2015), average temperatures in these places under the RCP 8.5 scenario (2071–2099) may increase by 8 °C for Cuiabá and Rio Branco, and 5 °C for Teresina.

With regards to publications on heat stress, few studies have explored future scenarios for Brazil. Global average temperature rise projections for the RCP 8.5 scenario show that the north and most of the central-west regions may reach maximum WBGT values above 40 °C, with a 7.7 °C increase in global average temperature (Kjellstrom et al., 2015) (Table 4.1). According to Sherwood and Huber (2010), an increase around 10 °C in global average temperature might lead to values above 35 °C in most of the North and Central-west regions, as well as in the state of

Table 4.1 WBGT projections for Brazil (approximate) under RCP 8.5 according to different temperature rise values

	Global average temperature rise			
	No ↑	↑ 4.4°C	↑ 7.7°C	↑ 10°C
North	28 – 31°C	34 – 38°C	39 – >40°C	35 – 40°C
Northeast	24 – 30°C	32 – 35 °C	33 – >40°C	32.5 – 35°C
Central-west	27 – 31°C	32 – 37°C	37 – >40°C	35 – 40°C (MT and MS) 32.5 – 37.5°C (GO)
Southeast	24 – 30°C	25 – 33°C	35 – 40°C	32.5 – 37.5°C (ES, SP and MG). 35 – 40°C (RJ)
South	24 – 27°C	25 – 32°C	30 – 35°C	32.5 – 37.5°C
Authors	*Kjellstrom et al., 2015a*	*Kjellstrom et al., 2015a*	*Kjellstrom et al.,*	*Sherwood, Huber, 2010*

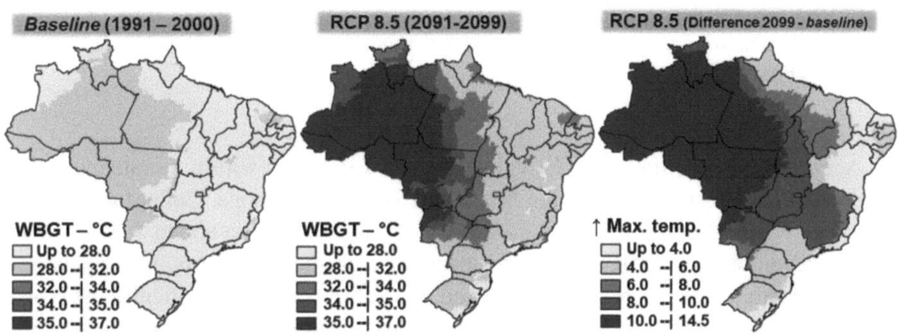

Fig. 4.3 Maximum Wet-Bulb Globe Temperature (WBGT) projections (warmest month) for 1990–2000 (baseline) and 2091–2099, and maximum temperature rise under the RCP 8.5 scenario for Brazil (Data source: Kjellstrom et al., 2016 (http://www.climatechip.org/). Map source: Oliveira, BFA)

Rio de Janeiro (Table 4.1). In terms of health, this extreme temperature rise means that, in these regions, human uninhabitability thresholds might be reached during part of the year, or all year long.

An analysis based on Hadley Global Environment Model (HadGEM-2ES) projections for the RCP 8.5 scenario from 2090 to 2099 show potential WBGT and UTCI values for Brazil, with their respective temperature rises (Figs. 4.3 and 4.4). In those places where maximum temperature rises range between 10 °C and 14 °C, maximum WBGT might exceed 35 °C, especially in the states of Acre, Amazonas and Mato Grosso. Thus, these estimates show that the threshold for human physiological adaptability to extreme heat may be exceeded for many hours, days or months in the year, since temperature projections have been toned down as a result of focusing on a 10-year average (Fig. 4.3).

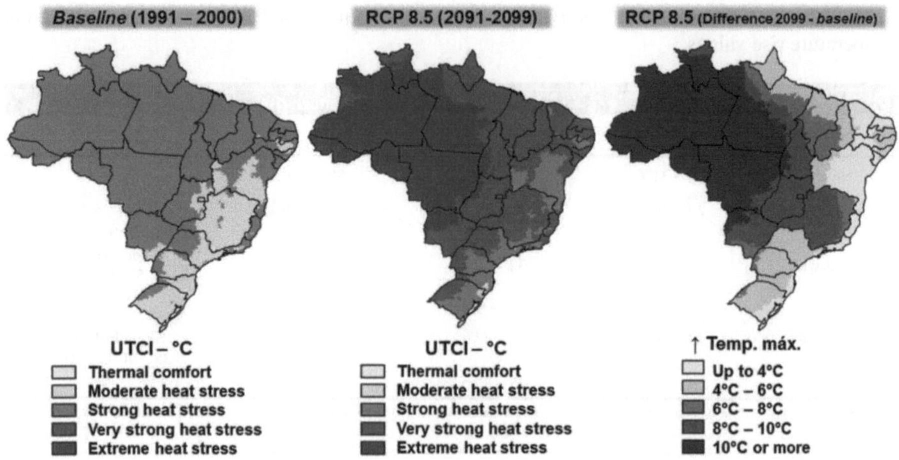

Fig. 4.4 Maximum Universal Thermal Climate Index (UTCI) projections (warmest month) for 1990–2000 (baseline) and 2091–2099, and maximum temperature rise under the RCP 8.5 scenario for Brazil (Data source: Kjellstrom et al., 2016 (http://www.climatechip.org/). Map source: Oliveira, BFA

According to maximum UTCI values, most of Brazil is within a scale classified as strong heat stress, except for the South and Southeast regions. Under the RCP 8.5 scenario, where the average temperature rise is predicted to be above 4 °C, most of the Brazilian territory will experience strong or extreme heat stress. In areas where the maximum temperature rise is predicted to be above 10 °C, specifically the North and Central-west regions, UTCI will exceed 46 °C, indicating extreme heat stress (Fig. 4.4).

Considering that the temperature rise for the end of the century in a pessimistic climate change scenario may lead to WBGT values above the adaptability limit for several periods, the distribution of this indicator and of UTCI for at least 1 capital in each of the country's main regions has been analysed (Fig. 4.5). For Porto Velho (RO) in the North, Cuiabá (MT) in the Central-west, and Teresina (PI) in the Northeast, where a maximum temperature rise between 8 °C and 5 °C has been observed, maximum WBGT values will average over 30 °C every month. These values will remain stable during all months of the year in Teresina, while peaks of approximately 35 °C will be observed in Cuiabá and Porto Velho in the month of October. For these places, UTCI values indicate intense heat stress during all months of the year, with extreme risks to human health.

An analysis based on HadGEM2 ES projections for WBGT and UTCI has been made for Brazilian capitals under the RCP 8.5 scenario (2071–2099) (Table 4.2). For the North region capitals, maximum WBGT values exceeded 32 °C, reaching 35 °C in Porto Velho (RO) and 34 °C in Rio Branco (AC) and Manaus (AM). In terms of temperature rise, it can exceed 10 °C in all three capitals above. With regards to exposure, estimates show that the total population in Rio Branco, Manaus

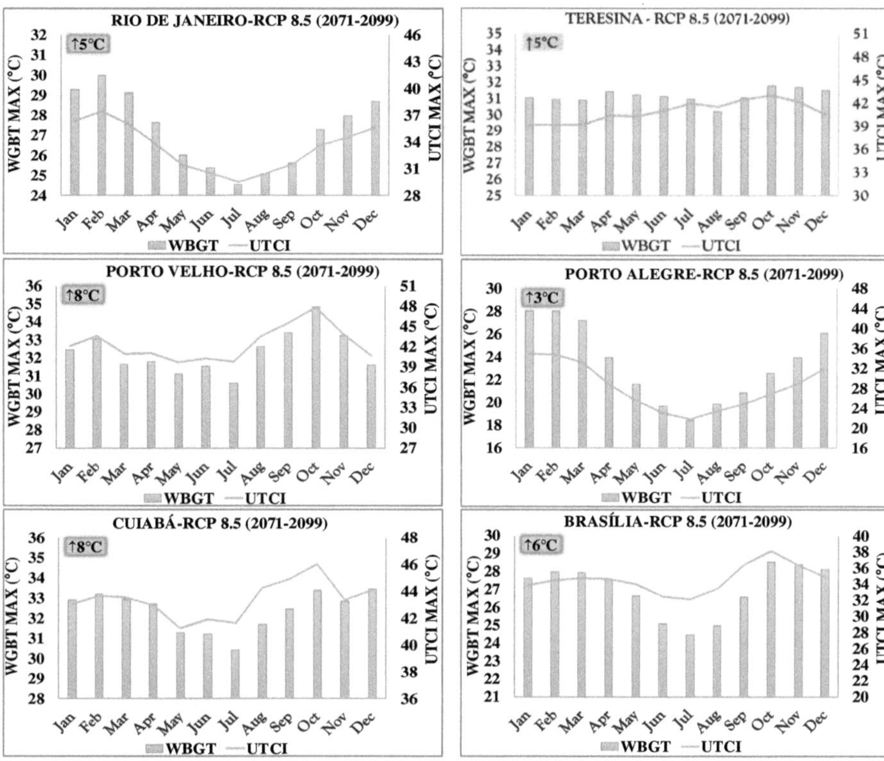

Fig. 4.5 Monthly projections for maximum Wet bulb globe temperature (WBGT) and Universal Thermal Climate Index (UTCI) for 2071–2099 under the Representative Concentration Pathway 8.5 (RCP 8.5) scenario for some Brazilian capitals (Data source for the charts: Kjellstrom et al., 2016 (http://www.climatechip.org/))

and Porto Velho will correspond, respectively, to 57, 49 and 24% of the state population. According to UTCI, these people will be exposed to extreme heat stress, totalling 4,890,572 inhabitants, 19% of whom (943,028) are children under the age of 14, and 7% (357,942) are elderly above the age of 60. These two groups are most vulnerable to extreme heat.

In addition to WBGT and UTCI, the apparent temperature has also been studied in Brazil, in particular in a case-crossover study in the city of São Paulo (Bell et al., 2008). The apparent temperature reflects individuals' physiological experience with regards to a combination of values including relative air humidity and temperature. In this study, a 4 °C apparent temperature rise on the same day of exposure was associated with a 6.2% increase (IC95% 2.9–9.0) in mortality by all causes, adjusted by the effects of particulate matter (PM_{10}). The results were 16% (IC95% 4.14–30.9) and 8.8% (IC95% 4.78–12.9) for deaths by respiratory diseases and among the elderly, respectively.

Table 4.2 Maximum WBGT and UTCI projections for hottest month from 2071 to 2099 under the RCP 8.5 scenario for Brazilian state capitals, with their respective populations

	Capitals	Population	% Exp. Population	ΔT (°C)	WBGT max	UTCI max
North	Macapá	990,362	56.7%	7	32.1	41.6
	Belém	2,597,322	20.1%	7	32.2	41.9
	Palmas	285,145	11.6%	9	32.1	44.3
	Rio Branco	661,060	56.7%	10	34.3	46.8
	Boa Vista	595,341	61.5%	11	33.3	46.0
	Porto Velho	651,927	23.9%	12	34.9	47.9
	Manaus	3,577,584	48.9%	13	34.4	48.0
North-east	Fortaleza	3,255,513	28.2%	4	30.9	38.5
	Natal	1,295,056	25.1%	4	31.1	38.8
	Salvador	3,157,848	18.1%	4	31.1	38.9
	João Pessoa	891,731	17.2%	4	30.9	38.7
	Recife	2,193,904	17.5%	4	30.3	37.7
	Aracajú	886,085	25.3%	4	30.9	38.7
	Maceió	1,096,429	27.5%	4	30.4	38.0
	Teresina	788,451	24.9%	6	31.8	43.0
	São Luís	1,294,474	15.0%	6	31.5	40.2
Southeast	Vitória	576,324	9.2%	3	30.9	39.1
	Rio de Janeiro	7,868,194	39.6%	3	30.0	37.5
	São Paulo	16,721,683	27.6%	4	27.2	33.4
	Belo Horizonte	3,189,014	12.3%	8	27.9	36.6
South	Florianópolis	764,778	6.3%	3	26.8	32.2
	Porto Alegre	730,663	6.1%	3	28.1	35.1
	Curitiba	2,399,481	16.4%	4	25.9	31.6
Central-west	Goiânia	2,460,373	21.5%	8	28.9	40.4
	Campo Grande	1,314,655	31.2%	10	31.8	42.9
	Cuiabá	1,000,087	18.9%	10	33.4	46.0
DF	Distrito Federal	6,545,453	97.3%	8	28.6	38.2

Data source for the table: Kjellstrom, T. et al., 2016 (http://www.climatechip.org/). Projected exposed population available at IBGE.

WGBTmax. (°C)
- >28.6
- 28.6 - 29.3
- 29.3 - 30.6
- 30.6 - 31.8
- 31.8 - 34.0
- 34.0 - 35.0

UTCI max. (°C)
- Moderate heat stress
- Strong heat stress
- Very strong heat stress
- Extreme heat stress

Concerning the heat load of people exposed at work, the average temperature rise in Brazil (around 4.5 °C) reduces productivity in several areas of the economy. HadGEM WBGT projections show that approximately 14% of annual daily working hours will be lost by Brazilian workers that perform heavy duties (e.g., in agriculture, construction and industry) (Kjellstrom et al., 2015) (Fig. 4.6). Under certain conditions, workers' health risks can be more extreme, as demonstrated by Bitencourt, Ruas, and Maia (2012). These authors analysed the atmospheric conditions on the day of death of 14 sugar cane cutters in the state of São Paulo and concluded that those deaths were associated with high air temperatures. In 50% of the cases, the temperatures recorded on the day of death exceeded 27 °C. This situation may be aggravated if we consider the expansion of sugar cane to other regions, such as the Central-west, where an 8 °C average temperature rise is predicted for the end of the century.

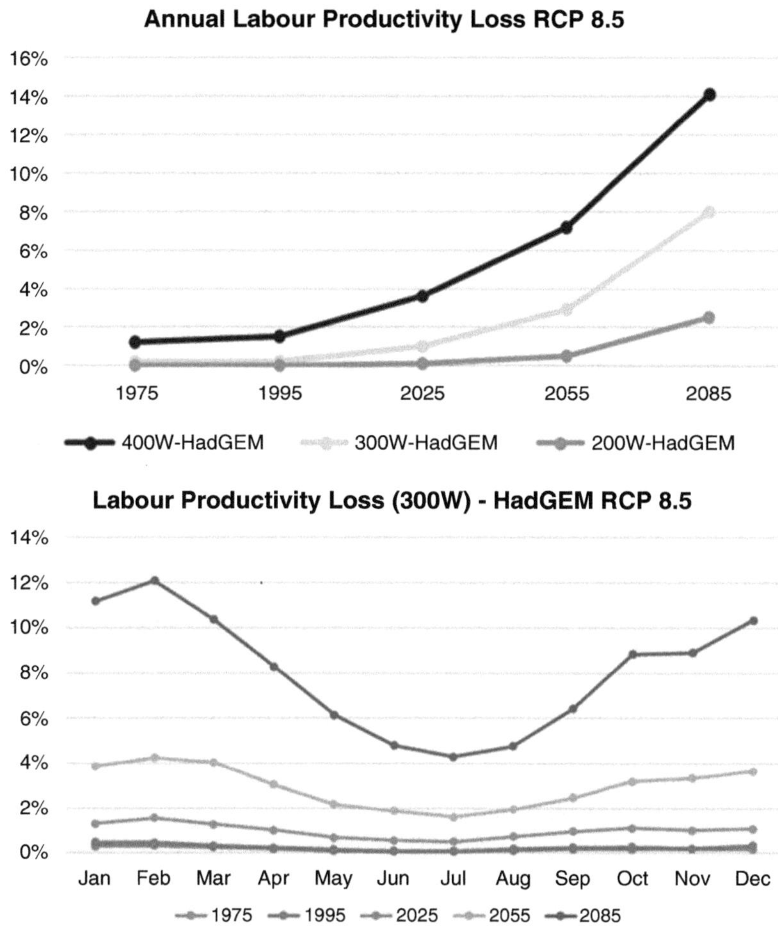

Fig. 4.6 Annual and monthly productivity loss for Brazil under the RCP 8.5 scenario between 1975 and 2085 (Data source for chart: Kjellstrom et al., 2015. Analysis for 'Climate and Health Country Profile – 2015' report)

4.5 Extreme Events

4.5.1 Mortality by Heat Waves

Human beings have an ideal temperature range for their survival. However, thanks to our ability to adapt to our environment, we can function adequately within a relatively wide temperature range. Nevertheless, recent sharp temperature variations have indicated that extreme temperature is associated with an increase in mortality. Among extreme temperature climate events, heat waves may impact heavily on human systems, causing the death of thousands, such as the 70,000 deaths in Europe in 2003 and the 15,000 deaths in Russia in 2010 (Matsueda, 2011; Robine et al., 2008). This is more evident among the vulnerable groups, as the elderly (Anderson & Bell, 2009; Son, Gouveia, Bravo, de Freitas, & Bell, 2016).

There is no universally acceptable definition for heat waves (Perkins & Alexander, 2013; Robinson, 2001), but these events are understood as periods of time with high temperatures (with or without a subtle start, and with an end), lasting for at least 3 days and having a perceptible impact on human activities (WHO, 2015b). An index that has been frequently used is the Warm Spell Duration Index (WSDI), which defines heat waves as the number of days per year when maximum temperatures exceed the 90th percentile, lasting for 5 or more consecutive days (Alexander & Arblaster, 2009; Costa et al., 2015; Meehl & Tebaldi, 2004; Sillmann, Kharin, Zwiers, Zhang, & Bronaugh, 2013; Zhang et al., 2011).

Regardless of which definition is adopted, heat waves can present different meteorological conditions, as they are based on local climate characteristics. Thus, some climate conditions may constitute a heat wave in some areas, but not in others. In addition, the spatial range, the length and the intensity of these events may vary considerably among different regions (Stefanon, D'Andrea, & Drobinski, 2012). Another important aspect of these events is the fact that they tend to be more exacerbated or accentuated in urban areas or mega-cities due to the formation of local heat islands. During this phenomenon, the solar heat trapped in the urban environment during the day is slowly released during the night, resulting in higher night temperatures.

A summary of the main studies on heat waves or exposure to extreme temperatures in Brazil is presented in Table 4.3. Son et al. (2016) have developed a study with the aim of exploring overall mortality deriving from the exposure to cold and warm temperatures, as well as during heat waves, over 14.5 years (1996–2010). The authors defined heat waves as average temperatures above the 96th percentile (26.5 °C) for more than 2 consecutive days. According to the results, the percentage increase in the risk of death during heat waves was 5.8 percent (IC95% 2,3. – 9.3) for all causes, 7.8% (IC95% 3.2–12.7) for cardiovascular diseases, and 10.8% (IC95% 2.8–19.3) for respiratory diseases. With a warming of 4 °C or more, heat waves are predicted to become longer and more intense, with a consequent increase in the risk of death caused by these events.

Guo et al. (2014) and Gasparrini et al. (2015) developed studies on the exposure to high temperatures in 18 Brazilian capitals, including the Federal District (*Distrito Federal*). The states not included in the study were Rio de Janeiro, Acre, Amapá, Roraima, Rondônia, Mato Grosso do Sul, Santa Catarina, Sergipe and Tocantins. In both studies, heat waves were not explicitly explored. The focus was on exposure to extreme temperatures (see definition in Table 4.3). Guo et al. (2014) compared the number of deaths by all causes, according to the average temperature variation in the 99th percentile compared with the 90th percentile, observing an 11% increase in mortality rates in Teresina, 8% in São Paulo and Belo Horizonte, and 7% in Curitiba.

Despite regional differences, in a recent report using results from Gasparrini et al. (2015), it was observed that mortality rates increase exponentially in all cities studied, including 18 Brazilian cities, starting at average temperatures above 30 °C,

Table 4.3 Publications on the effects of exposure to extreme temperatures in Brazil

Authors	Place	Perid	Outcome	Threshold/ΔT (°C)	%RR (IC95%)	Exposure/ Definition
Guo et al., 2014	Belém	1999–2011	Overall mortality rate	28.0–28.7 (ΔT 0.7)	0.94(0.88–1.01)	Exposure to temperatures above the 99th percentile compared with the 90th percentile
	Brasília	+	+	23.8–25.9 (ΔT 2.1)	1.04 (0.98–1.11)	
	Belo Horizonte	+	+	25.2–27.3 (ΔT 2.1)	**1.08 (1.03–1.13)**	
	Cuiabá	+	+	29.3–31.3 (ΔT 2.0)	1.02 (0.89–1.17)	
	Curitiba	+	+	22.5–24.7 (ΔT 2.2)	**1.07 (1.01–1.13)**	
	Fortaleza	+	+	28.1–28.8 (ΔT 0.7)	0.98 (0.93–1.03)	
	Goiânia	+	+	27.0–29.2 (ΔT 2.2)	**1.07 (1.02–1.13)**	
	João Pessoa	+	+	28.5–29.4 (ΔT 0.9)	1.07 (0.98–1.16)	
	Maceió	+	+	27.0–27.9 (ΔT 0.9)	1.03 (0.94–1.12)	
	Manaus	+	+	29.2–31.0 (ΔT 1.8)	1.06 (0.99–1.14)	
	Natal	+	+	25.6–28.4 (ΔT 2.8)	1.02 (0.95–1.09)	
	Porto Alegre	+	+	25.6–28.4 (ΔT 2.8)	**1.06 (1.02–1.11)**	
	Recife	+	+	27.7–28.7 (ΔT 1.0)	0.97 (0.93–1.01)	
	Salvador	+	+	27.6–28.6 (ΔT 1.0)	0.97 (0.93–1.02)	
	São Luís	+	+	28.0–28.7 (ΔT 0.7)	1.01 (0.95–1.08)	
	São Paulo	+	+	24.5–26.9 (ΔT 2.4)	**1.08 (1.05–1.12)**	
	Teresina	+	+	29.8–31.0 (ΔT 1.2)	**1.11 (1.01–1.21)**	
	Vitória	+	+	27.9–29.2 (ΔT 1.3)	1.06 (0.97–1.16)	

(continued)

Table 4.3 (continued)

Authors	Place	Perid	Outcome	Threshold/ΔT (°C)	%RR (IC95%)	Exposure/ Definition
Son et al., 2016	São Paulo	1996–2010	Overall mortality rate	26.5 (25.7–27.1)	2.2 (−0.7–5.5)* 7.4 (−1.5 – 16.9)**	2 or more consecutive days of exposure to average temperatures above the 96th percentile
		+	Cardiovascular	+	1.6 (−2.3–5.6)* 1.9 (−9.7 – 15.0)**	
		+	Respiratory	+	4.5 (−1.8–11.2)* 5.5 (−14.1–29.6)**	
Gasparrini et al., 2015	18 Brazilian cities (see Guo et al., 2014)	1999–2011	Overall mortality rate	Not specified in the article and supplementary material	Attributable fraction (presented in a chart and for Brazil)	Exposure indicator: exposure to extreme temperatures. Temperature values above the 97.5th percentile

[a]Considering a 1-day increase in the heat wave
[b]Considering a 1 °C increase in temperature
ΔT between average temperature in the 99th percentile vs. average temperature in the 90th percentile
Data source: development from the data publications on the impacts of temperature rise on mortality rates (Gasparrini et al., 2015; Guo et al., 2014; Son et al., 2016). Table source: Oliveira, BFA

and that values between 20 °C and 29 °C can be considered within a safety zone both for countries in temperate areas and for those with a warm weather (USB, 2016). According to this assessment, municipalities in the North and Central-west regions, as well as in the Northeast states of Maranhão and Piauí, might be in a high-risk zone for mortality caused by temperature rise under the RCP 8.5 scenario for 2071–2099. In addition, these places might present average temperature rises above 4 °C. Municipalities in the Central-west region, in particular, may see a 6 °C increase or more at the end of the century (Fig. 4.7).

With regards to age groups, extreme heat events had, and will probably continue to have, greater impact on the elderly. In São Paulo, a 9% increase was observed (IC95% 3.3–14.1) in mortality rates among the elderly as a result of heat waves (Son et al., 2016). In addition to the elderly, children are also considered biologically vulnerable to climate conditions, as they present higher risk of dehydration, and consequently diarrhoea. Considering the high-risk zone for heat related mortal-

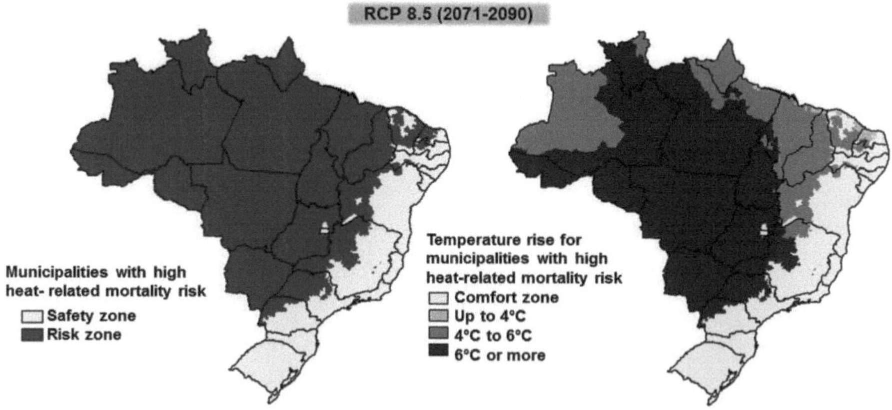

Fig. 4.7 Projections of municipalities with average temperatures in high mortality risk zones (temp >30 °C) under the RCP 8.5 scenario for 2071–2099 (Data source: USB, 2016 and regionalised Eta/HadGEM2-ES (Chou et al., 2014) with municipalised temperature projections (Costa et al., 2015). Map source: Oliveira, BFA

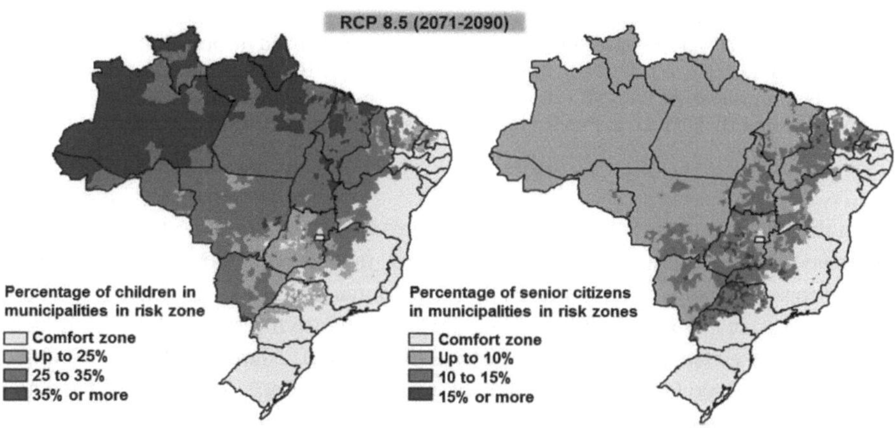

Fig. 4.8 Percentage of elderly and children in high mortality risk zone municipalities (temp >30 °C) under the RCP 8.5 scenario for 2071–2099 (Data source: USB, 2016 and regionalised Eta/HadGEM2-ES (Chou et al., 2014) with municipalised temperature projections (Costa et al., 2015). Map source: Oliveira, BFA

ity, the percentage of children exposed to this condition would range from 25 to 35% in most of the Central-west and North regions, and exceed 35% in some municipalities in the states of Acre, Amazonas and Pará (in a warming scenario of 4 °C or more). The percentage of elderly people in high mortality risk zones was 10%, except for some municipalities in the states of Mato Grosso, Mato Grosso do Sul, Goiás, Tocantins, Piauí and Maranhão, where results ranged from 10 to 15 percent (Fig. 4.8).

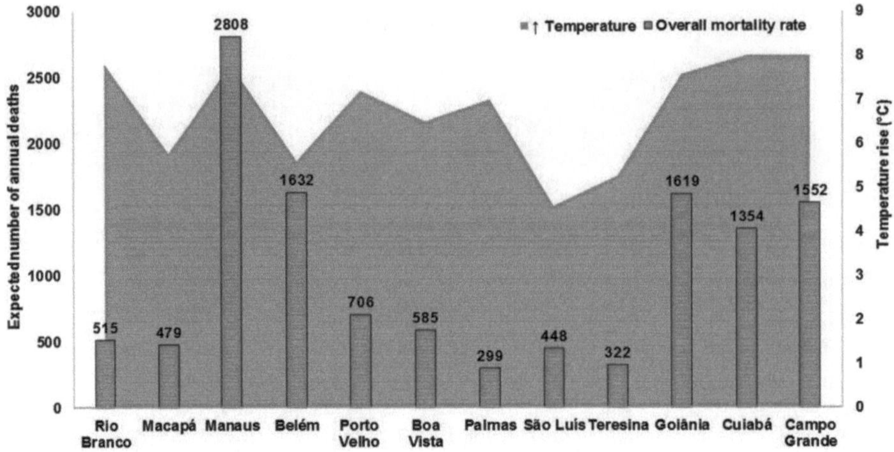

Fig. 4.9 Estimates for average expected deaths by all causes in capitals with average temperature above 30 °C under the RCP 8.5 scenario for 2071–2099 (*Estimates produced through the equation "EM = β x Standardised mortality rate x ΔT x P" (Ostro, 2004), where β refers to the variation in the number of deaths for every 1 °C increase in average temperature; the mortality rate was standardised for all ages and causes, excluding external causes in the baseline period (average 2000-–2010); ΔT average temperature difference in 2071-–2099 relative to the baseline period (1960-–1990), RCP 8.5 scenario with a 30 °C threshold. O corresponds to projected exposed population. The value of β was taken from the study by Son et al. (2016). Data source: the chart uses the temperature threshold set by USB (2016) based on the results of Gasparrini et al. (2015). Figure source: Oliveira, BFA. Ph.D. in Public Health and the environment, ENSP/FIOCRUZ)

In the health sector, the impacts of exposure to extreme temperatures have been estimated through quantitative methodologies, in particular due to excess deaths during these events (Anderson & Bell, 2009; Matsueda, 2011; Robine et al., 2008; Son et al., 2016). If we consider an average temperature threshold of 30 °C, as proposed recently (Gasparrini et al., 2015; USB, 2016), Manaus, Belém, Goiânia, Cuiabá and Campo Grande might present, respectively, 2808, 1632, 1619, 1354 and 1552 annual deaths for the RCP 8.5 scenario in 2070–2099 (Fig. 4.9). For these places, the estimated average temperature rise was around 8 °C, except for Goiânia, which presented a rise around 5 °C.

Considering specific causes for heat related mortality, the annual number of expected deaths by respiratory and cardiovascular diseases in high risk zones may be, respectively, 205 and 163 for Manaus, and 156 and 136 for Campo Grande, both with a predicted average temperature rise of 8 °C. It is worth noting that in a scenario with a warming above 4 °C, deaths by respiratory and cardiovascular causes may rise due to an expected increase in air pollutant concentrations, especially particulate matter and ozone (Fig. 4.10) in urban areas.

The estimated number of deaths was based on data recently published for Brazil, and some weaknesses must be mentioned: 1). When assuming the relative risk cal-

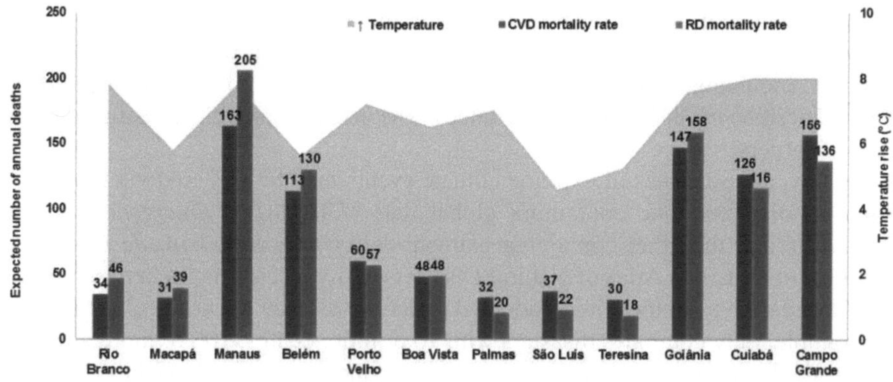

Fig. 4.10 Estimates for average expected deaths by all causes for capitals with an average temperature above 30 °C under the RCP 8.5 scenario for 2071–2099 (*Estimates produced through the equation ''EM = β x Standardised mortality rate x ΔT x P'' (Ostro, 2004), where β refers to the variation in the number of deaths for every 1 °C increase in average temperature; the mortality rate was standardised for all ages and causes, excluding external causes in the baseline period (average 2000--2010); ΔT average temperature difference in 2071--2099 relative to the baseline period (1960--1990), RCP 8.5 scenario with a 30 °C threshold. P corresponds to projected exposed population. The β values were taken from the study by Son et al. (2016). RD: Respiratory diseases, and CVD: Cardiovascular diseases. Data source: the chart uses the temperature threshold set by USB (2016) based on the results of Gasparrini et al. (2015). Figure source: Oliveira, BFA. Ph.D. in Public Health and the environment, ENSP/FIOCRUZ)

culated for the state of São Paulo and the 30 °C threshold, the estimates may be underestimated or overestimated; 2). The cities in high risk zones for heat related mortality already present, in general, high temperatures all over the year, and therefore the population are used to consecutive days of temperature rise, thus reducing the risk of deaths in these places. 3). The capitals have different socio-economic, demographic, health care and environmental characteristics that may influence the concentration-response function for each of them. 4). Potential adaptation processes have not been contemplated, as several heat wave events in the same season may lead the population to adopt practices that are able to minimise the effects of such events.

4.5.2 Natural Disasters and their Impacts on Human Health

Extreme climate events are particularly relevant for society and for ecosystems due to the gravity of their impacts, especially on the health sector (IPCC, 2013). In general, health impacts result from a combination of at least four factors, including the occurrence of a natural threat, the exposed population, socio-environmental vulnerability conditions and the scarcity or lack of capacity or measures to reduce potential health risks and damages (Freitas, Carvalho, Ximenes, Arraes, & Gomes, 2012;

Narváez, Lavell, & Ortega, 2009). For an extreme temperature rise scenario, with a consequent increase in natural disasters, the potential impacts on human health may be direct, including death or immediate hospitalisation due to trauma; or indirect, with an increase in vector-borne and water-borne diseases, as well as post-traumatic stress disorders.

Usually, the estimates of extreme climate events may be accessed via two databases on disasters: one containing global data *(EM-DAT – Emergency Events Database),* and the other containing national data (*Atlas Brasileiro de Desastres Naturais*, or Brazilian Atlas of Natural Disasters). Any discrepancies between these databases derive from the criteria adopted to define 'disaster occurrence'. EM-DAT, for example, only computes disasters with 10 or more deaths; 100 or more affected people; declaration of state of emergency; and request for international aid.

In Brazil, extreme climate events have historically caused a considerable number of deaths every year, in addition to expressive material losses. Considering EM-DAT's data, the number of extreme climate events was higher in 1998 and 2009 (when the phenomena El Niño and La Niña took place), with 10 occurrences and over 10 million people affected. In economic terms, events related to droughts and floods in 2014 resulted in approximately USD 5 million in damages (Fig. 4.11).

Despite the notorious advantage of EM-DAT's data due to its international coverage and standardisation of disaster-related concepts, the data available by the *Atlas Brasileiro de Desastres Naturais* provide a more detailed profile of extreme events in all regions of the country. Thus, the characterisation of these regions can qualitatively help to assess the impacts of a warming above 4 °C. Based on the Atlas's data, approximately 31,909 natural disasters were recorded in Brazil between 1991 and 2010, including extreme events such as flash floods and prolonged droughts.

Of all Brazilian regions, the Northeast suffered the worst impacts, with a total of 12,851 official reports and 45,827,366 people affected between 1991 and 2010. Droughts and floods accounted for 78% of the total events. In the North region, gradual floods, flash floods and droughts were the most frequent events, accounting for 39, 26 and 18 percent of the total, respectively. The Central-west region presents similar events to the North, with 38% of flash floods and inundations, followed by 27% of gradual floods and 21% of droughts. The Southeast region was mainly affected by flash floods (35%) and droughts (32%). The latter took place in the extreme north of Espírito Santo and the northwest of Minas Gerais. The South region, where there has been a history of atypical phenomena, such as Hurricane Catarina, presented 40% of droughts and 23% of flash floods (Universidade Federal de Santa Catarina, 2012) (Fig. 4.12; Table 4.4).

With regards to damages to the health sector, flash and gradual floods were responsible for most of the damages, with thousands of people affected between 1991 and 2010. In 2014, a situation analysis of the main extreme events that took place in Brazil from 1991 to 2010 showed, once again, that hydrological events, including flash and gradual floods, accounted for the highest morbidity averages (Freitas et al., 2014). Some specific extreme events were particularly serious, such as the 2008 floods in Santa Catarina, which resulted in 110 deaths and 121,000 directly exposed people, including the displaced; and the 2010 floods in Alagoas

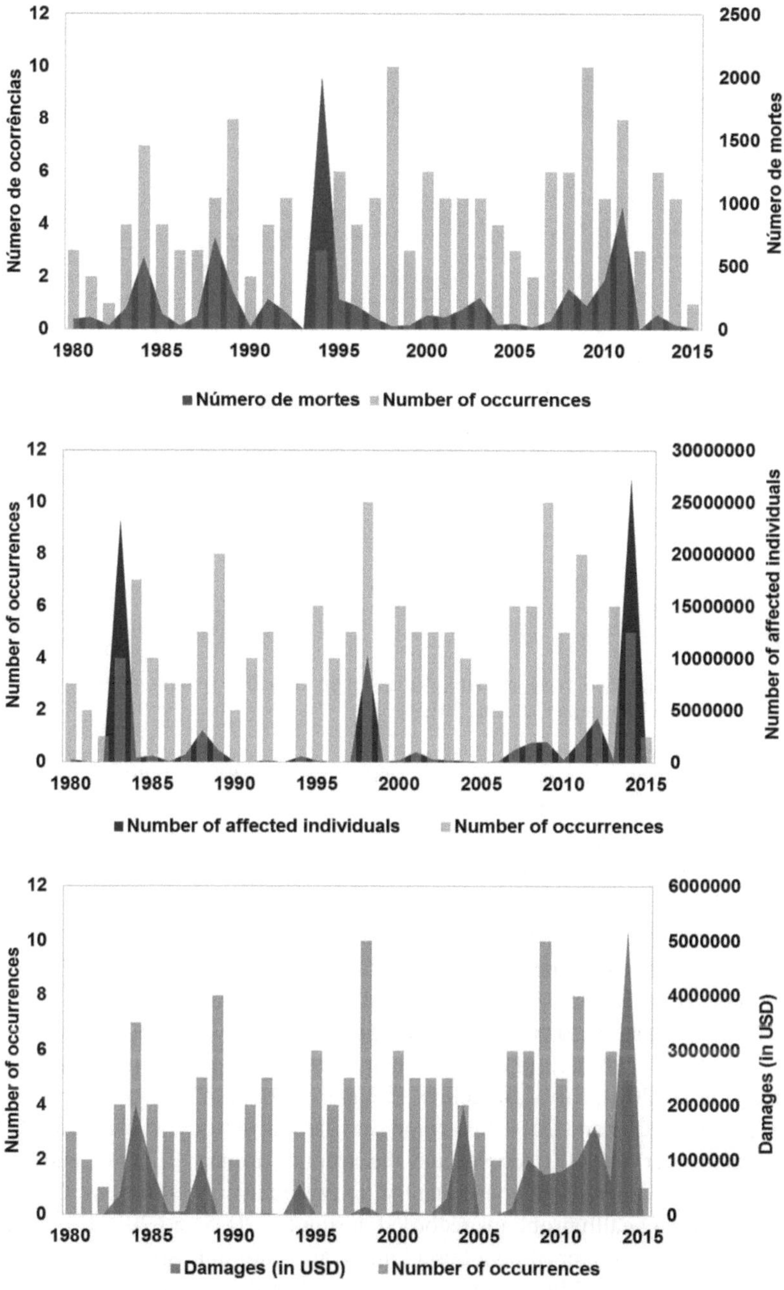

Fig. 4.11 Number of extreme climate events with their respective death records, affected population and economic damages caused to Brazil (1990–2015) (Data source: EM-DAT, 2015. Figure source: Oliveira, BFA

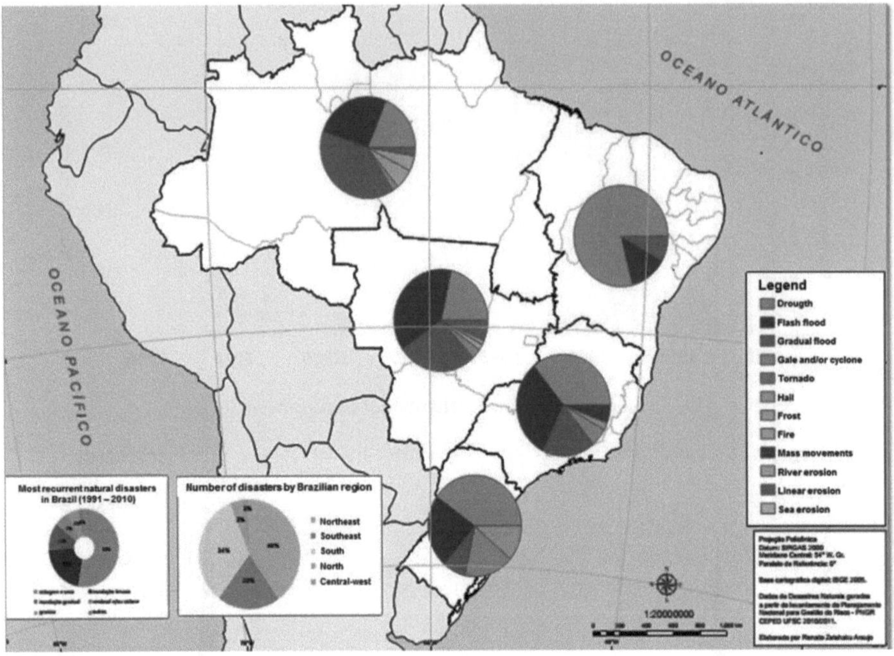

Fig. 4.12 Distribution and types of natural disaster in Brazilian regions (1991–2010)(Source: Universidade Federal de Santa Catarina, 2012)

and Pernambuco, which resulted in 56 deaths and 178,000 directly exposed people (World Bank, 2012). In some situations, floods affect populations living on riverbanks, but they can also cause landslides (another type of disaster), such as the events in the Serrana mountain region in Rio de Janeiro in 2011, which resulted in 947 officially documented deaths (Dourado, Arraes, & Silva, 2012).

In the North region, which can have a warming above 8 °C by the end of the century in a pessimistic climate change scenario, floods are gradual and are part of the river dynamics in the Amazon. In 2009 and 2012, floods in the state of Amazonas affected approximately 20,000 families living in precarious housing conditions. As a result, there were outbreaks of diseases such as diarrhoea, leptospirosis and dermatitis (Freitas et al., 2014; Silva, Barcellos, & Bacuri, 2010). In addition to hydrological events, between 2005 and 2010 this region saw the worst droughts ever recorded. The first affected over 167,000 people, and the second, around 122,000. During these events, forest fires increased; consequently, air pollutant concentrations also rose. As a result, the main health problems recorded during the 2005 and 2010 droughts were an increase in the number of people seeking health care in hospitals and clinics due to respiratory diseases (Mascarenhas et al., 2008; Smith et al., 2014).

For a warming scenario above 4 °C, it is hard to project the impacts on the health sector, as there is no way to predict the quantity and intensity of natural disasters for Brazil. However, based on the current situation, there is strong scientific evidence supporting increased occurrences of heat waves, floods, extreme rainfall and severe

Table 4.4 Number of events, affected population, recorded deaths or diseases per Brazilian region (1991–2010)

		No. of records	Affected	Deaths or diseases
North				
	Droughts	207	755,679	84,098
	Flash floods	292	951,875	13,149
	Gradual floods	433	1,424,117	60,083
North-east				
	Droughts	10,048	35,123,978	44,330
	Flash floods	1664	6,273,247	78,327
	Gradual floods	1033	4,190,308	68,901
Central-west				
	Droughts	173	825,471	1382
	Flash floods	303	3,295,464	664
	Gradual floods	219	288,894	920
Southeast				
	Droughts	2270	4,541,483	20,223
	Flash floods	2036	9,571,893	8930
	Gradual floods	1159	3,230,081	5941
South				
	Droughts	4246	7,189,994	1247
	Flash floods	2476	8,348,277	4831
	Gradual floods	832	1,096,840	1527

Data source: Universidade Federal de Santa Catarina, 2012

droughts all over the country (IPCC, 2013). Thus, these extreme events tend to be more qualitatively accentuated in terms of frequency and intensity over the years, which might inflate even further the number of deaths and people affected by natural disasters in Brazil.

In 2015, Rasch (2015) assessed the risk, exposure and vulnerability to floods in 1276 Brazilian urban municipalities. The vulnerability index was built including the following factors: age (the elderly and children are the most susceptible to the effects of floods); family income; quality and coverage of health care services; education; access to information; governance measures; degree of municipal autonomy in the decision making process; infrastructure; quality and size of buildings; housing; demographic density; and sanitation conditions. With regards to exposure, municipalities were classified according to their exposure to flood risks: 207 municipalities face a risk of coastal floods, 670 risk moderate floods (expected to flood at least twice in 100 years), and 129 are exposed to severe floods (expected to flood more than four times in 100 years).

The urban municipalities with very high vulnerability to floods are in the North region of the country, where in a pessimistic climate change scenario the temperature rise could exceed 8 °C at the end of the century (Fig. 4.13). However, if on one hand these municipalities are highly vulnerable to floods, on the other they are less populated when compared to other regions, such as the Southeast.

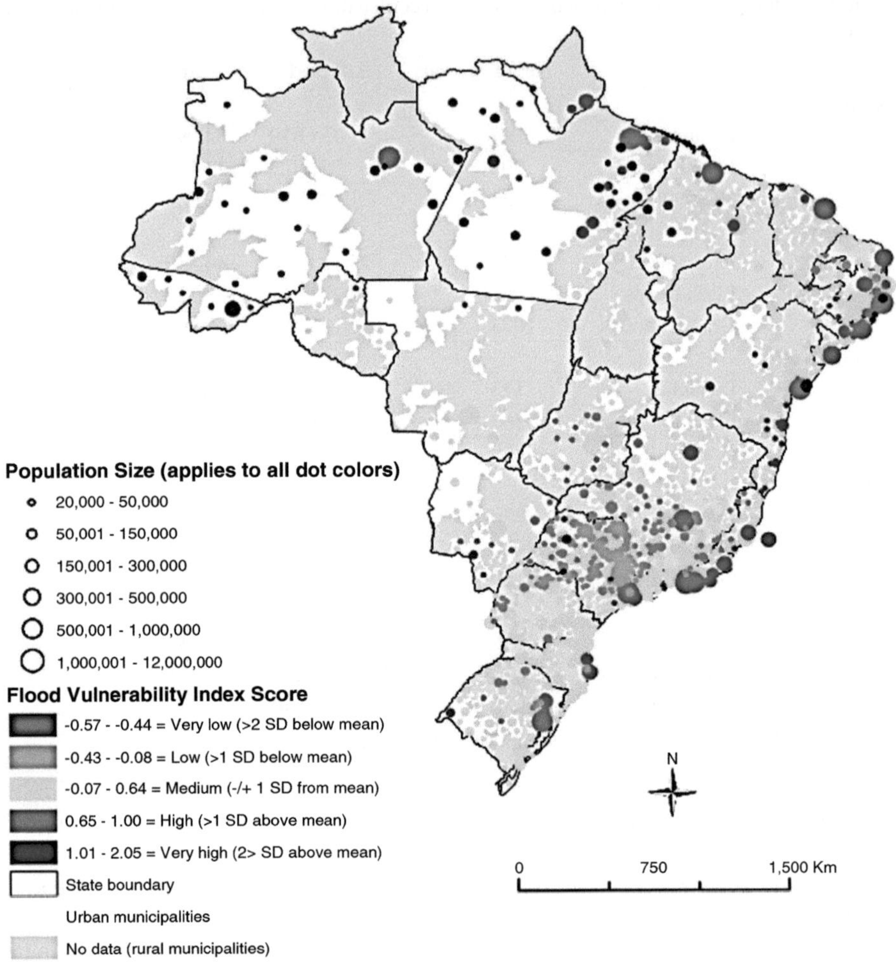

Fig. 4.13 Vulnerability to floods in 2010 for Brazilian municipalities, with their respective populations(Data source: Rasch, 2015)

Differently from the North region, most municipalities in the Southeast (where a 6 °C average temperature rise is predicted under RCP 8.5 for 2071–2099) presented low or medium vulnerability to floods, except for large cities such as Rio de Janeiro and São Paulo, which presented high vulnerability to floods and over 1 million people potentially exposed (Fig. 4.13).

Combining risk, exposure to floods and vulnerability indicators, 83 municipalities were classified as high risk, i.e., presented high exposure and vulnerability to floods (Fig. 4.14). Among the municipalities in the North region, three stand out: Feijó, Cruzeiro do Sul, and Sena Madureira, in the state of Acre. These towns may present an average temperature rise of 7 °C under the RCP 8.5 scenario for 2071–2099. São Paulo and Rio de Janeiro municipalities are among the 83 presenting high

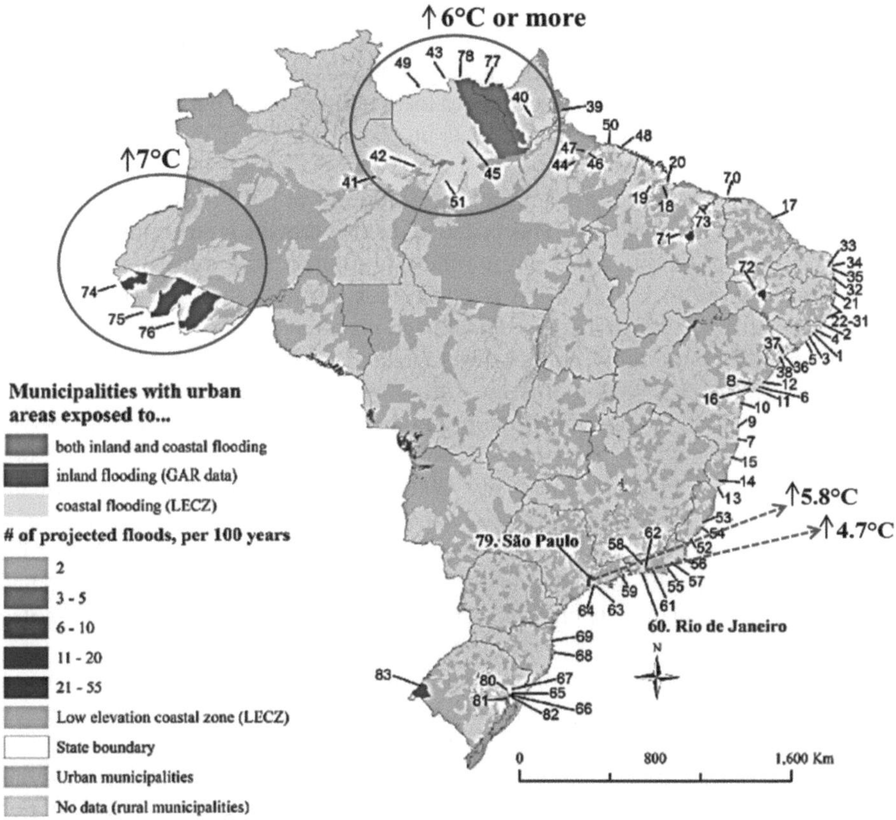

Fig. 4.14 Municipalities classified as high risk for floods (2010) and predicted average temperature rise under the RCP 8.5 scenario for 2071–2099 (Data source: Rasch, 2015 and regionalised Eta/HadGEM2-ES climate projections (Chou et al., 2014). Figure adapted by authors)

exposure and vulnerability to floods. They may face average temperature rises of 5.8 °C and 4.7 °C, respectively, under the RCP 8.5 scenario for 2071–2099, according to regionalised Eta-HadGEM projections (Chou et al., 2014) (Fig. 4.14).

4.6 Indirect Impacts of Climate Change on the Health Sector as a Response to Exposure to a Warming of 4 °C or More

4.6.1 Mortality by Heat

Future mortality projections associated with temperature rise show that this is one of the most probable impacts on the health sector, both by increased seasonal average temperatures, as well as the frequency and intensity of heat waves. Despite the focus on effects associated with extreme events, the potential effects of moderate heat are extremely relevant, as moderate heat days are more frequent than extreme heat ones (Gasparrini et al., 2015; Kinsgley et al., 2015).

In general, human populations adapt to local temperatures, but a sharp deviation of climate variability may contribute to an increased number of deaths. Thus, many studies have quantified the effects of heat and cold separately, using linear responses below and above temperature thresholds (Baccini et al., 2011; Hajat & Kosatky, 2014; McMichael et al., 2008). This is due to the fact that the temperature-mortality or temperature-morbidity ratios are normally non-linear with a distribution at U-, V- or in the form of J. Thus what is observed is an indication of a temperature threshold starting at which increased death and hospitalisation risks are observed.

With regards to mortality, studies developed in Brazil have mostly been descriptive and exploratory, showing only the effects of seasonal and spatial variations in the distribution of certain morbi-mortality outcomes (Botelho, Correia, Silva, Macedo, & Silva, 2003; Graudenz, Latorre, Tribess, Oliveira, & Kalil, 2006; Gonçalves, Braun, Dias, & Sharovsky, 2007; Pimentel, Grüdtner, & Zimerman, 2006). Recently, some studies have shown the relationship between temperature and mortality, and their potential health impacts in some Brazilian cities (Gasparrini et al., 2015; Guo et al., 2014; Son et al., 2016). However, future projection studies considering different climate change scenarios have not yet been carried out (Tables 5.5 and 5.6).

Old and recent publications show that the city of São Paulo has been the focus of most studies, and that several effects have been observed, according to different age groups and also specific causes of death. With regards to age groups, a 4 °C average temperature rise starting at 20 °C was associated with an 11% increase in overall mortality for children and the elderly (Gouveia, Shakoor, & Armstrong, 2003). For specific causes of death, mortality by respiratory diseases in São Paulo was higher when compared with all causes and cardiovascular diseases, with an 11% increase for each 4 °C average temperature rise in that city (Son et al., 2016).

With regards to temperature thresholds, the daily number of deaths was estimated for 18 municipalities in Brazil, with thresholds between 20.7 °C and 28.6 °C for Curitiba and Teresina, respectively (Guo et al., 2014) (Table 4.5). Subsequently, these values were redefined by Gasparrini et al. (2015), with thresholds of 21.3 °C for Curitiba and 28.1 °C for Cuiabá (Table 4.5).

With regards to the exposure-response function, the percentage mortality increase associated with temperature rises was significant only for Belo Horizonte, Curitiba, Goiânia, Porto Alegre, São Paulo and Teresina. For Belém, Fortaleza, Natal, Recife and Salvador, the results were negative, i.e., a drop in mortality was observed for each temperature rise. The absence of a positive exposure-response relationship in these places may suggest that the population have acclimatised or adapted to higher temperatures. In addition, higher temperature variations do not always present a higher death risk. For example, Porto Alegre and Teresina showed variations between the 99th percentile and the threshold of 4.9 °C and 2.4 °C, respectively, but the percentage death increase was 13% in Porto Alegre and 20% in Teresina.

Table 4.5 Publications on the impacts of temperature rise on mortality rates

Authors	Place	Period	Outcome	Lag	Threshold	ΔT	RR (IC95%)
Gouveia et al., 2003*	São Paulo	1991–1994	Overall mortality rate	0–1	20 °C	4 °C	**C: 1.11(1.07–1.15)**
				+	+	+	**A: 1.05 (1.04–1.08)**
				+	+	+	**E: 1.11 (1.08–1.12)**
			Cardiovascular	+	+	+	**A: 0.99 (0.97–1.01)**
				+	+	+	**E: 1.08 (1.07–1.09)**
			Respiratory	+	+	+	**A: 1.08 (1.04–1.12)**
				+	+	+	**E: 1.09 (1.07–1.12)**
Sharovsky et al., 2004	São Paulo	1996–1998	AMI mortality rate	MM2	22 °C	Up to 4.7 °C	**1.11 (1.06–1.16)**
McMichael et al., 2008*	Salvador	1996–1999	Overall mortality rate	0–1	23 °C	4 °C	**1.11 (1.04–1.18)**
	São Paulo	1991–1994	+	+	+	+	**1.10 (1.08–1.13)**
Guo et al., 2014**	Belém	1999–2011	Overall mortality rate	0–21	27.5 °C	1.2 °C	0.91 (0.81–1.02)
	Brasília	+	+	+	22.5 °C	3.4 °C	1.11 (0.99–1.24)
	Belo Horizonte	+	+	+	23.8 °C	3.5 °C	**1.18 (1.08–1.28)**
	Cuiabá	+	+	+	28.1 °C	3.2 °C	1.03 (0.79–1.34)
	Curitiba	+	+	+	**20.7 °C**	**4.0 °C**	**1.12 (1.00–1.24)**
	Fortaleza	+	+	+	27.7 °C	1.1 °C	0.98 (0.89–1.07)
	Goiânia	+	+	+	25.8 °C	3.4 °C	**1.14 (1.04–1.25)**
	João Pessoa	+	+	+	28.0 °C	1.4 °C	1.14 (0.97–1.35)
	Maceió	+	+	+	26.2 °C	1.7 °C	1.04 (0.88–1.23)
	Manaus	+	+	+	28.2 °C	3.1 °C	1.11 (0.97–1.26)
	Natal	+	+	+	**23.5 °C**	**4.9 °C**	0.98 (0.86–1.12)

(continued)

Table 4.5 (continued)

Authors	Place	Period	Outcome	Lag	Threshold	ΔT	RR (IC95%)
	Porto Alegre	+	+	+	**23.5 °C**	**4.9 °C**	**1.13 (1.04–1.23)**
	Recife	+	+	+	27.1 °C	1.6 °C	0.96 (0.89–1.04)
	Salvador	+	+	+	26.9 °C	1.7 °C	0.93 (0.86–1.01)
	São Luís	+	+	+	27.6 °C	1.1 °C	1.04 (0.93–1.17)
	São Paulo	+	+	+	**22.9 °C**	**4.0 °C**	**1.16 (1.09–1.24)**
	Teresina	+	+	+	28.6 °C	2.4 °C	**1.20 (1.01–1.44)**
	Vitória	+	+	+	26.8 °C	2.4 °C	1.14 (0.96–1.35)
Son et al., 2016	São Paulo	1996–2010	Overall mortalityrate	0–1	20.4 °C	≅ 4 °C	**1.05 (1.04–1.06)**
		+	Cardiovascular	+	+	+	**1.03 (1.02–1.04)**
		+	Respiratory	+	+	+	**1.11 (1.09–1.13)**

Legend: C: Children; A: Adults; E: Elderly; * In the article, the RR is considered for 1 °C increases in average temperature. In this text, it is transformed for every 3 °C increase. ** The threshold in the article is based on the percentile found for Brazil (P72). In the absence of this value for each city, the variation and threshold are approximate and based on P75. A recent publication by Gasparrini et al. (2015) reassessed the 18 Brazilian cities and reset the threshold. Data source: Publications on the impacts of temperature rise mortality rates (Gouveia et al., 2003; Guo et al., 2014; McMichael et al., 2008; Sharovsky et al., 2004; Son et al., 2016). Table source: Oliveira, BFA

In quantitative terms, Gasparrini et al. (2015) measured the fraction attributable to mortality by all causes deriving from exposure to heat and cold from 1997 to 2011 using data previously published by Guo et al. (2014). In Brazil, 2.8% and 0.7% of all deaths by all causes were attributable to cold and heat, respectively. Among Brazilian capitals, Fortaleza presented the highest percentage of deaths attributable to heat: 3075 deaths by all causes, or 1.5% of the total. In absolute numbers, the city of São Paulo presented the highest number of deaths attributable to heat, with 6322 deaths between 1991 and 2011. Based on these results and analysing regionalised Eta-HadGEM projections for the RCP 8.5 scenario in 2071–2099 as a snapshot of high Warming, some Brazilian cities may present a 6 °C average temperature rise in relation to their temperature threshold. This applies, for example, to Brasília, Cuiabá, Curitiba, Goiânia and São Paulo (Table 4.5).

Considering the temperature thresholds in these 18 Brazilian capitals (Table 4.6), the relative risk of mortality by all causes deriving from temperature variations was calculated for the RCP 4.5 and RCP 8.5 scenarios, which predict 3 °C and 6 °C average temperature rises in 2071–2099 (Fig. 4.15).

Table 4.6 Fraction attributable to heat related mortality by all causes between 1997 and 2011, and average temperature rise under RCP 8.5 (2071–2099)

Authors	Place	Threshold	%FR	Deaths/Heat	Difference (Ts – Tb) RCP 8.5 (2071–2099)
Gasparrini et al., 2015	Belém	26.7 °C	1.07	1423	5 °C
	Brasília	22.9 °C	0.31	356	6 °C
	Belo Horizonte	22.8 °C	0.47	2184	5 °C
	Cuiabá	28.1 °C	0.45	231	6 °C
	Curitiba	21.3 °C	0.31	481	7 °C
	Fortaleza	26.9 °C	1.52	3075	3 °C
	Goiânia	24.2 °C	1.14	1530	8 °C
	João Pessoa	27.1 °C	0.75	405	1 °C
	Maceió	25.6 °C	0.47	435	2 °C
	Manaus	27.4 °C	0.88	864	5 °C
	Natal	24.5 °C	0.30	242	5 °C
	Porto Alegre	24.2 °C	0.50	1079	1 °C
	Recife	25.7 °C	1.20	3161	2 °C
	Salvador	27.0 °C	0.23	514	1 °C
	São Luís	26.9 °C	0.80	655	4 °C
	São Paulo	21.5 °C	0.69	6322	8 °C
	Teresina	28.0 °C	0.62	417	3 °C
	Vitória	26.8 °C	0.38	197	0 °C

Data source: development from the data Gasparrini et al., 2015, and regionalised Eta/HadGEM2-ES (Chou et al., 2014) with municipalised projections (Costa et al., 2015). Table source: Oliveira, BFA

By the end of the century, under the RCP 8.5 scenario, all capitals will present a positive relative risk of mortality deriving from temperature variations above the threshold, except for Belém (PA), Fortaleza (CE), Natal (RN), Recife (PE) and Salvador (BA). At least a 20% increase is expected in mortality by all causes in Teresina (PI), Goiânia (GO), São Paulo (SP) and Curitiba (PR), respectively with 3 °C, 8 °C, 8 °C and 7 °C average temperature rises relative to the threshold. On the other hand, we have observed a death risk reduction in Belém (PA), Fortaleza (CE), Natal (RN), Recife (PE) and Salvador (BA) (Fig. 4.18).

Analysing a 4 °C temperature rise for the 18 capitals studied by Guo et al. (2014) and Gasparrini et al. (2015), we can observe that São Paulo already presents this temperature variation for the RCP 8.5 scenario in the first time window of the study. For the second time window in the same scenario, São Paulo, Curitiba, Goiânia and Brasília presented an average temperature rise above 4 °C relative to the optimal temperature, with 16,738, 1522, 1938 and 2862 expected deaths due to heat, respectively. For the third window, (2071–2099), all capitals included in the analysis faced impacts on mortality rates by all causes, except Belém, Salvador, Recife, Natal and

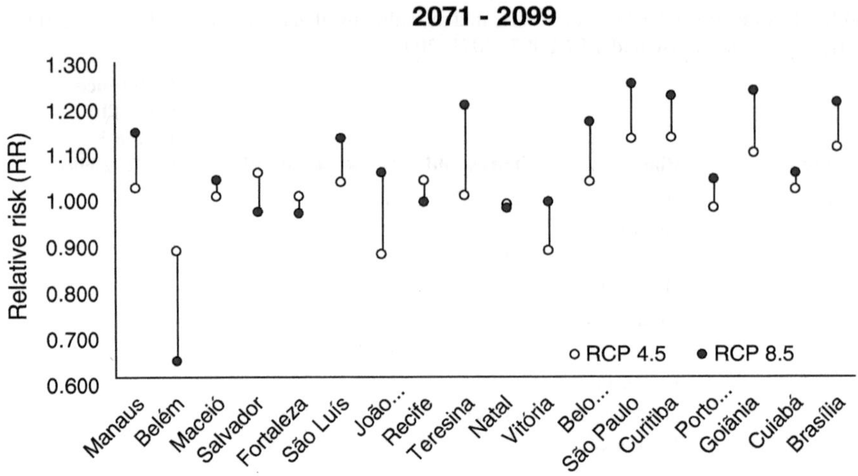

Fig. 4.15 Relative risk of death by all causes under the RCP 4.5 and RCP 8.5 scenarios for 2071–2099. Source: Guo et al. (2014) and Gasparrini et al. (2015). Data referring to state capitals. Relative risk increase given by RR = exp.$^{(\beta * \Delta T)}$

Vitória. In this period, the city of São Paulo presented an 8 °C difference between average and optimal temperatures, which might represent 20,000 additional deaths (Table 4.7).

In general, deaths resulting from temperature rises are associated with cardiovascular and respiratory diseases, and affect mainly the elderly and other individuals with pre-existing health conditions. In Brazil, a 4 °C increase in global average temperature (under scenario RCP 8.5) could lead to a daily mortality rate 7.5 times as high among the elderly, compared with the rates found in the baseline period (1995) (184 vs. 21.7 deaths per 100,000 people). In an optimistic scenario (RCP 2.6) with a predicted 1.4 °C increase in global average temperature, the heat related mortality rate among the elderly would reach 52 deaths per 100,000 people in 2085, a difference of 131 deaths per 100,000 elderly people when compared with the pessimistic scenario (Fig. 4.16).

Deaths resulting from air temperature rise, especially among the elderly, may be linked to a risk of dehydration due to immobility, depletion of body volume and reduction of compensatory cardiovascular mechanisms (Becker & Stewart, 2011). In addition, the health situation in this group can be aggravated by their susceptibility to pre-existing chronic conditions and by the continuous use of psychotropic medications, which affect their physical ability to adapt to climate change (Bouchama & Knochel, 2002).

Considering mortality by specific causes among the elderly, we have analysed the percentage increase of deaths by cardiovascular diseases due to average temperature rise under the RCP 8.5 scenario for the period from 2071 to 2099 in Brazil (Fig. 4.17). It is worth noting that an increase of up to 30% may be observed in mortality by cardiovascular diseases among the elderly, particularly in municipalities

Table 4.7 Estimated number of expected deaths by all causes and affecting all age groups as a result of temperature rises, according to different climate scenarios and time periods

	Number of expected deaths (all age groups) attributable to temperature rises											
	2011–2040				2041–2070				2071–2099			
	$T_S - T_{LM}$	RCP 4.5	$T_S - T_{LM}$	RCP 8.5	$T_S - T_{LM}$	RCP 4.5	$T_S - T_{LM}$	RCP 8.5	$T_S - T_{LM}$	RCP 4.5	$T_S - T_{LM}$	RCP 8.5
North												
Manaus	−0.34	−112	−0.05	−16	0.76	330	1.98	858	1.47	787	5.07	2716
Belém*	0.48	−285	0.86	−513	1.52	−1118	2.81	−2064	2.29	−1982	5.38	−4646
North-east												
Maceió	−0.94	−108	−0.74	−85	0.08	10	0.82	102	0.66	87	2.34	309
Salvador*§	−3.12	1690	−2.74	1483	−2.03	1167	−1.03	589	−1.40	845	0.63	−382
Fortaleza*§	−0.90	198	−0.58	127	0.00	0	1.07	−264	0.62	−167	2.69	−729
São Luís§	0.00	−1	0.38	57	0.95	156	2.27	372	1.70	302	4.36	775
João Pessoa§	−2.04	−679	−1.76	−587	−1.07	−400	−0.10	−39	−0.44	−182	1.49	615
Recife*§	−1.66	2326	−1.39	1952	−0.64	1020	0.25	−405	−0.03	62	1.80	−3201
Teresina	−1.04	−287	−0.45	−124	−0.07	−19	1.39	380	0.65	177	3.27	884
Natal*§	1.69	−29	1.91	−32	2.61	−53	3.48	−70	3.19	−74	5.03	−117
Southeast												
Vitória§	−3.60	−388	−3.40	−366	−2.69	−357	−1.92	−255	−2.15	−338	−0.19	−30
Belo Horizonte	0.18	115	0.53	336	1.07	748	2.33	1628	1.73	1308	4.54	3441
São Paulo	3.58	9425	3.87	10,183	4.24	12,777	5.55	16,738	4.85	16,369	8.15	27,489
South												
Curitiba	2.93	849	3.35	972	3.55	1157	4.67	1522	4.02	1443	7.24	2602
Porto Alegre	−2.49	−241	−2.16	−209	−1.88	−186	−1.23	−121	−1.54	−155	0.86	86

(continued)

Table 4.7 (continued)

	Number of expected deaths (all age groups) attributable to temperature rises											
Central-west												
Goiânia	2.50	733	3.14	920	3.33	1248	5.17	1938	4.16	1885	7.73	3503
Cuiabá[§]	0.43	13	1.07	31	1.30	47	3.00	108	2.02	86	5.87	251
Distrito Federal[§]												
Brasília	1.44	677	2.10	990	2.36	1628	4.15	2862	3.22	2903	6.47	5831

"–" Capitals that were not impacted by a temperature rise – average temperature (T_s) in relation threshold of temperature (T_{LM}) ($Ts-T_{LM} \leq 0$)
"*" Capitals that presented a relative risk below 1, i.e., a reduction in mortality resulting from temperature rise. Threshold of temperature (T_{LM}) and relative risks were taken from the cities studied by Guo et al. (2014) and Gasparrini et al. (2015), according to Tables 4.5 e 4.6
"§" Capitals where the relative risks were not statistically significant. Thus, they face a higher degree of uncertainty

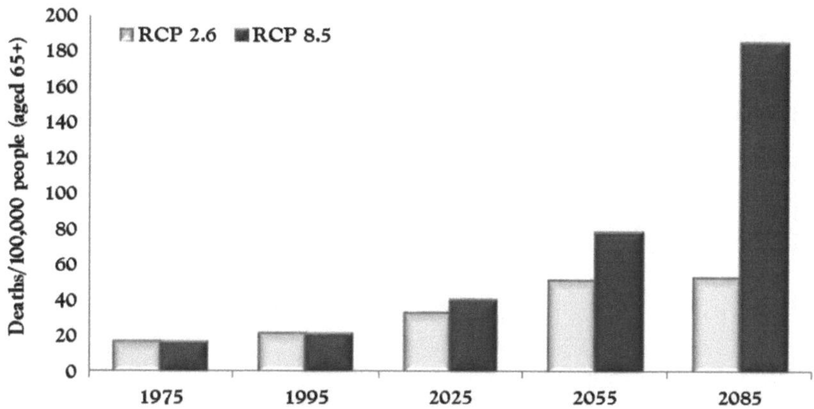

Fig. 4.16 Daily Mortality rate (per 100,000 people) among the elderly (65+) in Brazil according to climate scenarios for the years 1975, 1995, 2025, 2055 and 2085 (Data source: Kjellstrom et al., 2015 (WHO National Profiles on Climate Change and Human Health))

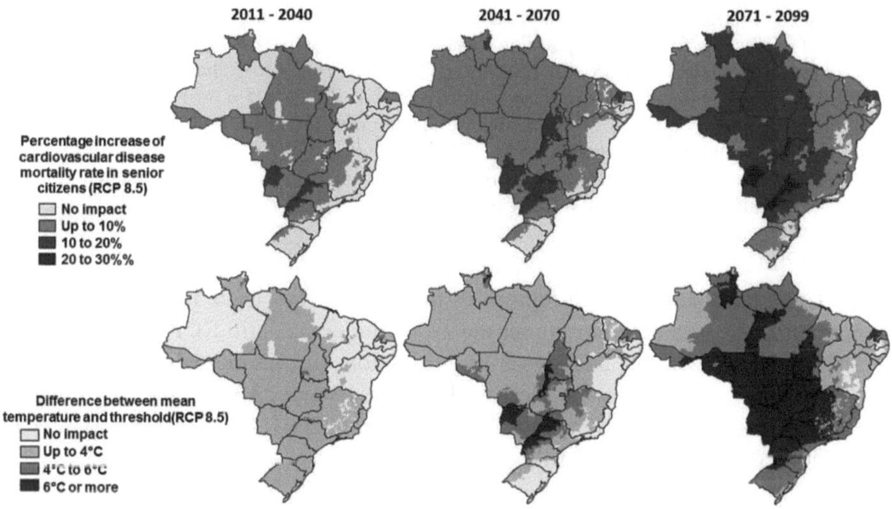

Fig. 4.17 Projections of percentage increase in mortality by circulatory diseases among the elderly (65+) and difference between average temperature and threshold under the RCP 8.5 scenario for the period 2071–2099 in Brazilian municipalities (For an estimate of the percentage increase in the number of deaths, the following expression was used: $\%RR = (\exp(\beta^*\Delta T) - 1)^*100$. The thresholds for the cities studied by Guo et al. (2014) and Gasparrini et al. (2015) were used as reference for the other municipalities in their respective states (Tables 4.5 and 4.6). The non-impacted municipalities correspond to those where the average temperature in the time windows was lower than the temperature thresholds ($Ts - TLM \leq 0$). In the absence of an exposure-response function between mortality by cardiovascular diseases and among the elderly, β was obtained from the study by Gouveia et al. (2003) for the city of São Paulo)

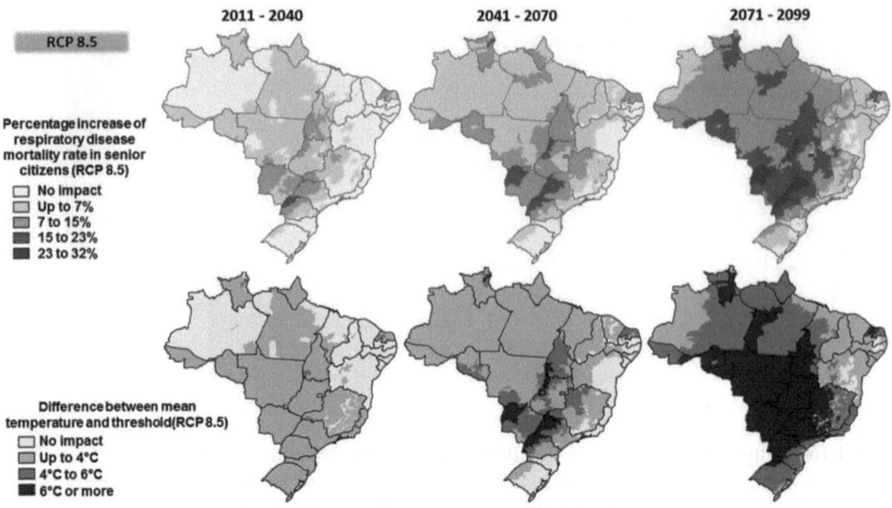

Fig. 4.18 Projection of percentage increase in mortality by respiratory diseases among the elderly (65+) and difference between average temperature and threshold under the RCP 8.5 scenario for the period 2071–2099 in Brazilian municipalities (For an estimate of the percentage increase in the number of deaths, the following expression was used: %RR = (exp (β*ΔT) – 1) *100. The thresholds for the cities studied by Guo et al. (2014) and Gasparrini et al. (2015) have been used as reference for the other municipalities in their respective states (Tables 4.5 e 4.6). The non-impacted municipalities correspond to those where the average temperature in the time windows was lower than the optimal minimum temperature (Ts – TLM ≤ 0). In the absence of an exposure-response function between mortality by respiratory diseases and among the elderly, β was obtained from the study by Gouveia et al. (2003) for the city of São Paulo)

in the Central-west and North regions of the country. For these municipalities, the temperature rise was equal or higher than 6 °C under scenario RCP 8.5.

In a scenario without adaptation or mitigation, the municipalities with a rise above 6 °C may experience an increase between 15 and 23% in mortality rates by respiratory diseases among the elderly in the period 2041–2071, and up to 32% in the period 2071–2099 (Fig. 4.18).

4.6.2 Vector-borne Diseases

Warming above 4 °C may have impacts on human health in different ways and intensities. Among the indirect impacts, those associated with vector-borne diseases represent a concern for the health sector, especially because climate variables may influence their respective transmission cycles, and therefore expand the current area of incidence of such diseases.

In Brazil, vector-borne diseases constitute an important cause of morbidity and mortality. In the current epidemiological scenario, dengue is considered the main re-emerging disease in the country, which saw, in 2015, the highest number of notifications since 1990, with over 1.5 million probable cases (Brasil, 2016a). In addition, the following are worth mentioning: malaria, Chagas disease, leishmaniasis, yellow fever, Oropouche fever, Mayaro fever, filariasis (bancroftian filariasis/elephantiasis and onchocerciasis/river blindness), West Nile fever, and encephalitis, among others. Some of these diseases are widely spread all over the country, such as dengue, while others are restricted to some regions, such as the Oropouche fever, in Pará. Adding to this scenario, the recent introduction of Zika and chikungunya has aggravated even further the epidemiological situation of these diseases in Brazil.

The introduction and expansion of these two new arboviruses – which have been known for a long time in Africa and Asia – have been cause of concern for Brazilian health authorities. The chikungunya virus was introduced in July/August 2014, after entering the Caribbean in December 2013; and the Zika virus was possibly introduced around the same time during the 2014 World Cup in Brazil. Despite Zika's apparent benignity, some more severe cases have been recorded and related to the disease, including some central nervous system compromise (Guillain-Barré Syndrome, transverse myelitis, and meningitis) (Campos, Bandeira, & Sardi, 2015; Oehler et al., 2014; Zanluca et al., 2015). In addition, the disease has been considered the possible cause for the alarming increase in microcephaly in 19 Brazilian states. Although this relationship has not been completely proved, the virus has been identified in the amniotic fluid of two babies born with this congenital malformation (Brasil, 2016a).

In general, multiple factors may influence the dynamics of vector-borne diseases, including environmental factors (vegetation, climate and hydrology), socio-demographic factors (migrations and demographic density), biological factors (life cycle of vectors), and health factors (population's susceptibility, and effectiveness of local health care systems and disease control programmes). Among the environmental components, climate variables play an important role in the vectors' temporal and spatial distribution, both limiting their propagation and influencing the dynamics of transmission.

Concerning climate variables, rainfall patterns may have a short and medium term effect, as they increase the number and quality of the vectors' breeding sites. Temperature constitutes an important ecological factor that may delay or accelerate the development and survival of vectors, as well as the extrinsic incubation period for some pathogens (Silveira-Neto, Nakano, Barbin, & Villa Nova, 1976). Therefore, environmental conditions that favour the development of vectors are relevant to better understand their population dynamics and predict the number of annual generations, as well as the periods with the most occurrences in infested areas (Beserra, Fernandes, Silva, Silva, & Santos, 2009).

In a warming scenario above 4 °C, the possibility of increased transmission of certain diseases (such as dengue) during El Niño events also becomes extremely relevant for the health sector, as these events may increase in a scenario with higher Warming. In the case of dengue, for example, the synchrony between its incidence at high temperatures during the 1997–1998 El Niño has been reported in some Southeast Asian countries (van Panhuis et al., 2015).

In this context, in view of the morbidity and mortality epidemiological situation, the impacts of the following diseases will be discussed in this report for scenarios with a warming equal or higher than 4 °C: dengue (Zika and chikungunya), malaria and leishmaniasis.

4.6.3 Dengue, Zika and Chikungunya

In Brazil, the dengue virus is numerically the most important human arbovirus, with over 1.5 million cases recorded in 2015 – a 176% increase relative to 2014 (Brasil, 2016a). One of the main characteristics of this disease is its high potential to spread to new areas. In Brazil, the dengue transmission area more than trebled between 2001 and 2011, growing from 2 to 7 million km^2 (Barcellos & Lowe, 2014), and consequently increasing the number of people at risk. These figures serve as a warning to health authorities and the population about the potential dissemination of other diseases transmitted by the same vector (the *Aedes aegypti*), such as Zika and Chikungunya.

Meteorological variables, such as temperature, play an important role in the life cycle of this vector, in the replication of viruses, and more generically in the cycle and vectorial capacity of the diseases above. In higher temperatures, larva development and mosquito bite rates increase, while the appearance of adult mosquitoes and virus replication time decrease. In addition, extreme temperatures may reduce the survival of mosquitoes, which could offset the positive effect of their abundance (Watts, Burke, Harrison, Whitmire, & Nisalak, 1987). In the specific case of dengue, for serotypes DEN-1 and DEN-4, *Aedes aegypti*'s feeding time and detection in salivary glands decreased from 9 days (at average temperatures between 26 °C and 28 °C) to 5 days (at average temperature of 30 °C) (Rohani, Wong, Zamre, Lee, & Zurainee, 2009).

Clearly, optimal values for the development of *Aedes aegypti* vary according to local characteristics. Based on studies published in Brazil, values between 21 °C and 29 °C are considered favourable for the development of the vector, and values between 22 °C and 30 °C are associated with higher longevity and fertility in adult mosquitoes (Beserra et al., 2009; Hii et al., 2009). Within this range, the optimal temperature for the development of the vector ranges from 24 °C to 26 °C (Gomes et al., 2012; Honório, Codeço, Alves, & Magalhães, 2009; Horta, Bruniera, Ker, Catita, & Ferreira, 2014). However, Beserra et al. (2009), while assessing the effects of temperature on the *Aedes aegypti*'s life cycle in the state of Paraíba, observed that the most favourable temperature is above 22 °C and below 32 °C, considering the development time and viability for the eggs, larvae and pupae phases, as well as in the fecundity of adults.

A study on dengue carried out in the state of Rio de Janeiro associated temperatures above 24 °C with an increase in the number of cases. For every 1 °C rise in the minimum temperature in a given month, a 45% increase was observed in the cases of dengue the following month. Considering this ratio as a parameter, a 480%

increase in the cases of dengue could be expected for a 4 °C increase in minimum temperature (Gomes et al., 2012).

Considering the RCP 8.5 scenario and the period between 2071 and 2099, municipalities in the Northeast, Southeast and South regions of the country (with an average temperature variation between 4 °C and 6 °C) would present favourable thermal conditions for the development of *Aedes aegypti*, with values between 22 °C and 32 °C (Beserra et al., 2009). According to the map in the study by Barcellos and Lowe (2014) about the expansion of this disease, we can expect an increased risk of transmission in the South region, a maintenance of endemic areas in the Northeast and Southeast, and a decrease in the North and Central-west. Therefore, a temperature rise above 6 °C in the North and Central-west regions would raise the average temperature above 32 °C, from which we could observe negative effects on the life cycle of *Aedes aegypti* (Fig. 4.19). Obviously, the occurrence of diseases transmitted by this vector in those regions is not explained by temperature alone, and therefore it is important to consider social, demographic and economic factors as well.

Although thermal conditions influence the development of certain vectors, vectorial capacity is commonly used as a transmission indicator for vector-borne infectious diseases. This indicator incorporates a set of physiological and behavioural characteristics of the insect, which, together with environmental conditions (such as temperature), favour the natural transmission of pathogens. According to the World Health Organization (WHO, 2015a), the average relative vectorial capacity for dengue transmission by *Aedes aegypti* should remain practically the same at the end of

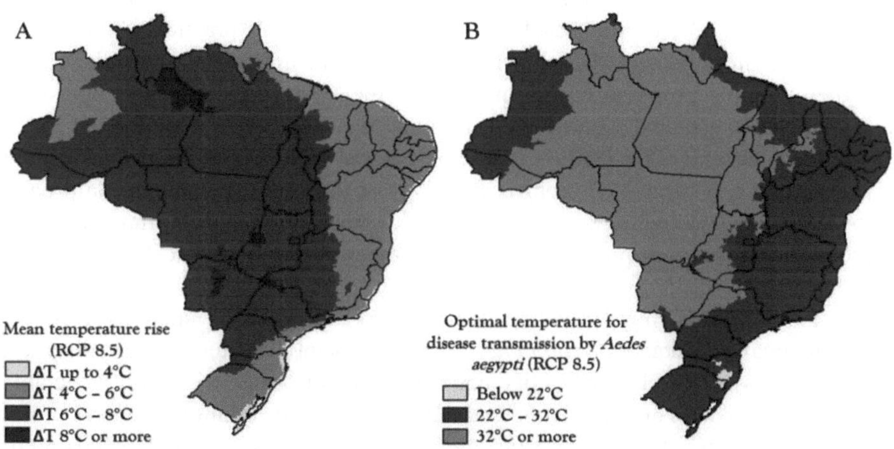

Fig. 4.19 Municipalities with favourable temperature (22 °C – 32 °C) for the transmission of Dengue, Zika and Chikungunya. ***Fig. A***: Average temperature rise under RCP 8.5 for the period 2071–2099 compared to the baseline period (1961–1990). *Fig. B*: Optimal or favourable temperature for the development of *Aedes aegypti* under the RCP 8.5 scenario for the period 2071–2099 (Data source: development from the data Regionalised Eta/HadGEM2-ES (Chou et al., 2014) with municipalised projections (Costa et al., 2015). Map source: Oliveira, BFA

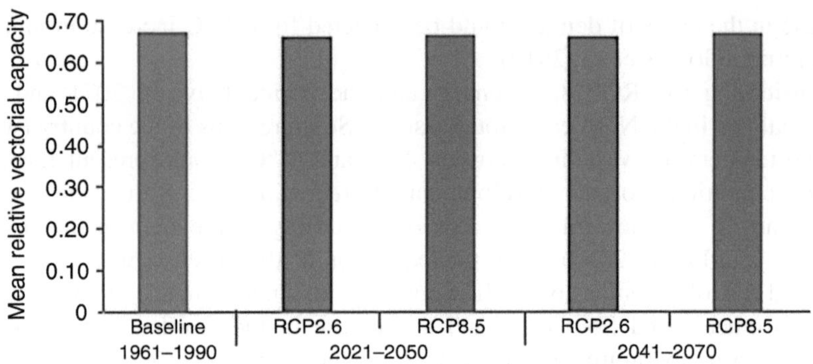

Fig. 4.20 *Aedes aegypti's* average relative vectorial capacity in Brazil under the RCP 2.6 and RCP 8.5 scenarios for the periods 2021–2050 and 2041–2070 (Source: Rocklöv et al. (2015). Specific analysis to characterise country profiles with regards to the health impacts of climate change. The results are based on health models described in 'Quantitative risk assessment of the effects of climate change on selected causes of death, 2030s and 2050s. Geneva: World Health Organization, 2014')

2070 as compared with the baseline period (1961–1990), with an approximate value of 0.67, which represents an endemic scenario for this disease in Brazil (Rocklöv, Quam, et al., 2015) (Fig. 4.20). It is worth noting that, for the period 2041–2070, considering the RCP 8.5 scenario and the average temperature for the municipalities, the temperature rise will be approximately 4 °C.

Data related to *Aedes aegypti's* vectorial capacity have also been used to estimate future dengue epidemic potential scenarios for the whole world, including Brazil, as shown in Fig. 4.21. In the baseline period (1980–2009), practically the whole Brazilian territory presented dengue epidemic potential areas (PED). A reduction in dengue epidemic potential based on *Aedes aegypti's* vectorial capacity for some areas in the North and Central-west regions was observed under the RCP 8.5 scenario for the end of the century. These places will present an average temperature rise between 6 °C and 7 °C, approximately. For the states of Espírito Santo and Rio de Janeiro, an increase is expected in dengue epidemic potential for 2071–2090. These states will present an average temperature rise of 4.6 °C and 5.1 °C, respectively.

Although *Aedes aegypti* is the main vector of diseases such as Dengue, Chikungunya and Zika, *Aedes albopictus* is also a potential vector for these diseases. In Brazil, this vector chooses to live and reproduce in places with more vegetable cover; it prefers wild environments, and therefore is more commonly found in rural and suburban areas. Differently from *Aedes aegypti*, *Aedes albopictus* feeds regularly on other vertebrate animals such as dogs, cats and cattle. These characteristics, together with its ability to move through forests and cities, turn this vector into a potential risk for the introduction of other pathological agents in urban areas. In some areas in Florida, for example, *Aedes aegypti* has been spatially and temporally replaced by *Aedes albopictus*, particularly in rural areas, suggesting that this species

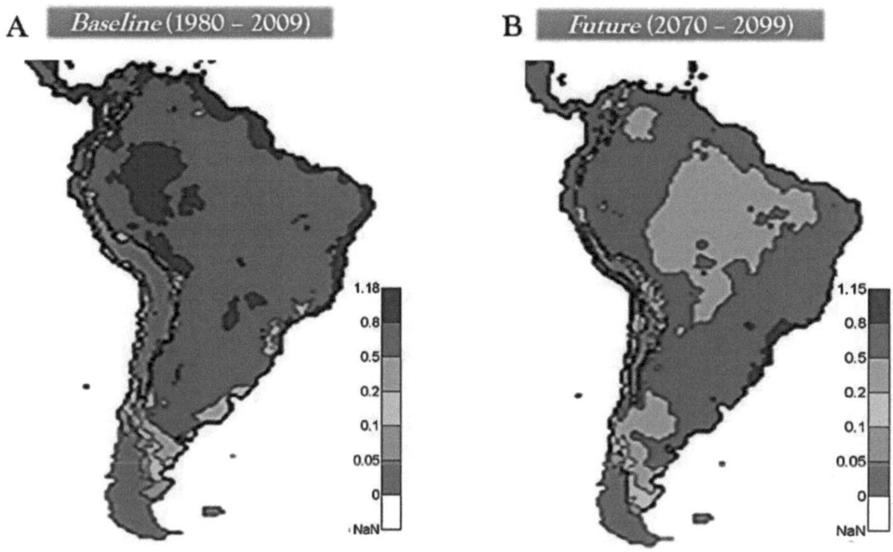

Fig. 4.21 Dengue epidemic potential based on *Aedes aegypti's* vectorial capacity in the baseline period (1980–2009) and for the RCP 8.5 scenario in Brazil, including day temperature variation data (Data source: Liu-Helmersson et al., 2014)

may be a better competitor for larval habitat, as well as more resistant to certain environmental conditions (Lounibos et al., 2010; Murrell, Damal, Lounibos, & Juliano, 2011). In 2015, about 272 Brazilian municipalities recorded the presence of *Aedes albopictus* (Brasil, 2015), but no cases of vector-borne diseases transmitted by this vector have been notified (Fig. 4.22).

4.6.4 Malaria

Malaria is a vector-borne infectious disease with a significant disease burden at global and regional levels (WHO, 2013). It is a serious public health burden, and its transmission dynamics is highly sensitive to climate conditions (Parham & Michael, 2010). Future temperature and ecosystem changes resulting from land use patterns may affect the transmission of this disease, accelerating the parasite-vector-host cycle. In Brazil, three *Plasmodium* species cause malaria: *Plasmodium malariae*, *Plasmodium vivax* and *Plasmodium falciparum*. The vector is a mosquito belonging to the order *Diptera*, family *Culicidae*, and genus *Anopheles*. The *Anopheles darlingi* is the main species that transmits this disease (Batista, Costa, & Silva, 2014; Molino-Cruz & Barillas-Mury, 2014). In terms of favourable climate conditions for malaria, studies suggest that, in general, there is risk of transmission at an optimal temperature of 31 °C; however, recent findings show that this value could be as low

Fig. 4.22 Brazilian municipalities recording the presence of *Aedes albopictus* (2015) (Data source: Brasil, 2015)

as 25 °C, with a risk reduction for temperatures above 28 °C (Blanford et al., 2013; Mordecai et al., 2012).

Currently, this disease has been showing a reduction in the number of cases, and its occurrence in Brazil is restricted to the Legal Amazon area, as we can see in Fig. 4.23. The high risk municipalities are primarily located in the state of Amazonas. Considering the average temperature variation map and using regionalised Eta/HadGEM2-ES (Chou et al., 2014), we can observe that high risk municipalities for malaria may reach average temperature variations between 4 °C and 6 °C under the RCP 8.5 scenario for 2071–2099.

In 2015, Laporta et al. (2015) studied the most favourable environmental conditions for the distribution of *Plasmodium falciparum* and its main vector, the *Anopheles darlingi*, as well as nine other potential vectors (species of the *Anopheles albitarsis* complex), according to two climate change scenarios for 2070 in South America. One of these two scenarios was extreme, with a predicted average temperature rise of 4 °C (RCP 8.5) for 2070. Under this scenario in Brazil, the *Plasmodium falciparum*'s distribution area would grow. The endemic areas in the North region of the country remained as such, but the disease expanded towards the Central-west (Fig. 4.24).

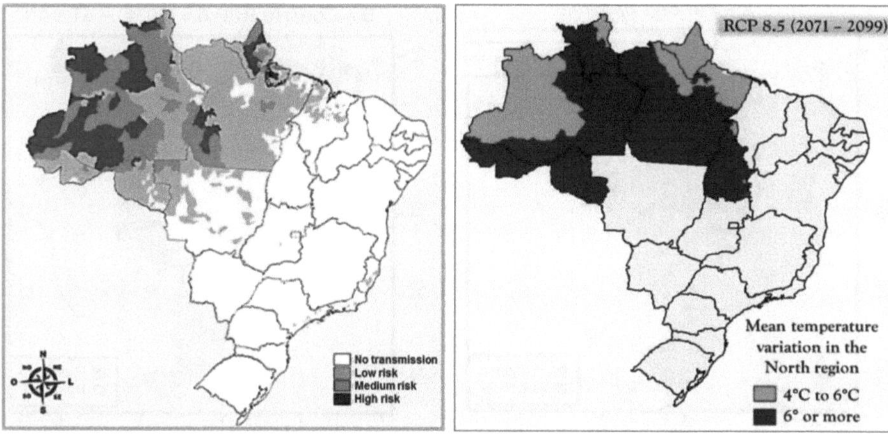

Fig. 4.23 Malaria risk map by municipality (2014), and average temperature variation for the North region under the RCP 8.5 scenario for the period 2071–2099 (Data source: development from the data Notifiable Diseases Information System (SINAN/SVS/SVS) and regionalised Eta/HadGEM2-ES (Chou et al., 2014) with municipalised projections (Costa et al., 2015). Map source for mean temperature in the North region: Oliveira, BFA. Ph.D. in Public Health and the environment, ENSP/FIOCRUZ)

With regards to its main vector, future projections for a temperature rise above 4 °C might significantly reduce adequate habitat conditions for the development of the *Anopheles darlingi,* with an impact on its distribution and abundance (Fig. 4.25). Thus, the considerable growth of the *Plasmodium falciparum*'s distribution potential (with projected temperature rises) highlights the importance of investigating other potential vectors for this disease (Fig. 4.26).

4.6.5 Leishmaniasis

Leishmaniasis is among the 6 major diseases that are neglected in the world, affecting men, women and children. These diseases are caused by protozoa of the genus *Leishmania*, and are transmitted to humans by hematophagous insects belonging to the genera *Phebotomus* and *Lutzomyia*. In the American continent, Brazil is the country with the highest incidence rates for visceral and tegumentary leishmaniasis (Alvar et al., 2012). The latter is particularly important, as it presents both chronicity and latency, and can develop metastasis leading to disfiguring clinical conditions.

In Brazil, tegumentary leishmaniasis is concentrated in the Brazilian Amazon region, and it is driven by environmental changes caused by humans, including land use and land cover changes in the Amazon forest, disorganised land occupation, and deforestation (Rangel, Costa, & Carvalho, 2014). Although 7 different species have been identified in Brazil, the main ones are *Leishmania amazonensis*, *Leishmania*

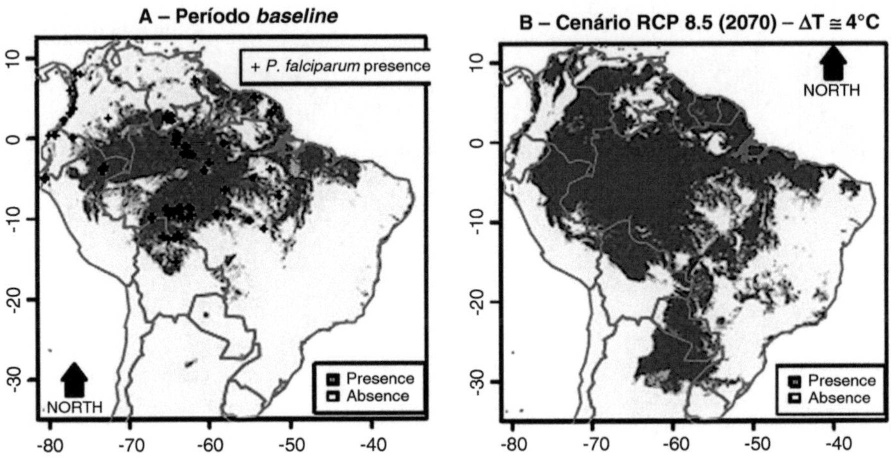

Fig. 4.24 Potential distribution area of *Plasmodium falciparum* in current environmental conditions (baseline) and for the RCP 8.5 scenario (2070) (Data source: Adapted from Laporta et al. (2015))

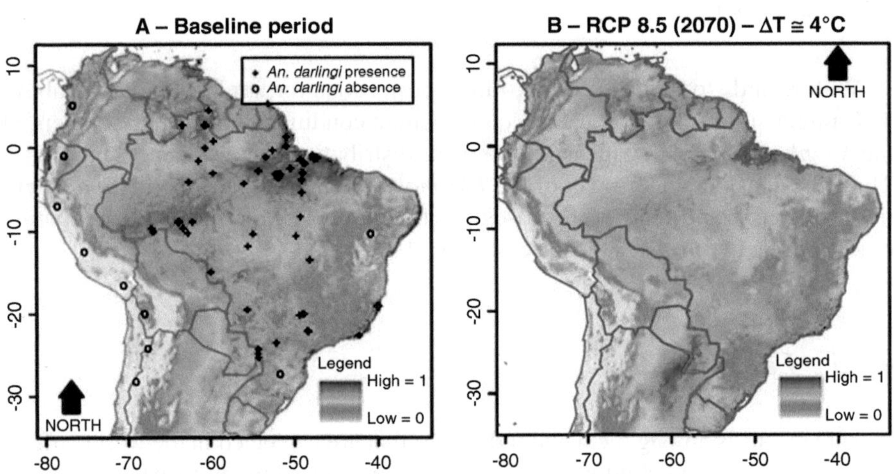

Fig. 4.25 Potential distribution of *Anopheles darlingi* in current environmental conditions (baseline) and for the RCP 8.5 scenario (2070) (Data source: Adapted from Laporta et al. (2015))

guyanensis and *Leishmania braziliensis*. The first is widely spread over the Amazon region. The *Lutzomyia flaviscutellata* has been reported as the main vector for this disease in the Amazon (Lainson & Shaw, 1968; Lainson et al., 1994).

As with other disease vectors, the distribution of the *Lutzomyia flaviscutellata* may be influenced by climate conditions (rainfall, in particular), and therefore it might expand to other areas with Warming above 4 °C. A temperature rise above

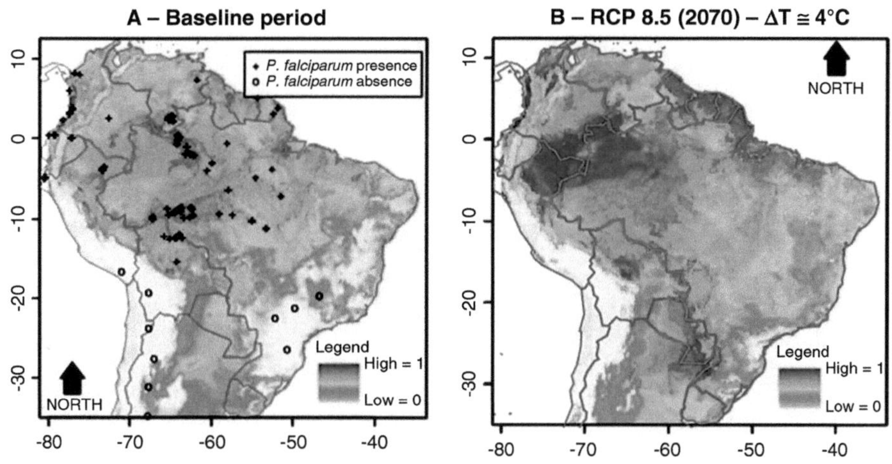

Fig. 4.26 Areas with high and low distribution potential for *Plasmodium falciparum* in current environmental conditions (baseline) and for the RCP 8.5 scenario (2070) (Data source: Adapted from Laporta et al. (2015))

4 °C is expected to intensify extreme events in the Amazon, such as prolonged droughts and changes in the rainfall regime, associated with a higher frequency of events such as El Niño (IPCC, 2013; IPCC, 2014). These factors may favour the expansion of the areas where tegumentary leishmaniasis occurs.

Future projections of favourable habitats for the development and expansion of *Lutzomyia flaviscutellata* according to different climate change scenarios in South America have been made by Carvalho, Rangel, Ready, and Vale (2015) for the year 2050 (median for 2041–2060). Based on the results of this study, there is a predicted expansion of the area considered climatically appropriate for the vector, including; in particular, the states of Minas Gerais, Mato Grosso do Sul and São Paulo. In Fig. 4.27, the highlighted areas in the south of Mato Grosso do Sul and Minas Gerais may present an average temperature rise around 5 °C and 4.5 °C, respectively, considering the RCP 8.5 scenario for a similar period to the one used for the study.

With regards to the number of cases, projections referring to variations in the number of hospitalisations by leishmaniasis have been made for three periods in Brazil (2010–2039; 2040–2079 and 2080–2100) under two different climate change scenarios (Mendes, Coelho, Féres, Souza, & Cunha, 2016). A total of 618 additional annual hospitalisations due to this disease are estimated for the end of the century in the A2 scenario, which predicts a 3.4 °C temperature rise, a percentage increase of 15% relative to the baseline period (1992–2002). Among the states that will be most affected, Rondônia, Pará, Amazonas, Bahia, Maranhão, Minas Gerais, São Paulo, Santa Catarina and Mato Grosso do Sul stand out.

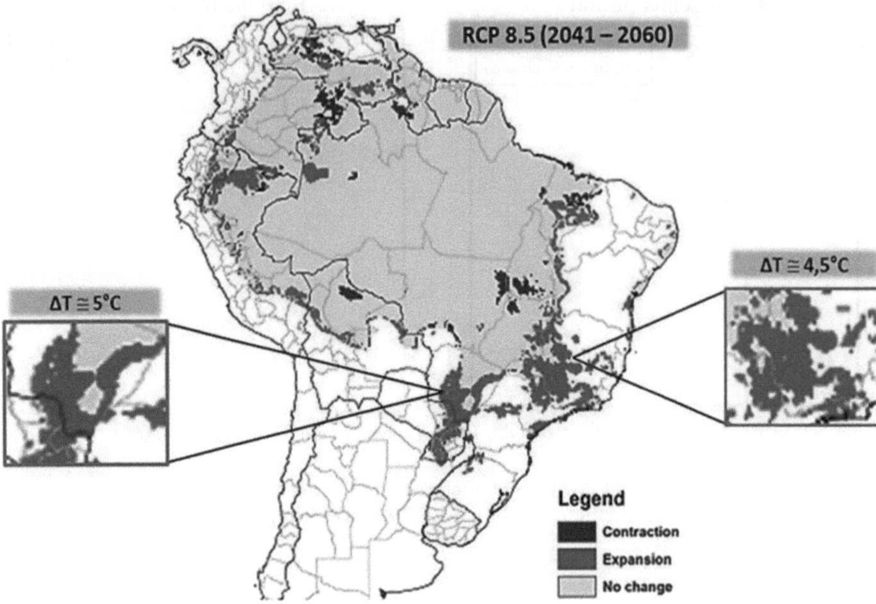

Fig. 4.27 Future projections of favourable climate conditions for the development of vector *Lutzomyia flaviscutellata,* under the RCP 8.5 scenario (2041–2060) (Data source: Adapted from Carvalho et al., 2015)

4.7 Water-borne and Food-borne Diseases

Water-borne diseases are caused by a variety of micro-organisms, biotoxins and toxic contaminants. Their epidemic features have been associated with the occurrence of extreme events, mainly in capitals and metropolitan areas, as well as other factors, such as floods, low-income population agglomeration, inadequate sanitation and high infestation of infected rodents.

In a 4 °C temperature rise scenario, with predictions of increasing extreme events, including in Brazil, people will be much more likely to contract water-borne infectious diseases, notably gastroenteritis, intestinal parasitosis, leptospirosis and viral hepatitis (Hep-A). In the specific case of gastroenteritis, outbreaks have been associated not only with post-flood periods (Ahern, Kovats, Wilkinson, Few, & Matthies, 2005; Heller, Colosimo, & Antunes, 2003), but also with prolonged droughts, food contamination and dehydration resulting from temperature rises (Fleury, Charron, Holt, Allen, & Maarouf, 2006; Naumova et al., 2007; Xu et al., 2014).

4.7.1 Leptospirosis

Leptospirosis is a worldwide disease, which is endemic in some Brazilian regions. It is caused by pathogenic *leptospiras* transmitted through contact with the urine of infected animals or with contaminated water, mud or soil. Urban epidemics are recorded every year, mainly in low income communities with no basic sanitation, and after floods, inundations or large natural disasters, such as in São Paulo, Rio de Janeiro, Salvador and Recife (Ko, Reis, Dourado, Johnson, & Riley, 1999). In general, flooded areas present a higher incidence of this disease as compared with non-flooded areas (Barcellos & Sabroza, 2001). The presence of a flood increases by 2.5 times the chance of contracting the disease (IC95%: 1.08–6.17) (Sarkar et al., 2002). With regards to the relationship between rainfall and leptospirosis in the city of São Paulo, for every 20 mm increase in rainfall, a 15.6% increase was observed in the number of cases (starting on the 14th day after the occurrence of a storm) (Coelho & Massad, 2012).

In Brazil, there are records of leptospirosis in all states, but there are more occurrences in the South and Southeast regions of the country. This epidemiological characterisation is extremely relevant for a temperature rise scenario. Under the RCP 8.5 scenario, rainfall is predicted to increase in the South region, in particular in the state of Rio Grande do Sul. This state has a high incidence of leptospirosis, with about 4.5 cases per 100,000 people – higher than the country rate (1.8 cases per 100,000 people) for the year 2015 (Brasil, 2016b).

Considering a 4 °C temperature rise scenario, the percentage increase in leptospirosis was analysed adapting the methodology described by Kolstad and Johansson (2011). The increased risk of leptospirosis was calculated with the expression $RR = ((\exp^{(\beta * \Delta P)}) - 1) * 100$, where β refers to the expected increase in the number of cases of leptospirosis for every 1 mm increase in rainfall (estimated from the results of Coelho & Massad, 2012); and ΔP refers to the difference between rainfall values for the RCP 8.5 scenario in the period 2071–2099 relative to the baseline period (1960–1990).

The results showed that, under the RCP 8.5 climate change scenario for 2071–2099, municipalities in the states of Rio Grande do Sul and Santa Catarina may experience a 150% increase in the cases of leptospirosis resulting from increased rainfall (Fig. 4.28). These municipalities are predicted to have an average temperature rise between 4 °C and 6 °C under the RCP 8.5 scenario (2071–2099) relative to the baseline period (1961–1990). It is noteworthy that these estimates are based on rainfall predictions, and therefore do not fully reflect the risk of floods. There might be heavy and intense rainfalls in other regions which will not necessarily represent an average rainfall increase at the end of the century, but which may cause floods and inundations (see list of municipalities risking floods in the Chapter on natural disasters). Furthermore, the incidence of leptospirosis depends on other factors in addition to climate change, such as lack of basic sanitation.

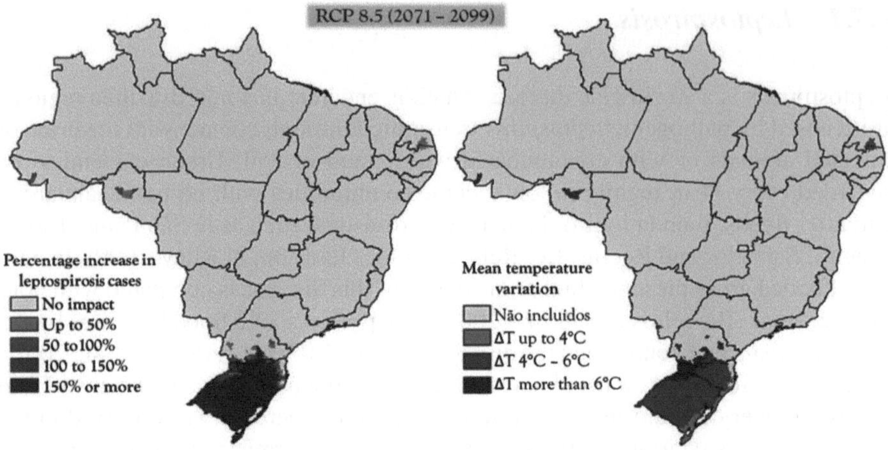

Fig. 4.28 Projections for percentage increase of the incidence of leptospirosis due to increased rainfall under the RCP 8.5 scenario for the period 2071–2099, and average temperature variation under the RCP 8.5 scenario for the period 2071–2099 relative to the baseline period (1961–1990) (Data source: development from the data Regionalised Eta/HadGEM2-ES (RCP 8.5 scenario) (Chou et al., 2014). The percentage increase was estimated adapting the methodology described by Kolstad and Johansson (2011) and using the results of Coelho and Massad (2012). Map source: Oliveira, BFA

4.7.2 *Diarrhoea*

Diarrhoeal diseases are orally or faecally-orally transmitted. The means of transmission varies according to the pathological agent. In general, diarrhoea's epidemiological chain is directly related to socio-economic factors, lack of basic sanitation (treated water supply and sewage network), protein-calorie malnutrition, food and water quality, and personal hygiene (Veronesi & Focaccia, 2004). Often, its mortality and hospitalisation indicators for children below the age of 5 – the most vulnerable group – are considered indicative of local socio-economic, health and environmental conditions (Bühler, Ignotti, Neves, & Hacon, 2014). In addition to socio-economic factors, rainfall, temperature and floods have been associated with the occurrence of diarrhoea in several studies (Ahern et al., 2005; Fleury et al., 2006; Heller et al., 2003; Naumova et al., 2007; Xu et al., 2014).

With regards to temperature, studies suggest that some etiological agents that cause diarrhoea tend to proliferate more intensely at higher temperatures (Carlton et al., 2016), and that inadequate food preparation, handling and packaging, especially at extreme temperatures, favour even further the proliferation of some pathogens (Bentham & Langford, 1995, 2001). For example, temperature rise has been positively associated with health risks caused by bacteria such as *Salmonellae*, *Campylobacter* and *Escherichia coli* (Carlton et al., 2016).

Favourable climate conditions for the growth of diarrhoea-causing pathogens vary according to the etiological agent. For bacterial proliferation, especially in food, the optimal temperature is above 35 °C for *Salmonellae*, *Escherichia coli*, *Shigella* and *Vibrio cholerae*, which respectively present a maximum temperature threshold equal to 45 °C – 47 °C, 46 °C, 45 °C – 47 °C and 43 °C (Lund & Baird-Parker, 2000). This is a risk for several areas in Brazil, considering that temperatures above 35 °C are easily reached for several days during the year in many regions of the country.

Concerning protozoa, the *Giardia lamblia* has maximum growth at 37 °C, with high mortality rates at temperatures above 39 °C (Baveja, Jyoti, Anand, & Agarwal, 1984). Low temperatures (4 °C to 20 °C) create the ideal environment for the survival of viral agents, such as the rotavirus. These agents tend to lose their infectivity when temperatures exceed 37 °C. However, even at environmental temperatures above 30 °C, rotavirus particles stored in faeces remain stable and can cause infections *in vitro* after 2 to 5 months in storage (D'Souza, Hall, & Becker, 2008).

Considering the RCP 8.5 scenario (2071–2099), an analysis showed that Brazilian municipalities with a temperature rise above 4 °C presented favourable conditions for the proliferation of pathogens associated with the occurrence of gastroenteritis. According to the data observed, almost the whole Brazilian territory will present favourable climate conditions for the proliferation of enterobacteria, with maximum average temperatures above 35 °C. With regards to the rotavirus, most municipalities in the Northeast and Southeast regions will present favourable environmental conditions for the transmissions of this disease, while there will be a reduction in the North and Northeast regions. In a strip of land between the North and Northeast regions, maximum average temperatures are expected to exceed 39 °C, which would result in an unfavourable environment for the proliferation of protozoa such as *Giardia lamblia* (Fig. 4.29).

With regards to the threshold temperature values above which we can observe temperature effects, in Dhaka, Bangladesh, a temporal series study showed a 40.2% increase in hospitalisations by diarrhoea due to rotavirus for every 1 °C temperature rise starting at 29 °C (Hashizume et al., 2008). In Brazil, in a study carried out in Rio Branco (AC), hospitalisations of children under the age of 5 by diarrhoea were associated with maximum temperature, with temperature effects starting at 32 °C. Thus, for a 4 °C maximum temperature rise, the number of notified cases of diarrhoea in children under the age of 5 might double in Rio Branco (Oliveira, Protázio, Jünger, & Hacon, 2016).

Although the analyses in Fig. 4.29 provide relevant data to assess the impact of temperature rise on gastroenteritis, the occurrence of this disease results from the interaction of several factors, including basic sanitation, hygiene habits, social strata and proper storage. Clearly, temperature variations due to climate change may (and possibly have already) affect the incidence of diarrhoea in several places, mainly among more socio-economically vulnerable populations. Based on the results of a recent meta-analysis and systematic review covering 26 studies, a 4 °C average temperature rise might raise the number of cases by 31% (Carlton et al., 2016); this increase could reach 41% in low and medium income countries, and 26% in high income countries (Carlton et al., 2016).

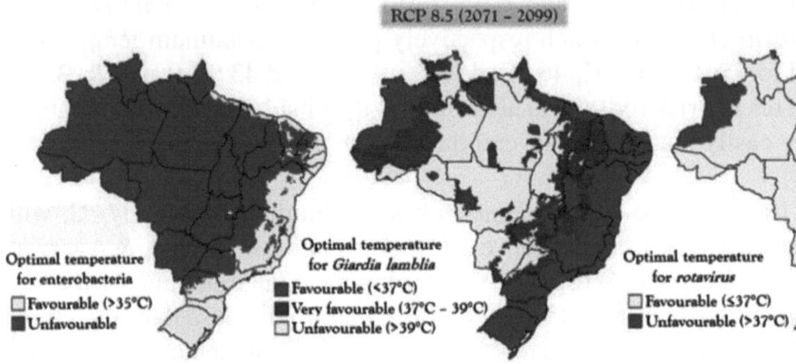

Fig. 4.29 Optimal temperature for the proliferation of pathogens associated with diarrhoeal diseases under the RCP scenario (2077–2099) (Data source: enterobacteria: Lund & Baird-Parker, 2000; Giardia lamblia: Baveja et al., 1984; rotavirus: D'Souza et al., 2008. For the optimal temperature, the maximum average temperature was used (2071–2099). Regionalised Eta/HadGEM2-ES (Chou et al., 2014) with municipalised projections (Costa et al., 2015). Map source: Oliveira, BFA

Assuming that the risks for diseases are different according to social strata, the potential risk of hospitalisation by diarrhoea deriving from an average temperature rise above 4 °C under the RCP 8.5 scenario (2071–2099) was analysed for all Brazilian municipalities, adjusting the analysis by socio-economic vulnerability indicators, as described by Hacon et al. (2016). The index was normalised, and municipalities with values between 0.000 and 0.401 were considered low vulnerability, whereas those above 0.401 were considered as medium and high socio-economic vulnerability. For the low and medium-high vulnerability municipalities, the concentration-response function was respectively 0.058 and 0.086, as presented by Carlton et al. (2016).

If socio-economic vulnerability conditions remain stable over the next few decades, most municipalities in the North, Central-west and Northeast regions may present a more than 45% increase in cases of diarrhoea. In the North region, an average temperature rise around 7 °C might increase by 76% the number of cases. This same percentage may be observed in the north of the Southeast region and the northwest of the Northeast region, where the states of Maranhão and Piauí are located (Fig. 4.30).

While analysing the municipalities that might present a 50% increase in the cases of diarrhoea under the RCP 8.5 scenario for the period 2071–2099, we observed that 34% of the municipalities and 64% of the child population might be at risk in the North region of the country, where the predicted average temperature rise is 7 °C. It is also noteworthy that 23% of the municipalities in the Northeast and Southeast regions might present an increase of more than 50% in the cases of diarrhoea (Fig. 4.31).

In general, these results may be underestimated, as other environmental factors, such as an increase in extreme events, may also favour the occurrence of diarrhoea in these places. Heat waves, prolonged droughts and floods have been strongly linked with the increase of diarrhoeal diseases, as they affect the dynamics of water-

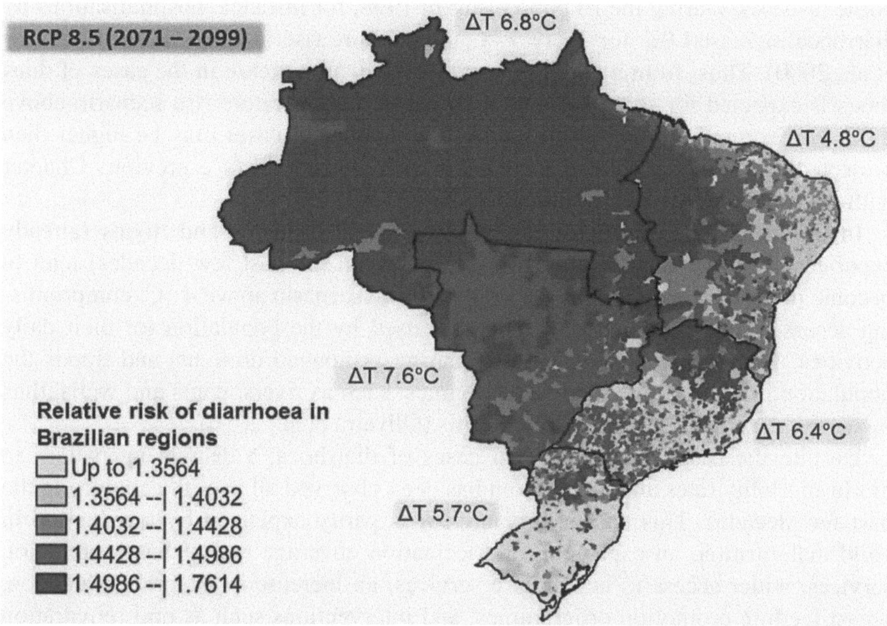

Fig. 4.30 Estimated risk projections for the occurrence of diarrhoea in Brazilian municipalities, according to temperature rise under the RCP 8.5 scenario for the period 2071–2099 (Data source: development from the data Eta/HadGEM2-ES (RCP 8.5 scenario) (Chou et al., 2014) with municipalised projections (Costa et al., 2015). The percentage increase was estimated adapting the methodology described by Kolstad and Johansson (2011) and using the results of Carlton et al. (2016), adjusted by the vulnerability index described by Hacon et al. (2016))

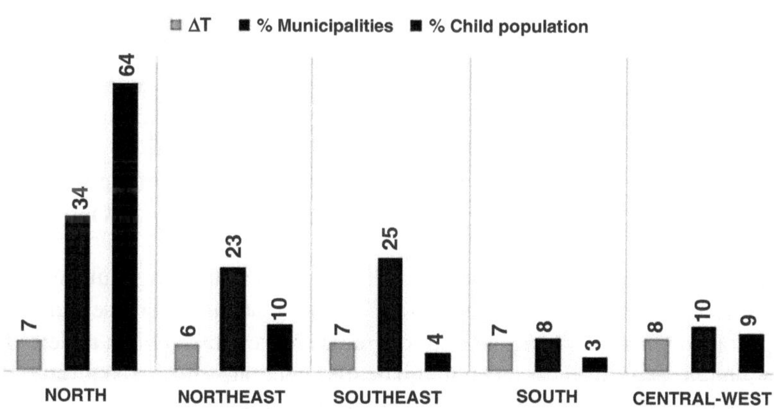

Fig. 4.31 Temperature variation, percentage of municipalities, and percentage of child population presenting a more than 50% risk of diarrhoea under the RCP 8.5 scenario for the period 2071–2099 (Data source: Eta/HadGEM2-ES (RCP 8.5 scenario) (Chou et al., 2014) with municipalised projections (Costa et al., 2015). Percentage increase estimated adapting the methodology described by Kolstad and Johansson (2011) and using the results of Carlton et al. (2016). Population distribution based on the 2010 Census population data. Chart source: Oliveira, BFA

borne diseases. During the El Niño event in 1998, for instance, hospitalisations by diarrhoea increased 8% for every 1 °C temperature rise in Lima, Peru (Checkley et al., 2000). Thus, from a qualitative perspective, an increase in the cases of diarrhoea is expected for some regions of Brazil in a temperature rise scenario above 4 °C. For example, in the South region, the number of cases may be higher than projected due to the predicted increase in average rainfall (see previous Chapter with regards to increased risk of leptospirosis).

In the North and Northeast regions, prolonged droughts and floods (already experienced by the inhabitants of those regions in the past few decades) tend to become more frequent and intense in a warming scenario above 4 °C, compromising access to and the quality of the water used by the population for their daily activities. This would happen because during prolonged droughts and floods the population may seek untreated water sources, such as rivers, dams and wells, thus favouring the occurrence of gastroenteritis (Oliveira et al., 2016).

Despite the projected increase in cases of diarrhoea, a decreasing pattern in morbi-mortality rates among children has been observed all over the country in the past few decades. This temporal behaviour is partly explained by a reduction in child malnutrition, an expansion of vaccination coverage and of basic sanitation services, wider access to health care services, an increase in maternal education, breast feeding promotion programmes, and interventions such as oral rehydration therapy and vaccination against the rotavirus (Benicio & Monteiro, 2000; Claeson & Waldman, 2000; Kosek, Bern, & Guerrant, 2003; Vanderlei, Silva, & Braga, 2003).

4.8 Socio-economic and Health Care Vulnerability for a Warming Above 4 °C

Numerous human responses to climate change are directly associated with individual and collective vulnerability aspects, such as age, health care services, physiological resistance and social determinants and constraints. Therefore, health impacts deriving from a warming above 4 °C will depend on the ability of the health system to adapt to the expected climate variability and to more intense and more frequent extreme events.

With regards to such ability, social conditions such as housing, nutrition, sanitation and access to health care services are factors that may increase vulnerability among those populations exposed to temperature rises, which, when added to other environmental exposures, may synergistically generate more risks, and consequently increase the demand for and the costs of health care.

Considering socio-economic, demographic and health care indicators, Hacon et al. (2016) compared spatial distribution of vulnerability indicators (including access to health care services) with the climate risk for each Brazilian municipality in different warming scenarios and different types of vulnerability. In the general analysis, the indicators highlighted precarious human development conditions in the North and Northeast regions, which suggests less ability to face health problems in the challenging context of climate change.

The socio-economic and health care vulnerability indices described by Hacon et al. (2016) have been classified as low, moderate, high and very high, according to quartiles, i.e., the low vulnerability municipalities correspond to the 25% with the lowest values. With regards to vulnerabilities associated with the conditions of local services, those municipalities classified as high and very high vulnerability are located in the North and Northeast regions. Considering municipalities with very high vulnerability to the conditions of health care services, the average temperature rise will be, under the RCP 8.5 scenario for the period 2071–2099, between 4 °C and 6 °C in most of the Northeast region, and above 6 °C in the North (Fig. 4.32). Thus, health authorities and managers in these places may consider implementing proactive actions to improve the response of health care services to a potential warming above the expected.

With regards to socio-economic conditions, a stronger pattern of municipalities with high and very high socio-economic vulnerability was observed in the North and Northeast regions, in the north of the Southeast region (specifically the north of the state of Minas Gerais), and in some municipalities in the state of Mato Grosso, in the Central-west region. For the very high vulnerability municipalities, a temperature rise between 4 °C and 6 °C was observed in the Northeast region, and above 6 °C for some high socio-economic vulnerability municipalities in the North region (RCP 8.5 scenario for 2071–2099). It is worth naming some isolated municipalities in some states, with their respective temperature rises: Xique-Xique (BA) with ↑ of

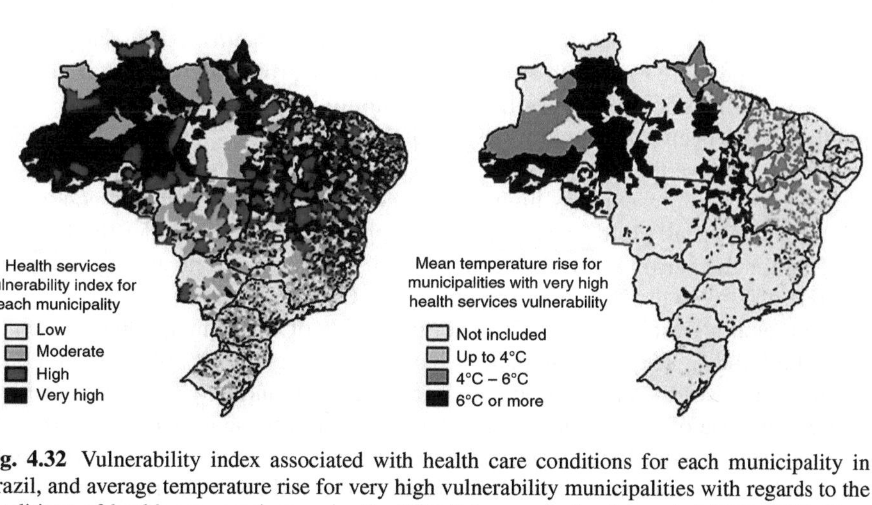

Fig. 4.32 Vulnerability index associated with health care conditions for each municipality in Brazil, and average temperature rise for very high vulnerability municipalities with regards to the conditions of health care services under the RCP 8.5 scenario for the period 2071–2099 (Data source: Hacon et al., 2016 and Eta/HadGEM2-ES (RCP 8.5 scenario) (Chou et al., 2014) with municipalised projections (Costa et al., 2015). Map source: Oliveira, BFA

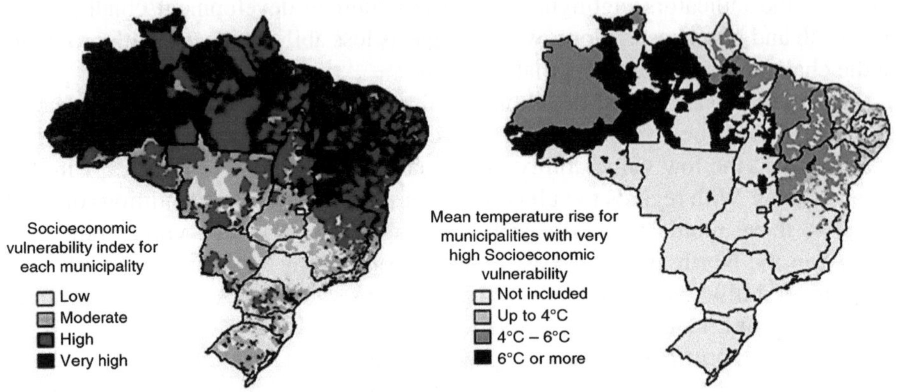

Fig. 4.33 Socio-economic vulnerability index for each municipality in Brazil, and average temperature rise for very high socio-economic vulnerability municipalities under RCP 8.5 for the period 2071–2099 (Data source: Hacon et al., 2016 and Eta/HadGEM2-ES (RCP 8.5 scenario) (Chou et al., 2014) with municipalised projections (Costa et al., 2015). Map source: Oliveira, BFA

8 °C; Bonito de Minas (MG) with ↑ of 7 °C; Campinápolis (MT) with ↑ of 8 °C; Campo Novo de Rondônia with ↑ of 7 °C; Seringueiras (RO) with ↑ of 8 °C; Nova União (RO) with ↑ of 8 °C; and Vale do Arani (RO) with ↑ of 7 °C (Fig. 4.33).

In addition to health care conditions and socio-economic factors, biological vulnerability is also an aspect to be considered in an extreme temperature scenario. Children, for example, are more sensitive to socio-economic conditions, and pathologies such as diarrhoea, dehydration and respiratory diseases may be aggravated during events such as floods and prolonged droughts. The elderly, on the other hand, are prone to thermal discomfort and cardiovascular problems, particularly during extreme temperature rises such as heat waves. Thus, a profile of the exposed population in municipalities with very high vulnerability to socio-economic and health conditions has been presented in Figs. 4.34 and 4.35.

We can observe that, in the North and Northeast regions, approximately 80% of all municipalities present very high vulnerability with regards to health conditions, affecting 40% of all children and elderly living in those areas. Under the RCP 8.5 scenario for the period 2071–2099, they would be exposed to an average temperature rise of 5 °C in the Northeast region, and 7 °C in the North (Fig. 4.35).

With regards to socio-economic conditions, 92% of the municipalities presented a very high index. These municipalities have approximately 70% of the child population and 63% of the elderly population in the region, and may present an average temperature rise of 7 °C under the RCP 8.5 scenario for the period 2071–2099. In the Northeast region, 98% of all municipalities have been classified as very high vulnerability, representing 77 and 75 percent of the child and elderly population in the region, respectively. These municipalities may experience an average temperature rise of 4.8 °C under the RCP 8.5 scenario at the end of the century. These estimates show that the North and Northeast regions, in particular, are priority concerns for an extreme temperature rise (Fig. 4.35).

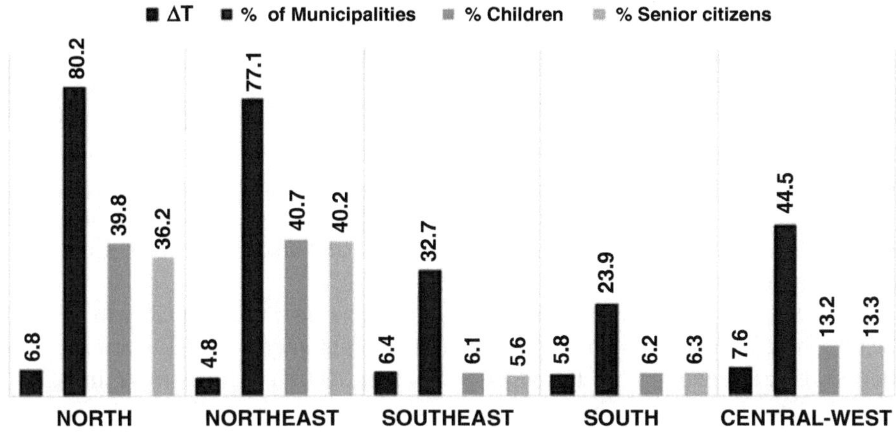

Fig. 4.34 Temperature variation, percentage of municipalities, and percentage of child and elderly population in municipalities with very high vulnerability to health conditions under the RCP 8.5 scenario for the period 2071–2099 (Data source: Hacon et al., 2016 and Eta/HadGEM2-ES (RCP 8.5 scenario) (Chou et al., 2014) with municipalised projections (Costa et al., 2015). Map source: Oliveira, BFA

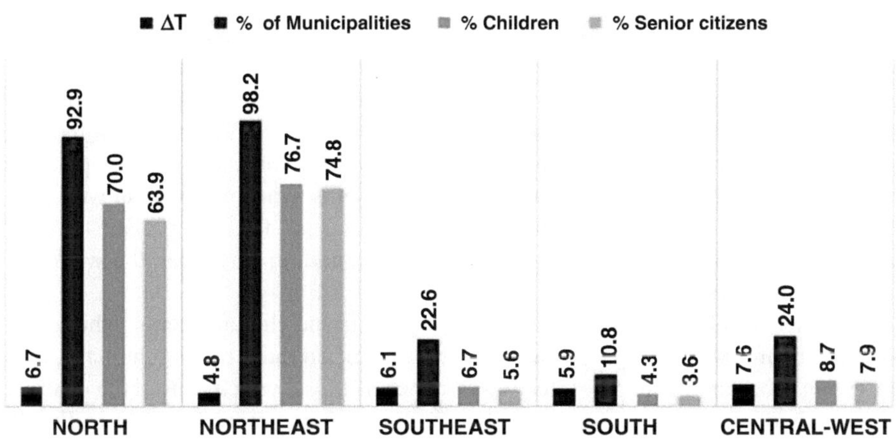

Fig. 4.35 Temperature variation, percentage of municipalities, and percentage of child and elderly population in municipalities with very high socio-economic vulnerability under the RCP 8.5 scenario for the period 2071–2099 (Data source: Hacon et al., 2016 and Eta/HadGEM2-ES (RCP 8.5 scenario) (Chou et al., 2014) with municipalised projections (Costa et al., 2015). Map source: Oliveira, BFA

4.9 Final Considerations on the Potential Impacts of a Warming Above 4 °C on the Health Sector

Climate change is one of the most likely threats to public health in the future, and an analysis of current and future consequences of a temperature rise above 4 °C represents a major challenge for the sector. This is also true because predicting the consequences of a temperature rise above the expected may improve the ability of the health sector to deal with emergency situations and with the extreme risks associated with climate change.

Despite the complexity and multiplicity of factors involved in quality of life and health, there is consensus among the scientific community that high warming may have a direct impact on the Brazilian population, including heat stress, increased risk of vector-borne, water-borne and food-borne diseases, as well as cardiovascular and respiratory diseases. In addition, the increase frequency of extreme climate events, such as heat waves, prolonged droughts, floods or inundations may also lead to a number of impacts on public health, particularly for the most vulnerable groups.

Notably, the climate change impacts on human health are complex, often indirect and with multiple factors. Therefore, they can vary from one place to another. In climate terms, the projections show that all regions will exceed the 4 °C threshold under the RCP 8.5 scenario, particularly in the North and Central-west regions, where the average warming may exceed 6 °C and 7.5 °C, respectively. Among the predicted impacts of these rises, heat stress is particularly important, as it allows a clear definition of human biological limits for thermal comfort and biological adaptability to extreme heat (Sherwood & Huber, 2010). The physical and thermodynamic limits of the human body were measured with the WBGT and UTCI indices. According to the results, in several municipalities in the North region the population will be exposed for many hours, days and even months to an environment with temperatures above 35 °C and high humidity, which suggests, for example, the need for acclimatisation and reduction in physical exercise, and, in particular, the workload of some activities.

Another widely highlighted impact refers to extreme climate events, which, differently from indirect impacts, can be abrupt and often intolerable to humans, thus affecting thousands of people. Heat waves, such as those, which caused the death of thousands of people in Europe and Russia, may become more frequent in Brazil, affecting especially the elderly, who are more vulnerable to extreme heat conditions (Anderson & Bell, 2009; Kinsgley et al., 2015; Matsueda, 2011; Robine et al., 2008; Son et al., 2016). Actually, with the current increase in Brazil's elderly population, an imminent risk of death and hospitalisation resulting from extreme heat within this group is expected by the health sector. Thus, adaptation and reduction plans for these vulnerabilities need to be developed and implemented, according to local characteristics.

Natural disasters, particularly floods and prolonged droughts, are other extreme events worthy of attention. In the case of floods, in addition to direct impacts, increased temperatures and rainfalls, together with socio-economic and demographic variables, are likely to maximise the increase of water-borne diseases, such as diarrhoea and leptospirosis. In the North and Northeast regions, prolonged droughts may affect the availability of and access to drinking water for people's daily needs, and indirectly reduce crop productivity, thus reducing food availability and consequently increasing malnutrition. Although current conditions qualitatively show the potential impacts of these events on human health, the discussion, albeit extremely relevant, is still very recent in the health sector.

Extreme climate and environmental conditions, with temperature rises above 4 °C, may influence the adoption of new human behaviours due to the direct or indirect need for populations to adapt to these extreme risks. Thus, new lifestyles, daily routines, eating habits, and means of accessing water may be modified, resulting in new nosological profiles. In certain situations, the inadaptability of some regions to high Warming may favour forced human migration, and, in addition to social, demographic and economic consequences, we may observe an increase in diseases sus Dengue, Malaria, and more recently, Zika and Chikungunya.

In the current epidemiological situation, vector-borne infectious diseases may be an aggravating factor in a temperature rise scenario. This is because climate alterations that are favourable to the spread of vectors and other diseases, as well as the growing number of international flights, which favour the movements of diseased or infected people during the incubation period, may, in addition to Chikungunya and Zika, favour the introduction of new arboviruses in the country, aggravating even further the current scenario. In view of that, local communities and all spheres of government should avoid extreme health risks through strategic plans and actions to improve health surveillance systems (not only for diseases themselves, but also for early signs, particularly of vector populations), and invest in systematic vaccination coverage, as possible.

In epidemiological alert situations, such as epidemics, heat waves and natural disasters, the health system needs to reinforce and quickly implement a set of tools to obtain timely data on the risks in order to launch effective interventions. Considering this situation, notification frequency or usual case definitions may be changed during an emergency period.

From a social perspective, the impacts of temperature rise on human health will be distributed asymmetrically among different regions of the country, with the least favoured groups suffering the strongest impacts. In this context, the best response the health sector can give to a warming above 4 °C is to work together with other sectors to reduce our vulnerability to such events, strongly improving social inequality, sanitation infrastructure, education, transport services and health care services.

In conclusion, the health sector must consider measures to reduce vulnerabilities and implement adaptation measures in order to reduce as much as possible the impacts of extreme events and temperature rise. In this context, the results of this review and the projections made for several regions of Brazil are extremely relevant for strategic planning and priority setting, as well as for the implementation of proactive actions that contribute to expand and enable health services, reduce and control social and environmental vulnerabilities, and implement adaptation plans.

References

Ahern, M., Kovats, R. S., Wilkinson, P., Few, R., & Matthies, F. (2005). Global health impacts of floods: Epidemiologic evidence. *Epidemiologic Reviews, 27*, 36–46.

Alexander, L. V., & Arblaster, J. M. (2009). Assessing trends in observed and modelled climate extremes over Australia in relation to future projections. *International Journal of Climatology, 29*, 417–435. https://doi.org/10.1002/joc.1730

Alvar, J., Vélez, I. D., Bern, C., Herrero, M., Desjeux, P., Cano, J., et al. (2012). Leishmaniasis worldwide and global estimates of its incidence. *PLoS One, 7*(5), e35671. https://doi.org/10.1371/journal.pone.0035671 PMID: 22693548

American Conference of Governmental Industrial Hygienists (ACGIH). (1996). Threshold Limit Values for chemical substances and physical agents. Biological Exposure Indices, American Conference of Governmental Industrial Hygienists, Cincinnati, OH.

Anderson, B. G., & Bell, M. L. (2009). Weather-related mortality: How heat, cold and heatwaves affect mortality in the United States. *Epidemiology, 20*(2), 205–2013.

Baccini, M., Kosatsky, T., Analitis, A., Anderson, H. R., D'Ovidio, M., Menne, B., et al. (2011). Impact of heat on mortality in 15 European cities: attributable deaths under different weather scenarios. *Journal of Epidemiology and Community Health, 65*(1), 64–70.

Barcellos, C., & Lowe, R. (2014). Expansion of the dengue transmission area in Brazil: the role of climate and cities. *Tropical Medicine & International Health, 19*(2), 159–168.

Barcellos, C., Monteiro, A. M. V., Corvalán, C., Gurgel, H. C., Carvalho, M. S., Artaxo, P., et al. (2009). Mudanças climáticas e ambientais e as doenças infecciosas: cenários e incertezas para o Brasil. *Epidemiol Serv Saúde, 18*(3), 285–304.

Barcellos, C., & Sabroza, P. C. (2001). The place behind the case: leptospirosis risks and associated environmental conditions in a flood-related outbreak in Rio de Janeiro. *Cadernos de Saúde Pública, 17*(Suppl), S59–S67.

Basagaña, X., Sartini, C., Barrera-Gómez, J., Dadvand, P., Cunillera, J., Ostro, B., et al. (2011). Heat waves and cause-specific mortality at all ages. *Epidemiology, 22*, 765–772.

Batista, E. P. A., Costa, E. F. M., & Silva, A. A. (2014). Anopheles darlingi (Diptera: Culicidae) displays increased attractiveness to infected individuals with Plasmodium vivax gametocytes. *Parasites & Vectors, 7*, 251.

Baveja, U. K., Jyoti, A. S., Anand, B. S., & Agarwal, D. S. (1984 Sep). Optimum conditions for growth of Giardia lamblia in vitro. *The Journal of Communicable Diseases, 16*(3), 184–189.

Becker, J. A., & Stewart, L. K. (2011). Heat-related illness. *American Family Physician, 83*(11), 1325–1330.

Bell, M. L., O'Neill, M. S., Ranjit, N., Borja-Aburto, V. H., Cifuentes, L. A., & Gouveia, N. C. (2008 Aug). Vulnerability to heat-related mortality in Latin America: a case-crossover study in Sao Paulo, Brazil, Santiago, Chile and Mexico City, Mexico. *International Journal of Epidemiology, 37*(4), 796–804.

Benicio, M. H., & Monteiro, C. A. (2000). Tendência secular da doença diarréica na infância na cidade de São Paulo (1984-1996). *Rev Saúde Pública, 34*, 83–90.

Bennet, H., Jones, R., Keating, G., Woodward, A., Hales, S., & Metcalfe, S. (2014). Health and equity impacts of climate change in Aotearoa-New Zealand, and health gains from climate action. *The New Zealand Medical Journal, 127*(1406), 16–31.

Bentham, G., & Langford, I. (1995). Climate change and the incidence of food poisoning in England and Wales. *International Journal of Biometeorology, 39*, 81–86.

Bentham, G., & Langford, I. (2001). Environmental temperatures and the incidence of food poisoning in England and Wales. *International Journal of Biometeorology, 45*, 22–26.

Beserra, E. B., Fernandes, C. R. M., Silva, S. A. O., Silva, L. A., & Santos, J. W. (2009). Efeitos da temperatura no ciclo de vida, exigências térmicas e estimativas do número de gerações anuais de Aedes aegypti (Diptera, Culicidae). *Ser Zool, 99*(2), 142–148.

Bitencourt, D. P., Ruas, A. C., & Maia, P. A. (2012). Análise da contribuição das variáveis meteorológicas no estresse térmico associada à morte de cortadores de cana-de-açúcar. *Cadernos de Saúde Pública, 28*(1), 65–74.

Blanford, J. I., Blanford, S., Crane, R. G., Mann, M. E., Paaijmans, K. P., Schreiber, K. V., et al. (2013). Implications of temperature variation for malaria parasite development across Africa. *Scientific Reports, 3*, 1300.

Botelho, C., Correia, A. L., Silva, A. M., Macedo, A. G., & Silva, C. O. (2003). Environmental factors and hospitalization of under-five children with acute respiratory infection. *Cadernos de Saúde Pública, 19*(6), 1771–1780.

Bouchama, A., & Knochel, J. P. (2002). Heat stroke. *The New England Journal of Medicine, 346*(25), 1978–1988.

Brasil. Ministério da Saúde. (2015). Secretaria de Vigilância em Saúde. Resultados LIRAa Nacional 2015. Brasília, 2015. Available at http://portalsaude.saude.gov.br/images/pdf/2015/dezembro/03/LIRAa-NACIONAL-2015-Verso-II.pdf

Brasil. Ministério da Saúde. (2016a). Secretaria de Vigilância em Saúde. Situação epidemiológica da dengue, 2016. [Accessed Jan 2016]. Available at: http://portalsaude.saude.gov.br/index.php/situacao-epidemiologica-dados-dengue

Brasil. Ministério da Saúde. (2016b). Secretaria de Vigilância em Saúde. Situação epidemiológica da leptospirose, 2015. [Accessed Jan 2016]. Available at: http://portalsaude.saude.gov.br/index.php/situacao-epidemiologica-dados

Bröde, O., Fiala, D., & Blazejczyk, K. (2011). Deriving the operational procedure for the universal thermal climate index UTCI. *International Journal of Biometeorology, special issue UTCI.*

Bühler, H. F., Ignotti, E., Neves, S. M. A. S., & Hacon, S. (2014). Spatial analysis of integrated determinant indicators of mortality from acute diarrhea in children under 1 year of age in geographical regions. *Ciênc. Saúde Coletiva., 19*(10), 4131–4140.

Caminade, C., Kovats, S., Rocklöv, J., Tompkins, A. M., Morse, A. P., Colón-González, F. J., et al. (2014). Impact of climate change on global malaria distribution. *Proceedings of the National Academy of Sciences of the United States of America, 111*, 3286–3291.

Campos, G. S., Bandeira, A. C., & Sardi, S. I. (2015 Oct). Zika virus outbreak, Bahia, Brazil. *Emerging Infectious Diseases, 21*(10), 5.

Carmargo, M. G., & Furlan, M. M. D. P. (2011). Resposta fisiológica do corpo às temperaturas elevadas: exercício, extremos de temperatura e doenças térmicas. *Saúde e Pesquisa, 4*(2), 278–288.

Carvalho, B. M., Rangel, E. F., Ready, P. D., & Vale, M. M. (2015). Ecological niche modeling predicts southward expansion of Lutzomyia (Nyssomya) flaviscutellata (Diptera: Psychodidae: Phlebotominae), Vector of Leishmania (Leishmania) amazonensis in South America, under climate change. *PLosONE, 10*(11), e0143282.

Carlton, E. J., Woster, A. P., DeWitt, P., Goldstein, R. S., & Levy, K. (2016). A systematic review and meta-analysis of ambient temperature and diarrhoeal diseases. *International Journal of Epidemiology, 45*(1), 117–130.

Checkley, W., Epstein, L. D., Gilman, R. H., Figueroa, D., Cama, R. I., Patz, J. A., et al. (2000). Effect of El Niño and ambient temperature on hospital admissions for diarrhoeal diseases in Peruvian children. *Lancet, 355*(9202), 442–450.

Chou, S. C., Lyra, A., Mourão, C., Dereczynski, C., Pilotto, I., Gomes, J., et al. (2014). Evaluation of the eta simulations nested in three global climate models. *American Journal of Climate Change, 3*(5), 438–454.

Claeson, M., & Waldman, R. J. (2000). The evolution of child health programmes in developing countries: from targeting diseases to targeting people. *Bulletin of the World Health Organization, 78*, 1234–1245.

Coelho, M. S., & Massad, E. (2012). The impact of climate on Leptospirosis in São Paulo, Brazil. *International Journal of Biometeorology, 56*(2), 233–241.

Costa, D., Hacon, S., Siqueira, A., Pinheiro, S., Gonçalves, K., Oliveira, A., et al. (2015). Municipal temperature and heatwave predictions as a tool for integrated socio-environmental impact analysis in Brazil. *American Journal of Climate Change, 4*, 385–396.

D'Ippoliti, D., et al. (2010). The impact of heat waves on mortality in 9 European cities: results from the EuroHEAT project. *Environmental Health, 9*, 37.

Doherty, R. M., Heal, M. R., Wilkinson, P., Pattenden, S., Vieno, M., Armstrong, B., et al. (2009). Current and future climate- and air pollution-mentioned impacts on human health. *Environmental Health, 8*(Suppl I), S8.

Dourado, F., Arraes, T. C., & Silva, F. e. (2012). O Megadesastre da Região Serrana do Rio de Janeiro – as causas dos moimentos de massa e a distribuição espacial dos investimentos de reconstrução no pós-desastre. *Anuário do Instituto de Geociências, 25*(2), 43–54.

D'Souza, R. M., Hall, G., & Becker, N. G. (2008). Climatic factors associated with hospitalizations for rotavirus diarrhoea in children under 5 years of age. *Epidemiology and Infection, 136*, 56–64.

EM-DAT. (2015). The International disasters database. Centre for Research on the Epidemiology of Disasters – CREA DISDAT [Accessed Nov 2015]. Available at http://www.emdat.be/disaster-data-collection-initiative-portal

Fleury, M., Charron, D., Holt, J., Allen, O., & Maarouf, A. (2006). A time series analysis of the relation of ambient temperature and common bacterial enteric infections in two Canadian provinces. *International Journal of Biometeorology, 50*, 385–391.

Freitas, C. M., Carvalho, M. L., Ximenes, E. F., Arraes, E. F., & Gomes, J. O. (2012). Vulnerabilidade socioambiental, redução do risco de desastres e construção da resiliência – lições do terremoto no Haiti e das chuvas fortes na Região Serrana, Brasil. *Ciência & Saúde Coletiva, 17*(6), 1577–1586.

Freitas, C. M., Silva, D. R. X., Sena, A. R. M., Silva, E. L., Sales, L. B. F., Carvalho, M. L., et al. (2014). Desastres naturais e saúde: uma análise da situação do Brasil. *Ciência & Saúde Coletiva, 19*(9), 3645–3656.

Gasparrini, A., Guo, Y., Hashizume, M., Lavigne, E., Zanobetti, A., Schwartz, J., et al. (2015). Mortality risk attributable to high and low ambient temperature: a multicountry observational study. *Lancet, 6736*(14), 62114–62110.

Giang, P. N., Dung, V., Bao Giang, K., Vinhc, H. V., & Rocklöv, J. (2014). The effect of temperature on cardiovascular disease hospital admissions among elderly people in Thai Nguyen Province, Vietnam. *Global Health Action, 8*, 23649.

Gomes, A. F., Nobre, A. A., & Cruz, O. G. (2012 Nov). Temporal analysis of the relationship between dengue and meteorological variables in the city of Rio de Janeiro, Brazil, 2001-2009. *Cadernos de Saúde Pública, 28*(11), 2189–2197.

Gonçalves, F. L., Braun, S., Dias, P. L., & Sharovsky, R. (2007). Influences of the weather and air pollutants on cardiovascular disease in the metropolitan area of São Paulo. *Environmental Research, 104*(2), 275–281.

Gouveia, N., & Fletcher, T. (2000). Time series analysis of air pollution and mortality: effects by cause, age and socioeconomic status. *Journal of Epidemiology and Community Health, 54*(10), 75–75.

Gouveia, N., Freitas, C. U., Martins, L. C., & Marcilio, I. O. (2006). Respiratory and cardiovascular hospitalizations associated with air pollution in city of São Paulo, Brazil. *Cadernos de Saúde Pública, 22*(12), 2669–2677.

Gouveia, N., Shakoor, H., & Armstrong, B. (2003). Socioeconomic differentials in the temperature-mortality relationship in São Paulo, Brazil. *International Journal of Epidemiology, 32*, 390–397.

Gracie, R., Barcellos, C., Magalhães, M., Souza-Santos, R., & Barrocas, P. R. (2014). Geographical scale effects on the analysis of leptospirosis determinants. *International Journal of Environmental Research and Public Health, 11*(10), 10366–10383.

Graudenz, G. S., Latorre, M. R., Tribess, A., Oliveira, C. H., & Kalil, J. (2006). Persistent allergic rhinitis and indoor air quality perception – an experimental approach. *Indoor Air, 16*(4), 313–319.

Guimarães, R. M., Cruz, O. G., Parreira, V. G., Mazoto, M. L., Vieira, J. D., & Asmus, C. I. (2014). Temporal analysis of the relationship between leptospirosis and the occurrence of flooding due to rainfall in the city of Rio de Janeiro, Brazil, 2007-2012. *Ciência & Saúde Coletiva, 19*(9), 3683–3692.

Guo, Y., Gasparrini, A. B., Li, S., Tawatsupa, B., Tobias, A., et al. (2014). Global variation in the effects of ambient temperature on mortality: a systematic evaluation. *Epidemiology, 25*(6), 781–709.

Hacon, et al. (2016). Vulnerabilidade, riscos e impactos das mudanças climáticas sobre a saúde do Brasil. In *Terceira Comunicação Nacional sobre Mudanças Climáticas à UNFCC (TCN-UNFCC).*

Hajat, S., & Kosatky, T. (2014). Heat-related mortality: a review and exploration of heterogeneity. *Journal of Epidemiology and Community Health, 64*, 753e760.

Hashizume, M., Armstrong, B., Wagatsuma, Y., Faruque, A. S. G., Hayashi, T., & Sack, D. A. (2008). Rotavirus infections and climate variability in Dhaka, Bangladesh: a time-series analysis. *Epidemiology and Infection, 136*(9), 1281–1289.

Heller, L., Colosimo, E. A., & Antunes, C. M. F. (2003). Environmental sanitation conditions and health impact: A case-control study. *Revista da Sociedade Brasileira de Medicina Tropical, 36*, 41 50.

Hii, Y. L., Rocklöv, J., Ng, N., Tang, C. S., Pang, F. Y., & Sauerborn, R. (2009). Climate variability and increase in intensity and magnitude of dengue incidence in Singapore. *Global Health Action, 2*, 2036.

Honório, N. A., Codeço, C. T., Alves, F. C., & Magalhães, M. (2009). Temporal distribution of Aedes aegypti in different districts of Rio de Janeiro, Brazil, measured by two types of traps. *Journal of Medical Entomology, 46*, 1001–1014.

Horta, M. A., Bruniera, R., Ker, F., Catita, C., & Ferreira, A. P. (2014). Temporal relationship between environmental factors and the occurrence of dengue fever. *International Journal of Environmental Health Research, 24*(5), 471–481.

Huynen, M., & Martens, P. C. (2015). Climate change effects on heat- and cold-related mortality in the Netherlands: A scenario-based integrated environmental health impact assessment. *International Journal of Environmental Research and Public Health, 12*(10), 13295–13320.

Ignotti, E., Hacon, S., Junger, W. L., Mourão, D., Longo, K., Freitas, S., et al. (2010 Apr). Air pollution and hospital admissions for respiratory diseases in the subequatorial Amazon: a time series approach. *Cadernos de Saúde Pública, 26*(4), 747–761.

Intergovernmental Panel on Climate Change. (2014). Climate Change 2014: Mitigation of climate change. New York: Cambridge University Press; 2014.

Intergovernmental Panel on Climate Change (IPCC). (2013). *Climate change: The physical science basis*: Summary for policymakers; Intergovernmental panel on climate change, Working group I contribution to the IPCC fifth assessment report; IPCC: Geneva, Switzerland, 2013.

ISO 7243. (1982). *Hot Environments—Estimation of the heat stress on working man, based on the WBGT-index (wet bulb globe temperature).* Geneva, Switzerland: International Standards Organization

Jacob, D. J., & Winner, D. A. (2009). Effect of climate change on air quality. *Atmospheric Environment, 43*(1), 51–63.

Kaiser, R., et al. (2007). The effect of the 1995 heat wave in Chicago on all-cause and cause-specific mortality. *American Journal of Public Health, 97*(Suppl 1), S158–S162.

Kinsgley, S. L., Eliot, M. N., Gold, J., Vanderslice, R. R., & Wellenius, G. A. (2015). Current and projected heat-related morbidity and mortality in Rhode Island. *Environmental Health Perspectives.*

Kjellstrom, T., Woodward, A., Gohar, L., Lowe, J., Lemke, B., Lines, L., et al. (2015). The risk of heat stress to people. In D. King, D. Schrag, Z. Dadi, Q. Ye, & A. Ghosh (Eds.), *Climate change: A risk assessment.* London: UK Foreign and Commonwealth Office.

Kjellstrom, T., et al. (2015). Hothaps analysis for 2015 WHO project. In *Who national profiles on climate change and human health – contribution on heat stress, occupational health and labor productivity*.

Kjellstrom T., et al. (2016). Climate change health impact profiles (ClimateCHIPs). Available: in http://www.climatechip.org/about-us

Ko, A. I., Reis, M. G., Dourado, C. M. R., Johnson, W. D., & Riley, L. W. (1999). Urban epidemic of severe leptospirosis in Brazil. *The Lancet, 354*(9181), 820–825.

Kolstad, E. W., & Johansson, K. A. (2011). Uncertainties associated with quantifying climate change impacts on human health: A case study for diarrhea. *Environmental Health Perspectives, 119*(3), 299–305.

Kosek, M., Bern, C., & Guerrant, R. L. (2003). The global burden of diarrhoeal disease, as estimated from studies published between 1992 and 2000. *Bulletin of the World Health Organization, 81*, 197–204 Epub 2003.

Lainson, R., & Shaw, J. J. (1968). Leishmaniasis in Brazil I. Observations on enzootic rodent leishmaniasis - Incrimination of Lutzomyia flaviscutellata (Mangabeira) as the vector in the lower Amazonian Basin. *Transactions of the Royal Society of Tropical Medicine and Hygiene, 62*, 385–395.

Lainson, R., Shaw, J. J., Silveira, F. T., de Souza, A. A., Braga, R. R., & Ishikawa, E. A. (1994). The dermal leishmaniases of Brazil, with special reference to the eco-epidemiology of the disease in Amazonia. *Memórias do Instituto Oswaldo Cruz, 89*(3), 435–443.

Laporta, G. Z., Linton, Y.-M., Wilkerson, R. C., Bergo, E. S., Nagaki, S. S., Sant'Ana, D. C., et al. (2015). Malaria vectors in South America: current and future scenarios. *Parasites & Vectors, 8*, 426.

Le Tertre, A., et al. (2006). Impact of the 2003 heatwave on all-cause mortality in 9 French cities. *Epidemiology, 17*, 75–79.

Liu-Helmersson, J., Stenlund, H., Wilder-Smith, A., & Rocklöv, J. (2014). Vectorial capacity of Aedes aegypti: Effects of temperature and implications for global dengue epidemic potential. *PLosOne, 9*(3), 289783.

Lounibos, L. P., O'Meara, G. F., Juliano, S. A., Nishimura, N., Escher, R. L., Reiskind, M. H., et al. (2010). Differential survivorship of invasive mosquito species in south Florida cemeteries: Do site-specific microclimates explain patterns of coexistence and exclusion? *Annals of the Entomological Society of America, 103*, 757–770.

Lund, T. C., & Baird-Parker, G. W. (2000). *The microbiological safety and quality of food* (pp. 101–121). Gaithersburg, MD: Aspen Publishers.

Maia, P. A., Ruas, A. C., & Bittencourt, D. P. (2015). Wet-bulb globe temperature index estimation using meteorological data from São Paulo State, Brasil. *International Journal of Biometeorology, 59*(10), 1395–1403.

Mascarenhas, M. D. M., Vieira, L. C., Lanzieri, T. M., Leal, A. P. P. R., Duarte, A. F., & Hatch, D. L. (2008). Poluição atmosférica devido à queima de biomassa florestal e atendimentos de emergência por doença respiratória em Rio Branco, Brasil - Setembro, 2008. *Jornal Brasileiro e Pneumologia, 34*(1), 46–46.

Matsueda, M. (2011). Predictability of Euro-Russian blocking in summer of 2010. *Geophysical Research Letters, 38*(6).

McMichael, A. J., Wilkinson, P., Kovats, R. S., Pattenden, S., Hajat, S., Armstrong, B., et al. (2008). International study of temperature, heat and urban mortality: the "ISOTHURM" project. *International Journal of Epidemiology, 37*, 1121–1131.

Meehl, G. A., & Tebaldi, C. (2004). More intense, more frequent, and longer lasting heat waves in the 21st century. *Science, 305*, 994–997.

Mendes, C. S., Coelho, A. B., Féres, J. G., Souza, E. C., & Cunha, D. A. (2016). Impacto das mudanças climáticas sobre a leishmaniose no Brasil. *Ciência & Saúde Coletiva, 21*(1), 263–272.

Molino-Cruz, A., & Barillas-Mury, C. (2014). The remarkable journey of adaptation of the Plasmodium falciparum malaria parasite to New World anopheline mosquitoes. *Memórias do Instituto Oswaldo Cruz, 109*, 1–6.

Mordecai, E. A., Paaijmans, K. P., Johnson, L. R., Blazer, C., Ben-Horin, T., Moor, E., et al. (2012). Optimal temperature for malaria transmission is dramatically lower than previously predicted. *Ecology Letters, 16*(1), 22–30.

Moreno, A. R. (2006). Climate change and human health in Latin America: drives, effects and policies. *Regional Environmental Change, 6*, 157–164.

Murrell, E. G., Damal, K., Lounibos, L. P., & Juliano, S. A. (2011). Distributions of competing container mosquitoes depend on detritus types, nutrient ratios, and food availability. *Annals of the Entomological Society of America, 104*, 688–698.

Narváez, L., Lavell, A., & Ortega, G. P. (2009). La gestión del riesgo de desastres: un enfoque basado en procesos. San Isidro: Secretaría General de la Comunidad Andina; 2009.

Nassis, G. P., Brito, J., Dvorak, J., Chalabi, H., & Racinais, S. (2015). The association of environmental heat stress with performance: analysis of the 2014 FIFA World Cup Brazil. *British Journal of Sports Medicine, 49*, 609–613.

Naumova, E. N., Jagai, J. S., Matyas, B., DeMaria, A., MacNeill, I. B., & Griffiths, J. K. (2007). Seasonality in six enterically transmitted diseases and ambient temperature. *Epidemiology and Infection, 135*, 281–292.

Oehler, E., Watrin, L., Larre, P., Leparc-Golfrt, I., Lestère, S., Valour, F., et al. (2014). Zika virus infection complicated by Guillain-Barré syndrome: case report, French Polynesia, December 2013. *Euro Surveillance, 19*(9), 20720.

Oliveira, B. F., Ignotti, E., Artaxo, P., Saldiva, P. H., Junger, W. L., & Hacon, S. (2012). Risk assessment of PM (2.5) to child residents in Brazilian Amazon region with biofuel production. *Environmental Health, 11*, 64.

Oliveira, B. F. A., Protázio, G., Jünger, W. L., Hacon, S. (2016). Tendência e a influência de variáveis climáticas nas internações por diarreia em menores de cinco anos, 2016 (no prelo).

Oliveira, T. V., Marinho, D. P., Costa Neto, C., & Kligerman, D. C. (2012). Climate variables, living conditions and the health of the population: leptospirosis in the city of Rio de Janeiro from 1996 to 2009. *Ciência & Saúde Coletiva, 17*(6), 1569–1576.

Ostro, B. (2004). Outdoor air pollution: Assessing the environmental burden of disease at national and local levels. Geneva, World Health Organization, 2004 (WHO Environmental Burden of Disease Series, No. 5).

Otto, et al. (2014). A tool for the estimation and analysis of local climate and population heat exposure. Available: www.climatechip.org, 2014.

Parham, P. E., & Michael, E. (2010). Modeling the effects of weather and climate change on malaria transmission. *Environmental HealthPerspectives, 118*, 620–626.

Patz, J. A., Frumkim, H., Holloway, T., Vimont, D. J., & Hainese, A. (2014). Climate change: challanges and opportunities for global health. *JAMA, 312*, 1565–2580.

Perkins, S. E., & Alexander, L. V. (2013). On the measurement of heatwaves. *Journal of Climate, 26*, 4500–4517.

Pimentel, M., Grüdtner, L., & Zimerman, L. I. (2006). Seasonal variation of ventricular tachycardia registered in 24-hour Holter monitoring. *Arquivos Brasileiros de Cardiologia, 87*(4), 403–406.

Rangel, E. F., Costa, S. M., & Carvalho, B. M. (2014). Environmental changes and the geographic spreading of American cutaneous leishmaniasis in Brazil. In D. Claborn (Ed.), *Leishmaniasis – Trends in epidemiology, diagnosis and treatment*. Rijeka, Croatia: InTech. https://doi.org/10.5772/57207

Rasch, R. J. (2015). Assessing urban vulnerability to flood hazard in Brazilian municipalities. *Environment and Urbanization, 28*, 145. https://doi.org/10.1177/0956247815620961

Robine, J.-M., Cheung, S. L. K., Le Roy, S., et al. (2008). Death toll exceeded 70000 in Europe during the summer of 2003. *Comptes Rendus Biologies, 331*(2), 171–178.

Robinson, P. J. (2001). On the definition of a heat wave. *Journal of Applied Meteorology, 40*, 762–765.
Rocklöv, J., Quam, M., et al. (2015). *Climate and health country profiles – Brazil – 2015*. Geneva, Switzerland: World Health Organization.
Rohani, A., Wong, Y. C., Zamre, I., Lee, H. L., & Zurainee, M. N. (2009). The effect of extrinsic incubation temperature on development of dengue serotype 2 and 4 viruses in Aedes aegypti (L.). *The Southeast Asian Journal of Tropical Medicine and Public Health, 40*, 942–950.
Rossi, F. A., Krüger, E. L., & Bröde, P. (2012). Definição de faixas de conforto e desconforto térmico para espaços abertos em Curitiba, PR, com índice UTCI. *Ambiente construído, 12*(1), 41–59.
Sarkar, U., Nascimento, S. F., Barbosa, R., Martins, R., Nuevo, H., Kalafanos, I., et al. (2002). Population-based case-control investigation of risk factors for leptospirosis during an urban epidemic. *American Journal of Tropical Medicine and Hygiene, 66*(5), 605–610.
Sherwood, S. (2014). What is wet bulb temperatures? [Accessed Oct 2014]. Available at http://web.science.unsw.edu.au/~stevensherwood/wetbulb.html
Sherwood, S. C., & Huber, M. (2010). An adaptability limit to climate change due to heat stress. *Proceedings of the National Academy of Sciences of the United States of America, 107*(21), 9552–9555.
Sillmann, J., Kharin, V. V., Zwiers, F. W., Zhang, X., & Bronaugh, D. (2013). Climate extremes indices in the CMIP5 multi-model ensemble: Part 2. Future climate projections. *Journal of Geophysical Research: Atmospheres, 118*, 2473–2493.
Silva, D. X., Barcellos, C., & Bacuri, R. (2010). Vulnerabilidade e efeitos das mudanças climáticas na saúde pública em Manaus. Project report on "Guidance for Conducting Assessments of Health Vulnerability and Public Health and Health Care Interventions to Address Climate Change", funded by the Pan-American Health Organisation (Contract BR/CNT/0991502.001 of October 2009), 2010. [interned based document] [Accessed 11 Dec 2013]. Available at: http://www.climasaude.icict.fiocruz.br/docs/vulnerabilidade _manaus_relat_final2_x_edit.pdf respiratórias
Silveira-Neto, S., Nakano, O., Barbin, D., & Villa Nova, N. (1976). *Manual de ecologia dos insetos* (p. 419). São Paulo, Brazil: Agronômica Ceres.
Sharovsky, R., César, L. A. M., & Ramires, J. A. F. (2004). Temperature, air pollution, and mortality from myocardial infarction in SÃ£o Paulo, Brazil. *Brazilian Journal of Medical and Biological Research, 37*(11), 1651–1657
Smith, L. T., Aragão, L. E., Sabel, C. E., & Nakaya, T. (2014). Drought impacts on children's respiratory health in the Brazilian Amazon. *Scientific Reports, 16*(4), 3726.
Son, J. Y., Gouveia, N., Bravo, M. A., de Freitas, C. U., & Bell, M. L. (2016). The impact of temperature on mortality in a subtropical city: effects of cold, heat, and heat waves in São Paulo, Brazil. *International Journal of Biometeorology, 60*(1), 113–121.
Stefanon, M., D'Andrea, F., & Drobinski, P. (2012). Heatwave classification over Europe and the Mediterranean region. *Environmental Research Letters, 7*(1), 014023.
Universidade Federal de Santa Catarina (UFSC- CEPED). (2012). Centro Universitário de Estudos e Pesquisas sobre Desastres. Atlas brasileiro de desastres naturais 1991 a 2010: volume Brasil. Florianópolis: CEPED UFSC; 2012.
USB and Society Report. (2016). Middle-class exposure: Locating and characterizing the global middle class. In *Climate change: A risk to the global middle class. Exposure, vulnerability & economic impact* Available: https://www.ubs.com/climatechange
van Panhuis, W. G., Choisy, M., Xiong, X., Chok, N. S., Akarasewi, P., Iamsirithaworn, S., et al. (2015). Region-wide synchrony and traveling waves of dengue across eight countries in Southeast Asia. *Proceedings of the National Academy of Sciences of the United States of America, 112*(42), 13069–13074.
Vanderlei, L. C. M., Silva, G. A. P., & Braga, J. U. (2003). Fatores de risco para internamento por diarreia aguda em menores de dois anos: estudo de caso-controle. *Cadernos de Saúde Pública, 19*(2), 455–463.

Veronesi, R., & Focaccia, R. (2004). *Tratado de Infectologia* (2nd ed.). São Paulo, Brazil: Atheneu.
Watts, D. M., Burke, D. S., Harrison, B. A., Whitmire, R. E., & Nisalak, A. (1987). Effect of temperature on the vector efficiency of Aedes aegypti for dengue virus. *The American Journal of Tropical Medicine and Hygiene, 36*, 143–152.
World Bank. (2012). 4C: Turn down the heat. A report for the World Bank by the Potsdam Institute for Climate Impact Research and Climate Analytics, November 2012, Washington DC 20433, 2012.
World Health Organization(WHO). (2013). Global malaria programme. In: World Malaria Report. Geneva, 2013.
World Health Organization (WHO). (2014). Quantitative risk assessment of the effects of climate change on selected causes of death, 2030s and 2050s. Geneva, World Health Organization, 2014.
World Organization Health (WHO). (2015a). Climate and health country profiles – Brazil – . United Nations: Framework convention on climate change, 2015a.
World Organization Health (WHO). (2015b). Heatwaves and heath: guidance on warning-system development. WMO-No. 1142, 2015b.
Wu, J., Zhou, Y., Gao, Y., Fu, J. S., Johnson, B. A., Huang, C., et al. (2014). Estimation and uncertainty analysis of impacts of future heat waves on mortality in the eastern United States. *Environmental Health Perspectives, 122*(1), 10–16.
Xu, Z., Liu, Y., Ma, Z., Toloo, G. S., Hu, W., & Tong, S. (2014). Assessment of the temperature effect on childhood diarrhea using satellite imagery. *Scientific Reports, 4*, 5389.
Zanluca, C., Melo, V. C. A., Mosimann, A. L. P., Santos, G. I. V., Santos, C. N. D., & Luz, K. (2015 Jun). The first report of autochthonous transmission of Zika virus in Brazil. *Memórias do Instituto Oswaldo Cruz, 110*(4), 569–572.
Zhang, X., Alexander, L., Hegerl, G. C., Jones, P., Tank, A. K., Peterson, T. C., et al. (2011). Indices for monitoring changes in extremes based on daily temperature and precipitation data. *Wiley Interdisciplinary Reviews: Climate Change, 2*, 851–870.

Chapter 5
Biodiversity Sector: Risks of Temperature Increase to Biodiversity and Ecosystems

Fabio Rubio Scarano

5.1 Introduction

The Intergovernmental Panel on Climate Change (IPCC, 2014) states that there is more abundant and comprehensive evidence of climate change impacts for natural systems than for human systems. Moreover, the report shows that the most vulnerable natural systems are those that lost a significant portion of their life-supporting mechanisms. In parallel, for human systems, the IPCC recognizes that the poor people are the most vulnerable to climate impacts (Fisher et al., 2014; Magrin et al., 2014). Indeed, one of the main conclusions of the Working Group II in the fifth assessment report of the IPCC (2014) is that practices that promote sustainable development in the present – by combining social justice, environmental health and economic productivity – reduce future risks imposed by climate change and are thus adaptive.

Since biodiversity is essential for the good flow of ecosystem services for people, which ensure vital water and food security, protection of cultural values and identity, among other services (McNeely et al., 2009), one can conclude that human and natural systems are coupled, although disciplinary science often treats them separately. Therefore, risks that climate change impose on biodiversity have direct and indirect consequences to human wellbeing and livelihoods.

Nevertheless, global biodiversity is declining. Estimates are that species extinction is at least one thousand times higher than historic rates (Mace et al., 2005). Diamond (1989) proposed that this is due to an "Evil Quartet" that comprises habitat loss, overkill, invasive species and extinction chains. Later, Thomas et al. (2004) argued that climate change turned the quartet into a quintet, although there is no

F. R. Scarano (✉)
Federal University of Rio de Janeiro (UFRJ) and Brazilian Foundation for Sustainable Development (FBDS), Rio de Janeiro, Brazil
e-mail: fscarano@fbds.org.br

C. A. Nobre et al. (eds.), *Climate Change Risks in Brazil*,
https://doi.org/10.1007/978-3-319-92881-4_5

doubt that at this point in time habitat destruction remains as the main causal factor behind species extinctions (Baillie et al., 2004). Indeed, it has been claimed that climate change and biosphere integrity are two core planetary boundaries that, once transgressed, may drive the Earth system into a new undesirable state for humankind (Steffen et al., 2015).

Brazil is the country with the largest number of species worldwide. From the 17 countries that host 70% of the species on the planet, it is the most diverse (Mittermeier et al., 1997). Furthermore, it holds the greatest proportion of superficial freshwater, and ranks among the top food producers in the world (Scarano et al., 2012). Thus, it is a key premise of this Chapter that if biodiversity and ecosystems in Brazil were at risk, the potential impacts that might derive from it would have planetary consequences. This Chapter revises the available literature on impacts and vulnerability of biodiversity and ecosystems in Brazil published since the latest reports from the IPCC (Magrin et al., 2014) and the Brazilian Panel on Climate Change – PBMC (Souza-Filho et al., 2014).

5.2 Vulnerability

The latest IPCC (Magrin et al., 2014) and PBMC reports (Souza-Filho et al., 2014) refer to climate change vulnerability in Brazil at three different organisation levels, based on the best science available at that moment: biomes, ecosystems and species. The literature published after the reports endorses these panels' findings and adds new discussion (Table 5.1).

In relation to biomes, there are more reports referring to climate change vulnerability in biodiversity hotspots (Atlantic Forest and Cerrado), in the Amazon, and in the Caatinga. The Atlantic Forest is one of the three hotspots most vulnerable to climate change in the world. This is due to a combination of high risk related to the emergence of new climate and the disappearance of the current one, as well as susceptibility to invasive species and expansion of pasture areas for cattle raising (Béllard et al., 2014). In the case of the Cerrado, high replacement rates of original vegetation cover and other land uses made the biome extremely vulnerable (Sawyer, 2008; Strassburg et al., 2017). For the Amazon, there is the potential for savannisation based on drought and rising temperatures brought about by deforestation and climate change, despite the existence of uncertainties around timing for a supposed tipping point to be reached (see review in Magrin et al., 2014). In the Caatinga, vulnerability refers to the low percentage of protected areas, persistence of poverty and desertification process due to extreme droughts (Oliveira et al., 2012; Tabarelli et al. 2017). In the most recent study on the subject, Seddon et al. (2016) propose that the Amazon and Caatinga are among the world's biomes that present intrinsic vulnerability to climate variations and would have low response or resilience to possible changes. In the case of the Amazon, sensitivity is related to changes in temperature and cloud cover, and in the case of the Caatinga, water is the pivotal component of sensitivity.

Table 5.1 Brazilian biomes, ecosystems and species more vulnerable to climate change

Level of organisation	Types	Case	References
Biomes	Atlantic forest	Biodiversity hotspot	Béllard et al. (2014); Joly et al. (2014); Scarano and Ceotto (2015)
	Cerrado	Biodiversity hotspot	Sawyer (2008), Strassburg et al. (2017)
	Caatinga	Desertification	Seddon et al. (2016); Oliveira et al. (2012)
	Amazon	Savannisation	Anadón et al. (2014); Balch et al. (2015); Seddon et al. (2016)
Ecosystems	High altitude	High thermal sensitivity	Laurance (2015)
	Coastal and marine	Rising sea level	Godoy and Lacerda (2015)
	Urban	Heat waves	Lucena et al. (2012); Rosenzweig et al. (2015)
Species	Endangered species	High ecological sensitivity	Keith et al. (2015); Urban (2015)
	Amphibians	High ecological sensitivity	Loyola et al. (2014)
	Corals	High ecological sensitivity	Descombes et al. (2015)
	Invertebrates	High ecological sensitivity and diminishing abundance	Faleiro et al. (2018); Ferro et al. (2014); Giannini et al. (2015)
	Plants, mammals, birds	Moving south	Giannini et al. (2015); Hoffmann et al. (2015); Oliveira et al. (2015)

In relation to ecosystems, the ones at high altitudes, on the coast and in urban areas are particularly vulnerable. The IPCC reports (Magrin et al., 2014) at the global level and the one produced by PBMC (Souza-Filho et al., 2014) for Brazil, already pointed out this trend and subsequent studies confirmed it. Laurance (2015) argues that high altitude ecosystems are made up of species with high thermal specialisation, making them vulnerable to climate change. The effects of climate change on coastal ecosystems, include impacts on coastal dynamics like floods and erosion. In Brazil, a recent example is the reported increase of the mangrove area in the state of Ceará by 24 hectares inland, which happened between 1992 and 2003 (Godoy and Lacerda, 2015). Urban ecosystems are among the most vulnerable to the general reduction in natural cover (Rosenzweig et al., 2015), both inside urban centers and in their periphery, and the recent water supply related impacts in big cities like São Paulo and Rio de Janeiro, are possibly related to a synergy between land use change and climate change (Cunningham et al., 2017). In the case of the Rio de Janeiro metropolitan region, the increase of heat islands in the last few decades also reflects this vulnerability (Lucena et al., 2012).

Recent studies on taxonomic groups and species also reinforce patterns already described in IPCC and PBMC about the vulnerability of corals (Descombes et al.,

2015), invertebrates (i.e. moths and bees, Faleiro et al., 2018, Ferro et al., 2014, Giannini et al., 2015) amphibians (Loyola et al., 2014), mammals (Ribeiro et al., 2016), and endangered species in general (Keith et al., 2015; Urban, 2015). They also predict that mammals, birds and plants will migrate south due to the increase in temperature (Giannini et al., 2015; Hoffmann et al., 2015; Oliveira et al., 2015).

5.3 Risk Assessment for Temperature Limits Exceeding 2–4 °C

Specific projections for biomes, ecosystems and Brazilian species in a warmer world, from 2 to 4 °C temperature rise, or over 4 °C are still hard to find in the literature, particularly using the IPCC (2013)'s RCP scenarios. However, the general patterns of the studies reviewed here are similar to the general vulnerability patterns described in Table 5.1.

By 2070, there is a 90% probability of temperature rise between 2 to 3 °C in scenario RCP 8.5 in Brazil (Figs. 2.4d and 2.5d in Chap. 2 of this book). In this case, Anadón et al. (2014) predict impoverishment and reduction of the Atlantic Forest and the Amazon. They also predict the expansion of the savanna (including the Caatinga, and in the direction of the Amazon) and of forest in the Pampas, all together with changes in species distribution and impoverishment. This projection is similar to the one provided by Yu et al., (2014), who also used scenario RCP 8.5, as well as Leadley et al. (2014), predicting a 3 °C rise for 2075, leading to the savannisation of Brazilian tropical forests and impoverishment of the Cerrado, in IPCC scenario A2 (2007). More recently, Zanin et al., (2017) found similar projections of forest contraction, expansion of open vegetation, and overall vegetation impoverishment.

In addition to the vulnerability of biomes, South America is the continent with the highest risk of species extinction (23%) that may be attributed to climate change. Extinction risk goes from 5.2% at 2 °C to 15.7% when temperatures reach over 4 °C (Urban, 2015). Visconti et al. (2015) proposed that in 2050, for a business as usual scenario with temperature increase higher than 2 °C (A1B, IPCC, 2007), species richness will decline rather sharply, particularly for mammals. In the Atlantic Forest, the endemic species from the top of the mountain in Brazil's east, the grey-backed tachuri bird (*Polystictus superciliaris*) would be endangered by 2080 (Hoffmann et al., 2015), following the IPCC A2 and B2 scenarios (2007). In the Cerrado, Aguiar et al., (2016) predicted that bat species would move south in scenario RCP 4.5 for 2050, when there is an 80% probability of temperature rising above 2 °C. Under these types of future scenarios, there is already evidence that usual conservation instruments, such as protected areas, may become inefficient for given species and ecosystems (Ferro et al., 2014; Feeley & Silman, 2016).

Two projections for Brazilian species under different climate change scenarios draw attention due to the impact on socio-economics. Oliveira et al. (2015) predicted the occurrence and distribution of 16 edible plant species in the Cerrado, which are important for local and traditional communities – among them the souari nut (*Caryocar brasiliense* Camb.), mouriche palm (*Mauritia flexuosa* L.f.) and locust berries (*Byrsonima verbascifolia* L.DC.). The projection was based on RCP 8.5 for 2080, when there is a 75% chance of temperatures rising above 4.5 °C in the Brazilian territory. The results point to the extinction of several of these species and a shift in the distribution of some of them to the south, in the direction of what is the Atlantic Forest today. The study suggests that the southeast area of the Cerrado needs conservation measures, as for these 16 species it is the most irreplaceable region. In another example, Giannini et al. (2015) predicted a shift in the distribution of *Meliponina quadrifasciata* – a stingless bee, native of the Atlantic Forest and key pollinator of native species and farming plants – from the north to the south, and from the coast to inland areas. However, the scenario used was A1B for 2030, 2050 and 2080. Results point to a deficit in coffee pollinators in the states of São Paulo and Minas Gerais in 2030 already, with potential impacts on the economy.

Another worrying and less known component of the effect of climate change on biodiversity refers to the impacts on plant growth. This component is important because plant productivity affects the functioning of ecosystems, food chains, oxygen supply, food, fibres and fuel for humanity. Mora et al. (2015) found that tropical areas, amongst them Brazil, may lose up to 200 good plant growing days a year by 2100, in a RCP 8.5 scenario (over 4.5 °C). The impact on the human population and economy would be devastating. In scenarios RCP 2.6 and 4.5, the impact would be much smaller. In evergreen perennial forests like the Amazon and Atlantic Forest, the impact on plant growth would be higher for all plant groups in scenario RCP 8.5 (Mora et al., 2015).

Segan et al. (2016) analysed the interaction between habitat loss and fragmentation with climate change. They argue that in order to avoid a bigger impact in Brazilian biomes and ecosystems by 2090, in a RCP 8.5 scenario (over 4.5 °C), the western part of the Amazon and the Pantanal should be protected, in addition to prioritising the restoration of areas in the Caatinga, Cerrado, Atlantic Forest and Pampa. Another recommendation is to develop schemes to account for climate change in spatial prioritisation studies for conservation (e.g., Faleiro et al., 2013; Jones et al., 2016; Lemes and Loyola, 2013; Zwiener et al., 2017).

The marine panorama is also reason for concern, especially in more severe climate change scenarios (RCP 6.0 and 8.5). Beaugrand et al., (2015) state that changes of this magnitude will affect the marine pelagic biodiversity more than changes in temperature that took place between the last glacial maximum and current times, 50% (RCP 6.0) and 70% (RCP 8.5) of the global ocean surface by 2100. Species loss will be particularly severe in the tropics, including the Brazilian coast.

5.4 Adaptation

The projections of the impacts of future climate trends on Brazilian biodiversity and ecosystems call for immediate action. Carbon mitigation alone will continue to be relevant and must speed up, but alone will not suffice to halt or circumvent ongoing climate trends. Thus, adaptation strategies are needed to boost resilience of vulnerable socio-ecological systems, always bearing in mind that there are limits for adaptation (Scarano, 2017), as the planetary boundaries framework suggests (Steffen et al., 2015). Whenever adaptation avoids or reduces climate risks without negatively impacting coupled human-natural systems, it becomes key to the sustainable development agenda (Juhola et al., 2016; Pant et al., 2015; Scarano, 2017). Therefore, although adaptation and sustainable development are not synonyms, an alignment between policy actions on these two fronts is desirable (Kasecker et al., 2017).

Although this Chapter examines the potential impacts of climate change, especially temperature rise, on biodiversity and ecosystems, the other side of the same coin is that biodiversity and ecosystems can be adaptive to climate change. In developing countries such as Brazil, reduction of the climate change vulnerability of local societies and its associated risks requires a combination of policy instruments related to biodiversity and ecosystem services conservation (e.g., establishment and effective management of protected areas, community management of natural areas, and ecological restoration) and socio-economic policies that foster livelihood diversification, income generation and poverty reduction. We call this type of action 'ecosystem-based adaptation to climate change' (EbA; Jones et al., 2012; Scarano, 2017).

Kasecker et al. (2017) found 397 Brazilian municipalities that combine high natural vegetation cover, high human poverty, and high exposure to climate change. These municipalities are mainly located in the Amazon, the Cerrado and the Caatinga biomes. This paper proposes that an EbA approach that conciliates biodiversity and ecosystem conservation actions with improvement of socioeconomic conditions is essential to reducing climate vulnerability in these municipalities. In parallel, Brazil also has municipalities that have lost a significant part of their vegetation cover and have high human poverty, as seen by examples in the Atlantic forest (Pires et al., 2017; Rezende et al., 2018) and in the Amazon (Silva and Prasad, 2017). This restoration gap is echoed by Brazil's commitment to the Paris Agreement of the climate convention to restore 12 million hectares by 2030, and backed up by national environmental legislation that turn mandatory the restoration of existing environmental debts within private rural properties (Scarano, 2017). However, restoration is often expensive. Both for conservation and for restoration, incentive mechanisms will be necessary to reach the necessary scale and to cover some of the implementation costs (Kasecker et al., 2017; Strassburg et al., 2017; Vieira et al., 2017).

5.5 Conclusions and Recommendations

Based on the reduced literature on the impact of an increase of 4°C or above in the average temperature, it is clear that in this scenario the effects would significantly amplify projections already made for settings with a 2 °C rise. The following projections stand out:

(a) Savannisation and vegetation impoverishment will be likely seen in an above 2 °C rise in 2070, which might be aggravated with the increase in temperature;
(b) Species' extinction risk may increase from 5.2% at 2 °C, to 15.7%, if the 4 °C barrier is broken - South America will be the continent most susceptible to species extinction;
(c) Extinction and changes in the distribution pattern of native species with edible and cultural value in the Cerrado would lead to socio-economic issues by 2080, when the temperature could break the 4 °C barrier;
(d) Impact on biodiversity might also negatively affect agriculture, such as in the case of the reduction in the native bee species in the Atlantic Forest that are important for coffee pollination. This reduction would be seen in 2030 and would get worse until extinction in 2050 and 2080, with temperature rising between 2 and 4 °C;
(e) In 2100, with an increase above 4.5 °C, Brazil would lose 200 days a year in suitable plant growing days, causing great impacts on biodiversity, productivity of ecosystems and the economy. For forest species in the Amazon and Atlantic Forest, the scenario would be particularly severe;
(f) In 2100, with a likely increase above 4.5 °C, biodiversity loss on tropical coasts, including Brazil's, would have a significant negative impact on food and economy.

The main practical recommendation found in this review are:

(a) To prioritise conservation (especially in the western Amazon, Pantanal and Cerrado) and restoration areas (particularly in the Caatinga, Cerrado, Atlantic Forest and Pampa), considering climate change scenarios, in order to promote ecosystem-based adaptation to climate change;
(b) To design and implement biodiversity and ecosystem conservation and restoration in a way that these actions foster livelihood improvement and poverty reduction in areas vulnerable to climate change to increase local adaptive capacity and resilience.

The main recommendations from the scientific point of view are:

(a) Due to the shortage of literature with predicted scenarios for Brazilian biodiversity in temperatures over 4 °C, it is critical for research to be encouraged in this area;

(b) Prioritisation schemes for conservation and restoration areas (used to guide public policies such as SNUC, Planaveg and the New Forest Code[1]) should incorporate the climate component to their analyses, considering scenarios with temperature increase.

In conclusion, while climate change poses obvious risks to biodiversity and the socioecological systems dependent on it in Brazil, the country's natural wealth, its biodiversity and ecosystems, are simultaneously the main source of alternatives for mitigation and adaptation (Scarano, 2017). To bring biodiversity and ecosystems to the centre of the development process of the country, rather than treating them as an obstacle to this process, will be strategic both to fight climate change and to promote a sustainable and inclusive development.

Acknowledgements I am very thankful to Prof. José Maria Cardoso da Silva (University of Miami, US) and Prof. Rafael Loyola (UFG, Brazil) for excellent suggestion for improving the manuscript. I am also grateful to the Brazilian Platform on Biodiversity and Ecosystem Services (BPBES) for support (CNPq; project: 405593/2015-5).

References

Aguiar, L. M. S., Bernard, E., Ribeiro, V., Machado, R. B., & Jones, G. (2016). Should I stay or should I go? Climate change effects on the future of Neotropical savannah bats. *Global Ecology and Conservation, 5*, 22–33.

Anadón, J. D., Sala, O. E., & Maestre, F. T. (2014). Climate change will increase savannas at the expense of forests and treeless vegetation in tropical and subtropical Americas. *Journal of Ecology, 102*, 1363–1373.

Baillie, J. E. M., Hilton-Taylor, C., & Stuart, S. N. (Eds.). (2004). *2004 IUCN red list of threatened species. A global species assessment*. Gland, Switzerland: IUCN.

Balch, J. K., Brando, P. M., Nepstad, D. C., Coe, M. T., Silvério, D., Massad, T. J., et al. (2015). The susceptibility of southeastern Amazon forests to fire: Insights from a large-scale burn experiment. *Bioscience, 65*(9), 893–905.

Beaugrand, G., Edwards, M., Raybaud, V., Goberville, E., & Kirby, R. R. (2015). Future vulnerability of marine biodiversity compared with contemporary and past changes. *Nature Climate Change, 5*, 695–701.

Béllard, C., Leclerc, C., Leroy, B., Bakkenes, M., Veloz, S., Thuiller, W., et al. (2014). Vulnerability of biodiversity hotspots to global change. *Global Ecology and Biogeography, 23*, 1376–1386.

Cunningham, C., Cunha, A. P., Brito, S., Marengo, J., & Coutinho, M. (2017). Climate change and drought in Brazil. In B. Marchezini, B. Wisner, L. R. Londe, & S. M. Saito (Eds.), *Reduction of vulnerability to disaters: From knowledge to action* (pp. 361–375). São Carlos, Brazil: Editora RiMa.

Descombes, P., Wisz, M. S., Leprieur, F., Parravicini, V., Heine, C., Olsen, S. M., et al. (2015). Forecasted coral reef decline in marine biodiversity hotspots under climate change. *Global Change Biology, 21*(7), 2479–2487.

Diamond, J. M. (1989). Overview of recent extinctions. In D. Western & M. C. Pearl (Eds.), *Conservation for the twenty-first century* (pp. 37–41). Oxford, UK: Oxford University Press.

[1] SNUC = National Conservation Units System; Planaveg = National Native Vegetation Restoration Plan; New Forest Code = Native Vegetation Protection Law.

Faleiro, F. A. V. M., Nemésio, A., & Loyola, R. (2018). Climate change likely to reduce orchid bee abundance even in climatic suitable sites. *Global Change Biology*. https://doi.org/10.1111/gcb.14112

Faleiro, F. V., Machado, R. B., & Loyola, R. D. (2013). Defining spatial conservation priorities in the face of land-use and climate change. *Biological Conservation, 158*, 248–257.

Feeley, K. J., & Silman, M. (2016). Disappearing climates may limit the efficacy of Amazonian protected areas in a warming world. *Diversity and Distributions, 22*, 1081–1084.

Ferro, V. G., Lemes, P., Melo, A. S., & Loyola, R. (2014). The reduced effectiveness of protected areas under climate change threatens Atlantic forest tiger moths. *PLoS One, 9*, e107792.

Fisher, J. A., Patenaude, G., Kalpana, G., et al. (2014). Understanding the relationships between ecosystem services and poverty alleviation: A conceptual framework. *Ecosystem Services, 7*, 34–45.

Giannini, T. C., Tambosi, L. R., Acosta, J. R., Saraiva, A. M., Imperatriz-Fonseca, V. L., & Metzger, J. P. (2015). Safeguarding ecosystem services: A methodological framework to buffer the joint effect of habitat configuration and climate change. *PLoS One, 10*(6), e0129225.

Godoy, M. D. P., & Lacerda, L. D. (2015). Mangroves response to climate change: A review of recent findings on mangrove extension and distribution. *Anais da Gym Brasileira de Ciências, 87*(2), 651–667.

Hoffmann, D., Vasconcelos, M. F., & Martins, R. P. (2015). How climate change can affect the distribution range and conservation status of an endemic bird from the highlands of eastern Brazil: The case of the gray-backed Tachuri, *Polystictus superciliaris* (Aves, Tyrannidae). *Biota Neotropica, 15*(2), 1–12.

IPCC. (2007). *Climate change 2007: The physical science basis. Contribution of working group I to the fourth assessment report of the intergovernmental panel on climate change*. Cambridge, UK: Cambridge University Press.

IPCC. (2013). *Climate change 2013: The physical science basis. Contribution of working group I to the fifth assessment report of the intergovernmental panel on climate change*. Cambridge, UK: Cambridge University Press.

IPCC. (2014). *Climate change 2014: Impacts, adaptation, and vulnerability. Part A: Global and sectoral aspects. Contribution of working group II to the fifth assessment report of the intergovernmental panel on climate change*. Cambridge, UK: Cambridge University Press.

Joly, C., Metzger, J. P., & Tabarelli, M. (2014). Experiences from the Brazilian Atlantic Forest: Ecological findings and conservation initiatives. *New Phytologist, 204*, 459–473.

Jones, H. P., Hole, D. G., & Zavaleta, E. S. (2012). Harnessing nature to help people adapt to climate change. *Nature Climate Change, 2*(7), 504–509.

Jones, K. R., Watson, J. E. M., Possingham, H. P., & Klein, C. J. (2016). Incorporating climate change into spatial conservation prioritisation: A review. *Biological Conservation, 194*, 121–130.

Juhola, S., Glaas, E., Linnér, B.-O., & Neset, T.-S. (2016). Redefining maladaptation. *Environmental Science and Policy, 55*, 135–140.

Kasecker, T. P., Ramos-Neto, M. B., Silva, J. M. C., & Scarano, F. R. (2017). Ecosystem-based adaptation to climate change: Defining hotspot municipalities for policy design and implementation in Brazil. *Mitigation and Adaptation Strategies to Global Change*. https://doi.org/10.1007/s11027-017-9768-6

Keith, D. A., Mahony, M., Hines, H., Elith, J., Regan, T. J., Baumgartner, J. B., et al. (2015). Detecting extinction risk from climate change by IUCN red list criteria. *Conservation, 28*(3–130), 810–819.

Laurance, W. F. (2015). Emerging threats to tropical forests. *Annals of the Missouri Botanical Garden, 100*(3), 159–169.

Leadley, P., Proença, V., Fernández-Manjarrés, P. H. M., Alkemade, R., Biggs, R., Bruley, E., et al. (2014). Interacting regional-scale regime shifts for biodiversity and ecosystem services. *Bioscience 64*(8):893–905: 665–679.

Lemes, P., & Loyola, R. D. (2013). Accommodating species climate-forced dispersal and uncertainties in spatial conservation planning. *PLoS One, 8*, e54323.

Loyola, R. D., Lemes, P., Brum, F. T., Provete, D. B., & Duarte, L. D. S. (2014). Clade-specific consequences of climate change to amphibians in Atlantic Forest protected areas. *Ecography, 37*, 65–72.

Lucena, A. J., Rotunno Filho, O. C., Peres, L. F., & France, J. R. A. (2012). A evolução da ilha de calor na region metropolitana do Rio de Janeiro. Revista Geonorte, 2(5) Edição Especial 2:8–21.

Mace, G., Masundire, H., & Baillie, J. E. M. (2005). Biodiversity. In R. Hassan, R. Scholes, & N. Ash (Eds.), *Ecosystems and human well-being: Current state and trends: Findings of the condition and trends working group* (pp. 77–122). Washington, DC: Island Press.

Magrin, G. O., Marengo, J. A., Boulanger, J.-P., Buckeridge, M. S., Castellanos, E., Poveda, G., et al. (2014). Central and South America. In V. R. Barros, C. B. Field, D. J. Dokken, M. D. Mastrandrea, K. J. Mach, T. E. Bilir, M. Chatterjee, K. L. Ebi, Y. O. Estrada, R. C. Genova, B. Girma, E. S. Kissel, A. N. Levy, S. MacCracken, P. R. Mastrandrea, & L. L. White (Eds.), *Climate change 2014: Impacts, adaptation, and vulnerability. Part B: Regional aspects. Contribution of working group II to the fifth assessment report of the intergovernmental panel on climate change* (pp. 1499–1566). Cambridge, UK: Cambridge University Press.

McNeely, J. A., Mittermeier, R. A., Brooks, T. M., Boltz, F., & Ash, N. (2009). *The wealth of nature: Ecosystem services, biodiversity, and human well-being*. Arlington, TX: CEMEX, Conservation International.

Mittermeier, R. A., Robles-Gil, P., & Mittermeier, C. G. (1997). *Megadiversity: Earth's biologically wealthiest nations*. Arlington, TX: CEMEX, Conservation International.

Mora, C., Caldwell, I. R., Caldwell, J. M., Fisher, M. R., Genco, B. M., & Running, S. W. (2015). Suitable days for plant growth disappear under projected climate change: potential human and biotic vulnerability. *PLOS Biology*. https://doi.org/10.1371/journal.pbio.1002167

Oliveira, G., Araújo, M. B., Rangel, T. F., Alagador, D., & Diniz-Filho, J. A. F. (2012). Conserving the Brazilian semiarid (Caatinga) biome under climate change. *Biodiversity and Conservation, 21*(11), 2913–2916.

Oliveira, G., Lima-Ribeiro, M. S., Terribile, L. C., Dobrovolski, R., Telles, M. P. C., & Diniz-Filho, J. A. F. (2015). Conservation biogeography of the cerrado's wild edible plants under climate change: Linking biotic stability with agricultural expansion. *Journal of Botany, 102*, 6–1.

Pant, L. P., Adhikari, B., & Bhattarai, K. K. (2015). Adaptive transition for transformations to sustainability in developing countries. *Current Opinion in Environmental Sustainability, 14*, 206–212.

Pires, A. P. F., Rezende, C. L., Assad, E. D., Loyola, R., & Scarano, F. R. (2017). Forest restoration can increase the Rio Doce watershed resilience. *Perspectives in Ecology and Conservation, 15*, 187–193.

Rezende, C. L., Fraga, J. S., Sessa, J. C., Souza, G. V. P., Assad, E. D., & Scarano, F. R. (2018). Land use policy as a driver for climate change adaptation: A case in the domain of the Brazilian Atlantic forest. *Land Use Policy, 72*, 563–569.

Ribeiro, B. R., Sales, L. P., De Marco, P., & Loyola, R. (2016). Assessing mammal exposure to climate change in the Brazilian Amazon. *PLoS One, 11*, e0165073.

Rosenzweig, C., Solecki, W., Romero-Lankao, P., Mehrotra, S., Dhakal, S., Bowman, T., et al. (2015). *ARC3.2 Summary for city leaders. Urban climate change research network*. New York: Columbia University.

Sawyer, D. (2008). Climate change, biofuels and eco-social impacts in the Brazilian Amazon and Cerrado. *Philosophical Transactions of the Royal Society B, 363*, 1747–1752.

Scarano, F. R. (2017). Ecosystem-based adaptation to climate change: Concept, scalability and a role for conservation science. *Perspectives in Ecology and Conservation, 15*, 65–73.

Scarano, F. R., & Ceotto, P. (2015). Brazilian Atlantic forest: Impact, vulnerability and adaptation to climate change. *Biodiversity and Conservation 24*(11), 2913–2916: 2319–2331.

Scarano, F. R., Guimarães, A., & Silva, J. M. (2012). Lead by example. *Nature, 486*, 25–26.

Seddon, A. W. R., Macias-Fauria, M., Long, P. R., Benz, D., & Willis, K. J. (2016). Sensitivity of global terrestrial ecosystems to climate variability. *Nature*. https://doi.org/10.1038/nature16986

Segan, D. B., Murray, K. A., & Watson, J. E. M. (2016). A global assessment of current and future biodiversity vulnerability to habitat loss–climate change interactions. *Global Ecology and Conservation, 5*, 12–21.

Silva, J. M. C., & Prasad, S. (2017). Green and socioeconomic infrastructures in the Brazilian Amazon: Implications for a changing climate. *Climate and Development*. https://doi.org/10.1080/17565529.2017.1411242

Souza-Filho, F. A., Scarano, F. R., Nicolodi, J. L., Vital, H., AHF, K., PEPF, T., et al. (2014). Recursos naturais, manejo e uso de ecosystems. In E. D. Assad & A. R. Magalhães (Eds.), *Impactos, vulnerabilidades e adaptação às mudanças climáticas. Contribuição do Grupo de Trabalho 2 do Painel Brasileiro de Mudanças Climáticas ao Primeiro Relatório da Avaliação Nacional sobre Mudanças Climáticas* (pp. 43–200). Rio de Janeiro, Brazil: COPPE, Rio de Janeiro Federal University.

Steffen, W., Richardson, K., Röckstrom, J., et al. (2015). Planetary boundaries: Guiding human development on a changing planet. *Science, 347*, 1259855. https://doi.org/10.1126/science.1259855

Strassburg, B. B. N., Brooks, T., Feltran-Barbieri, R., Iribarrem, A., Crouzeilles, R., Loyola, R., et al. (2017). Moment of truth for the Cerrado hotspot. *Nature Ecology and Evolution, 1*, 0099. https://doi.org/10.1038/s41559-017-0099

Tabarelli, M., Leal, I. R., Scarano, F. R., & Silva, J. M. C. (2017). The future of the Caatinga. In S. JMC, I. R. Leal, & M. Tabarelli (Eds.), *Caatinga* (pp. 461–474). Cham, Switzerland: Springer.

Thomas, C. D., Cameron, A., Green, R. E., Bakkenes, M., Beaumont, L. J., Collingham, Y. C., et al. (2004). Extinction risk from climate change. *Nature, 427*, 145–148.

Urban, M. C. (2015). Accelerating extinction risk from climate change. *Science, 348*, 571–573.

Vieira, R. R. S., Ribeiro, B. R., Resende, F. M., Brum, F. T., Machado, N., Sales, L. P., et al. (2017). Compliance to Brazil's forest code will not protect biodiversity and ecosystem services. *Diversity and Distributions, 24*, 434. https://doi.org/10.1111/ddi.12700

Visconti, P., Bakkenes, M., Baisero, D., Brooks, T., Butchart, S. H. M., Joppa, L., et al. (2015). Projecting global biodiversity indicators under future development scenarios. *Conservation Letters*. https://doi.org/10.1111/conl.12159

Yu, M., Wang, G., Parr, D., & Ahmed, K. F. (2014). Future changes of the terrestrial ecosystem based on a dynamic vegetation model driven with RCP8.5 climate projections from 19 GCMs. *Climatic Change, 127*, 257–271.

Zanin, M., Tessarolo, G., Machado, N., & Albernaz, A. L. M. (2017). Climatically-mediated land-cover change: Impacts on Brazilian territory. *Anais da Academia Brasileira de Ciências, 89*, 939–952.

Zwiener, V. P., Padial, A. A., Marques, M. C. M., Faleiro, F. V., Loyola, R., & Peterson, A. T. (2017). Planning for conservation and restoration under climate and land use change in the Brazilian Atlantic Forest. *Diversity and Distributions, 23*, 955–966.

Chapter 6
Climate Change and the Energy Sector in Brazil

Roberto Schaeffer, André F. P. Lucena, Isabella V. L. Costa, Eveline Vásquez, Cindy Viviescas, and Vanessa Huback

6.1 Introduction

Changes induced by global warming have led to the increase in frequency and intensity of extreme climate events like heat waves, hurricanes, floods and rise in sea level. This leads to significant alterations to water availability, flooding and duration of dry periods, among others. These changes in climate bring with them a long list of environmental and socio-economic impacts in different areas like biodiversity, coastal zones, water supply, agriculture, food security, health and energy, leading to warnings in relation to the future of the population and the planet's sustainability (Lelis, Calijuri, da Fonseca, & de Lima, 2011; Marengo et al., 2004; Streck & Alberto, 2006).

The energy sector stands out as a major climate change driving force. At the same time, the sector is subject to the adverse effects of climate change, particularly renewable energy sources that show much higher vulnerability vis-à-vis non-renewable sources like coal, oil and gas. Changes in the rainfall patterns may affect the production of energy from different sources, as well as other adverse impacts, such as the melting of glaciers, water supply problems, flooding, increasing potential soil loss and aggravating erosive processes, among others. (Lelis et al., 2011).

Additionally, climate change affects different crops used in the production of biofuels (Pinto & Assad, 2008; Schaeffer et al., 2008). Wind regimes define the generation of wind power (Pryor & Barthelmie, 2010), cloud formation has an impact on how much solar radiation is available for generating energy (Bull et al., 2007), as well as hydroelectricity (Gundry & Whittington, 1998), among other sources.

The study of these climate impacts in energy systems with a high share of renewables is ever more important. This is very much the case of Brazil, specifically for

R. Schaeffer (✉) · A. F. P. Lucena · I. V. L. Costa · E. Vásquez · C. Viviescas · V. Huback
COPPE/Federal University of Rio de Janeiro (COPPE/UFRJ), Rio de Janeiro, Brazil
e-mail: roberto@ppe.ufrj.br

C. A. Nobre et al. (eds.), *Climate Change Risks in Brazil*,
https://doi.org/10.1007/978-3-319-92881-4_6

the generation of hydroelectricity. Between the late 90s and 2011, hydraulic power represented 70% of the electricity generated in the country (ONS, 2015a). Together with other renewable sources in the electricity matrix, like wind and biomass, renewable sources reached nearly 90% in this period (Martins & Pereira, 2011).

Between 2012 and 2014, hydroelectricity underwent a sharp fall in its share, despite keeping its importance – 60% of the electricity generated (ONS, 2015b). This was very much a consequence of a period of low rainfall, which indicates the high vulnerability of hydropower to climate conditions. In the same way, the growing share of other renewables like wind and solar, or the higher number of biomass plants, leads to the need to understand the impacts that climate change may have on a electricity matrix, so as to assess its vulnerability.

Similarly, fossil fuel electricity generation plants need water for cooling. This happens in thermoelectric, as well as thermonuclear power plants, making these technologies susceptible to hydrological changes in water availability and temperature (Bull et al., 2007; Feeley et al., 2008); air temperature and humidity have an effect on the performance of natural gas turbines (Schaeffer et al., 2008); extreme climate events, like hurricanes, may affect oil production on off-shore platforms (Bull et al., 2007). In relation to electricity transmission, higher temperatures increase ohmic losses in electricity transmission lines (Perez, 2009).

The exposure of an energy system to impacts of climate change depends not only on the set of energy resources used and technology options, but also on the magnitude of the climate variation. In a world with higher GHG emissions, more severe increases in temperature are expected, causing greater impacts on the energy sector.

The magnitude of climate change depends on the joint international effort to reduce GHG emissions. Therefore, regardless of national mitigation efforts based on a more renewable energy system, it is possible for Brazil to end up facing impacts of great magnitude, generated by other countries' emissions.

Taking into account severe impacts resulting from a temperature increase of over 4 °C, it is important to assess the vulnerability of the Brazilian energy system. In this case, two possibilities are suggested. First, the country strives to put in place mitigation efforts, including a more climate dependent renewable energy matrix and consequently, becomes more vulnerable to climate change. Secondly and paradoxically, a lack of concern in relation to reducing emissions would lead the country to a more fossil based energy matrix, on one hand leading to higher emissions, while exposing it less to climate effects. Despite both paths being exposed to climate effects of great magnitude, this study does not intend on identifying which could be considered the most probable.

The aim of this chapter is to assess what would be the effects of extreme climate change, in other words, with average temperature increase of over 4°C, on the Brazilian energy system in different settings. In order for this objective to be met, a long bibliographical review was done of studies related to possible impacts of global climate change on the Brazilian electricity sector, as well as suitable adaptation options.

It is not the scope of this study to provide climate simulations and innovative impact assessments. The aim here is to supply a meta-analysis of already published studies, looking at them qualitatively. Notwithstanding, this approach may provide precious information for decision and policy makers, particularly those linked with climate negotiations and engaged in the drafting of mitigation and adaptation policies.

This Chapter is organised to initially present the possible impacts climate change may have on the energy system. Then, a bibliographical review on the impacts of climate change on the Brazilian electricity sector is provided, leading to an analysis of the possible effects that such big changes may have. After that, the adaptation options for Brazil are discussed, before the study's final conclusions are presented.

6.2 Climate Change and the Energy Sector in Brazil: Summary of Impacts

In relation to impacts on resource allocation, hydroelectric power depends directly on water and consequently, on the hydrological cycle. Therefore, 'the allocation of hydroelectric resources is the result of water surplus that turns into flow'" (Schaeffer et al., 2012). Hydro power plants also depend on the seasonality of hydrological cycles. In relation to the impacts on the energy supply, the amount of electricity that may be generated by hydro power plants depends on the installed generation capacity, as well as the water intake at the plants' reservoirs. Climate variations have a great influence on this type of system, as well as geographical dispersion and level of integration.

The availability and reliability of wind power depends on weather conditions. According to Pryor and Barthelmie (2010), the main mechanism through which climate change may have an impact on the allocation of wind power is geographical distribution and wind speed variability. Different from Hydroelectric power, wind energy cannot be regulated and thus, wind speed variations have a great impact on the energy produced by wind turbines.[1] Therefore, an impact assessment of climate change on the supply of energy may be done using the frequency distribution of wind speed and not just the value average.

Liquid biofuels are vulnerable to the effects of variable climate like temperature, rain and CO_2 levels. In terms of available natural resources, according to Schaeffer et al. (2012), the main impact is related to possible losses in fertile and appropriate areas for energy crops due to climate change. Also, according to Siqueira, Steinmetz, Salles, and Fernandes (2001), climate change has a direct effect on factors like crop yield, agricultural distribution zones, incidence of pests and availability of appropriate land for some crops. The increase in temperature may modify soil conditions,

[1] The amount of energy in the wind is proportionate to the wind speed elevated to the cube, which indicates how speed variations may have significant effects on the volume of available power (Pryor & Barthelmie, 2010).

reflecting on the fertility of plantations and their yield, which could be compensated by the increase in photosynthesis activity in some cases. Hydrological regime is also a very important factor, as well as CO_2 levels and extreme climate conditions.

It is worth mentioning that sugar cane productivity may benefit from temperature increase, although irrigation may become necessary in arid regions. Also, in the case of CO_2 fertilization, the productivity of some cultures may increase. However, many cultures may have their productivity decreased, impacting food availability. Furthermore, as high productivity land becomes more scarce, competition between energy and food crops may be exacerbated.

In addition, according to a Feres et al. (2010) study on the impact of climate change on land use pattern, water deficiency should become a reality in the Northeast forcing the migration of oilseeds like soybean and sunflower to regions in the south of the country, more suited to farming.

According to Cutforth and Judiesch (2007), climate change may affect solar power generation by altering the content of atmospheric vapour, cloud cover and characteristics that have an effect on atmospheric transmissivity. According to Bull et al. (2007), this may affect the generation of electricity by photovoltaic panels and concentrated solar power plants (CSP). In addition to impacts due to extreme climate change, the supply of energy may be affected by an increase in air temperature, which may modify the efficiency of photovoltaic cells and reduce electricity generation by these panels. As the CSP is a thermal machine, its efficiency is altered when there are changes in room temperature.

Wave power is the most used ocean energy source around the world, although it has not yet been developed and disseminated like other renewables. According to Schaeffer et al. (2012), climate change may affect wind, which causes direct impacts in wave formation. The average condition of waves in a place presents different long term trends throughout the world, following wind's climate effects in wave generation in a non-linear relationship.

Despite climate change not affecting the current amount of oil and gas, they may have an influence on our knowledge about these resources and access to them, in addition to affecting contingent resources. Offshore oil and natural gas supply or in low latitude easy access facilities may be disturbed by extreme climate events like strong hurricanes, which could lead to interruptions in production in order to avoid environmental damage or risk to human life. According to Schaeffer et al. (2012), supply may also be affected by structural damage caused by other extreme events like floods, which occur as a result of the rise in sea level and storms that may lead to erosion and other problems in the offshore oil production in Brazil.

Possible increase in the frequency and intensity of rainfall could lead to changing levels in rivers and floods, which could alter the quality and management of coal. This could lead to an increase in maintenance and operational costs for coal plants in coal preparation. Therefore, increasing coal management costs may interfere negatively with quantified reserves, which could affect their economic viability (Bull et al., 2007).

According to CCPS (2007), as global climate change may affect thermoelectric production by interfering with the generation cycle efficiency and water related needs for cooling plants. The technologies that could be affected by this are coal, natural gas, nuclear, CSP and biomass waste plants. The effects of room temperature changes on electricity generation efficiency in coal and nuclear plants are similar, as both operate under a Rankine cycle. According to Tolmasquim, Szklo, and Soares (2003), gas plants and those operating under a Brayton or combined cycle may see their turbine's power and efficiency affected by temperature and humidity variations. An increase in temperature due to climate change has an impact on gas turbine performance, leading to a reduction in generation or a rise in fuel consumption, according to Kehlhofer, Hannemann, Rukes, and Stirnimann (2009).

Thermal plants need large amounts of water and there are projections showing a drop in the availability of this resource in some areas in the future. Therefore, 'these plants may experience increased competition in relation to water with other uses like agriculture, in areas where there is water scarcity (Feeley et al., 2008). Changes in water quantity and quality affect mostly the generation of nuclear power.

On the demand side, the increase in temperature implies in a reduction in the demand for heating and a rise in the demand for cooling. In addition, according to Schaeffer et al. (2012), in order for developing countries to reach their development goals, they tend to have a higher demand for energy because of higher levels of urbanisation, electrification and standard of living. Furthermore, changes to temperature will affect the use of air conditioning not only in buildings, but also in vehicles, altering fuel consumption. According to Schaeffer et al. (2012), in the agricultural sector, temperature increase has led to a rise in demand for irrigation, thus, increasing the use of power for pumping water.

This Chapter is organised to present the possible impacts that climate change may have on energy systems. A general overview of Table 6.1 shows that several impacts and sectors have not yet been looked at formally, showing a lag in areas where good research may be developed.

6.3 Impact of Global Climate Change on the Brazilian Electricity Sector

Several energy sector studies are based on climate models (General Circulation Models – GCMs) in order to establish how these possible climate variations may have a direct or indirect impact on energy supply and demand. According to Lucena (2010), the impacts of climate change on several sectors have been studied since the 1980s, however, literature on the effects on the energy sector, particularly electricity is relatively new and limited. The energy and electricity systems are subject to alterations to global and regional climate conditions. Despite this, most of the current discussion in literature on the relationship between global climate change and energy focuses on the role played by the latter in GHG emissions, as well as mitigation alternatives.

Table 6.1 Summary of possible impacts on the energy sector related to climate variations

Energy sector	Climate variables	Related impacts	Energy sector studies
Thermoelectric (natural gas, coal and nuclear)	Air/water temperature.	Quantity and quality of water for cooling.	Kopytko and Perkins (2011)
	Air/water temperature, wind and humidity.	Cooling efficiency and operational turbine efficiency.	Schaeffer et al. (2008)
			Sathaye et al. (2013)
	Extreme climate events.	Erosion in surface mining.	
		Sea extraction interruptions.	
Oil and natural gas	Extreme climate events.	Sea extraction interruptions.	Harsem, Eide, and Heen (2011)
	Extreme climate events., air/water temperature, floods.	Land extraction interruptions.	Burkett (2011)
		Production and transport transfer interruptions.	
	Extreme climate events, floods, air temperature.	Import operation interruptions.	
	Extreme climate events.	Closing down of refineries.	
	Flood, extreme climate events, air/water temperature, floods.	Quantity and quality of water for cooling in oil refineries.	
Biomass	Air temperature, rainfall, humidity.	Desertification.	Fenger (2007)
	Extreme climate events.	Yield of bioenergy crops.	Siqueira et al. (2001)
		Land availability and distribution with appropriate edaphoclimatic conditions (agricultural zoning).	Pinto and Assad (2008)
	Carbon dioxide fertilization.		Brown, Rosenberg, Hays, Easterling, and Mearns (2000)
Hydroelectric	Air temperature, rainfall, extreme climate events.	Total and seasonal water availability (inflow at the plant's reservoir).	Fenger (2007)
			VicuñA, Leonardson, Dracup, Hanemann, and Dale (2005)
		Dry periods.	Lehner, Czisch, and Vassolo (2005)
		Changes in the operation of the hydroelectric system.	Lucena et al. (2009)
		Evaporation of reservoirs.	Hamlet, Lee, Mickelson, and Elsner (2010)
			Harrison and Whittington (2002)
			Whittington and Gundry (1996)
			Vicuna et al. (2007)
			See Table 6.2

(continued)

Table 6.1 (continued)

Energy sector	Climate variables	Related impacts	Energy sector studies
Demand	Air temperature, rainfall.	Increase in the demand for air conditioning in the summer.	See Table 6.1 in Schaeffer et al. (2012)
		Reduction in demand for heating in winter.	
		Increase in the demand for power in irrigation.	
Wind power	Wind and extreme climate events.	Changes to wind (intensity and duration), wind gradient, damage caused by extreme meteorological conditions.	Pryor and Barthelmie (2010)
			Fenger (2007)
			Lucena, Szklo, Schaeffer, and Dutra (2010)
			Harrison and Wallace (2005)
			Sailor, Hu, Li, and Rosen (2000)
			Sailor, Smith, and Hart (2008)
			Lucena, Szklo, and Schaeffer (2009b)
			Pereira, Martins, Pes, da Cruz Segundo, and Lyra (2013)
			Garreaud and Falvey (2009)
Solar energy	Air temperature, insolation level, humidity and rainfall.	Changes in insolation level (cloud formation).	Fenger (2007)
		Changes in efficiency due to radiation variation.	Fenger (2007)
		Efficiency reduction due to environmental conditions.	Crook, Jones, Forster, and Crook (2011)
			Wild, Folini, Henschel, Fischer, and Müller (2015)
Geothermal	Air/water temperature.	Cooling efficiency.	–
Waver power	Wind and extreme climate events.	Changes in waves.	Harrison and Wallace (2005)

Source: Schaeffer et al. (2012) (translation), Crook et al. (2011), Fidje and Martinson (2007), Garreaud and Falvey (2009), Lucena, Szklo, and Schaeffer (2009b), Pereira et al. (2013) and Wild et al. (2015)

In this Chapter, a scientific literature review on climate change impacts is made, taking into consideration more vulnerable sources. It focuses more on hydroelectricity because of its importance and the higher number of studies on the matter. The intention was to assess the published scientific literature on observed and projected impacts of climate change in the electricity sector, as a response to a 4°C or higher temperature increase. However, the scientific literature presents results using the nomenclature of IPCC's CO_2 concentration scenarios.

6.4 Review of Studies on Impacts on Electricity Production: The Case of Wind Power and Solar Energy

Studies on the impacts of climate change on energy production commonly approach a specific segment, generally a renewable source. Hydraulic and wind generation sources were the main object of scientific studies and papers, which built scenarios based on GCM climate projections. This Chapter will assess the literature on the impacts on wind power and solar energy. The following Chapter will emphasise studies on the impacts on hydroelectricity.

After the implementation of the Alternative Energy Sources Incentive Programme (PROINFA) in 2004, wind power has been growing significantly in the Brazilian energy matrix. The programme is aimed at encouraging the use of other renewable sources like wind, biomass and small hydro power plants (SHP). Consequently, the installed capacity went from 22 MW in 2003 to 7609 MW in 2010 (Martins & Pereira, 2011). Currently, installed wind power stands at 12509 MW, corresponding to 7,9% of the Brazilian electrical installed capacity (ANEEL, 2016b). The sector's expansion made it necessary for the wind potential of 143 GW to be reviewed, which had been established by the first Brazilian Atlas of Wind Potential (CEPEL, 2001), where towers of up to 50 metres in height were taken into consideration. New assessments now consider turbine of 120 metres or taller (MME, 2014).

Wind speed varies significantly with height and little is known on future projections of velocity at the height of a turbine shaft (over 50 metres): these projections are normally available at the relevant height. However, there are some methods that may be used to extrapolate wind speed for different heights (e.g. by a logarithmic rule, Dutra & szklo, 2008). Some key parameters have to be taken into account like rough terrain, which could vary depending on the type of vegetation cover (Dutra & szklo, 2008; Schaeffer et al., 2012). In this context, climate change may bring impacts to the vegetation cover (Nobre 2007) and thus, affect wind potential.

On the other hand, the relationship between wind speed and its energy density is not linear. Therefore, changes to the wind speed frequency distribution may affect the optimal point between the availability of natural resources and the turbine's power curve. A difficulty when assessing climate impacts in wind power generation is the fact that the results of the global GCM modelling do not provide enough information about speed variability in a fine spatial scale, increasing the need for downscaling techniques (Lucena, Szklo, & Schaeffer, 2009a). The natural time, day or seasonal variability of wind speed has a significant impact on the energy produced by turbines, making the operation more subject to changes in wind patterns resulting from climate change. This implies that the climate change analysis for wind generation has to be done using wind speed distribution and not just average values (Schaeffer et al., 2012). Pryor and Barthelmie (2010) show how simplified assumptions employed in similar studies may present diametrical results.

In relation to Brazilian generation history, wind power is a new source, which explains the limited literature on the impacts temperature rises may have on it. Lucena et al. (2010) used the 'delta method' to assess climate change impacts on wind generation potential in Brazil. The results of this study show that the wind potential will probably not suffer major negative impacts. On the contrary, for scenarios A2 and B2 results showed an increase in Brazil's wind potential as time goes by. The Brazilian Northeast, as well as the coast of the North and Northeast regions are areas that have shown to be particularly attractive for wind power exploration. These scenarios (A2 and B2) were dynamically downscaled in regional climate projections for Brazil by an expert team on Brazilian weather from CPTEC/INPE, who used the PRECIS (Providing Regional Climates for Impacts Studies) model (United Nations, 2016). This is a regional climate model developed by the Hadley Centre, which downscales the results from the general circulation model (GCM) HadCM3. Future projections for Brazilian wind potential were based on wind's average annual speed in 50 km by 50 km squares, for the time intervals considered by PRECIS.

Lucena et al. (2010) indicate that the wind average speeds would increase considerably in coastal regions in general, particularly in the country's North and Northeast regions. This study points to a greater frequency of wind with speeds over 8.5 m/s at the coast, which raises the possibility of including different turbine designs that may generate more power at higher speeds in future analyses. The results based on climate projections show that wind power generation could increase threefold in Brazil in scenario B2 and fourfold in scenario A2, when compared with the 2010 reference situation. However, these results are not determinant due to climate projection uncertainties and assumptions made in the study. In sum, this study indicates that wind power generation in Brazil will not be hindered by climate change.

Lucena, Szklo, and Schaeffer (2009b) provided a theoretical analysis on issues relevant to climate change impacts on wind power generation, such as the downscaling in speed distribution frequency, transposition of wind speed measuring height and possible alterations in vegetation cover. In addition, Pryor and Barthelmie (2010) conducted a review of studies focused on climate change impacts (Global Climate Change) in wind power generation. They analysed the mechanism through which climate change may influence wind resources and its operational conditions, as well as the tools that have been employed to quantify these effects and uncertainties related to them.

In Pereira et al. (2013), climate change impacts on wind power are assessed by simulating future scenarios on the country's gross potential, taking into account climate scenarios of the IPCC SRES A1B emissions. The analysis was done for Brazil's South and Northeast regions. Ground stations' data trend was studied like wind forecasts based on the global circulation model HadCM3. The Eta model was used to downscale a 40 km by 40 km resolution and 38 vertical layers. The Eta model was updated every 6 h with the boundary conditions of the HadCM3 outputs.

In the study of Pereira et al. (2013) the search for climate time series trends were not conclusive. On the other hand, Eta model predictions - HadCM3 for A1B scenario indicates an average growth trend from 15 to 30% for onshore wind power density for most of Brazil's Northeast region. Indeed, some regions showed an over 100% increase, particularly the Northeast. In addition, with the exception of the country's North and Northeast regions, the study pointed to a fall in future offshore wind power density, particularly off the coast of the state of Bahia.

Nevertheless, the same study pointed to a small increase in wind power density in Brazil's South region, when compared with results for the Northeast. This means an average increase of 10%, reaching over 20% in some areas. The central region of the Rio Grande do Sul state, which extends to the south of Uruguay, showed a small decreasing trend in wind power. This region also showed the highest seasonal variability, with a global minimum in the austral summer (December–February) and an increase in the rest of year, in relation to the baseline period. Therefore, according to Pereira et al. (2013), it is possible to expect that the impact of global climate change on wind power in Brazil's Northeast and South may be favourable to existing and future projects in both regions.

In addition, studies on coastal wind change were conducted, particularly for the South American west coast in Garreaud and Falvey (2009), using 15 GCMs and the PRECIS model, with 25 km horizontal resolution nested in the Hadley Centre Atmospheric Global Model - HadAM3. This study showed a large increase in wind near the surface, up to 15% in average speed for the A2 scenario. In the B2 scenario, results show seasonal wind patterns similar to the A2 scenario, but with up to 25% increase.

However, it must be remembered that changes in vegetation pattern may have significant impacts on wind speeds, as they are affected by friction with the soil surface. Wind progression at different heights is influenced by irregularities and characteristics of land biomes. Projections made by INPE for the 2070–2099 period, using global climate models, show more humid biomes (such as tropical forest) being replace by biomes adapted to less availability of water like the Cerrado, desert and semiarid (INPE, 2007). Such alterations could also influence wind potential in climate change scenarios.

Finally, it should be mentioned that the integration of wind power depends on the availability of generation sources that have the flexibility to cope with the variability and intermittency of wind power. For more information on the integration of wind power, see Acker et al. (2012). In this sense, hydropower is a good option, especially in the case of Brazil. However, there is limited literature on the impacts of climate change on renewable energy integration.

In the case of concentrated solar power plants (CSP), the direct influence of climate change in the production of electricity is mainly due to alterations in air temperature and radiation (Fenger, 2007). In addition, CSP plants may be indirectly affected by water availability for Rankine cycle cooling.

No solar power production studies were identified specifically for Brazil. Not much research on possible climate change impacts are available in the international literature for solar thermal or photovoltaic. Crook et al. (2011), Fenger (2007) and Fidje and Martinson (2007) based themselves on SRES emission scenarios. The study by Crook et al. (2011) used the A1B emission scenario and included several regions of countries belonging to different climate cones like the United States of America (USA), Saudi Arabia, Spain, Australia, Germany, China and Algeria. They used two climate models, HadGEM1 and HadCM3, and air temperature and solar radiation parameters (total and direct). The results indicate that the possible variations in the efficiency of solar panels depend very much on location. However, the study showed that energy production by solar panels is likely to increase in regions in Europe and China, with a small change in Algeria and Australia and a few percentage points fall in western USA and Saudi Arabia. On the other hand, the study also registered a possible alteration in energy generation efficiency in CSP, due to change in direct radiation. CSP generation should increase in over 10% in Europe and several percentage points in China, Algeria and Australia. In addition, in western USA and Saudi Arabia, the trend would be of a few percentage points reduction.

On the other hand, a recent study by Wild et al. (2015) assessed the possible impact of solar panel generation, based on RCP 8.5 scenarios, using 39 climate models and mainly parameters like air temperature and solar radiation. The study focused on regions in the following countries: United States of America, South Africa, Spain, India, Australia, Germany, China and Algeria. The results indicate a reduction in efficiency in several studied regions. Nevertheless, regions in Europe, in the southeast of North America and China showed an increase in generation output.

6.5 Review of Studies on Impacts on Electricity Production: The Case of Hydro Power

According to IPCC (2014), in many regions, changes in rainfall or the melting of snow and ice are altering hydrological systems, affecting water resources in terms of quantity and quality. Glaciers are shrinking almost all over the world and permafrost is melting in high altitude and latitude regions due to climate change, affecting the flow of available water resources.

Global and regional alterations have led to climate and hydrological changes in the Brazilian Amazon region. For example, changes in the use of soil, such as the conversion of over 700,000 km^2 of forest into pasture land and global warming, which has registered an increase in the average temperature from 0.6 to 0.9 °C in the last hundred years, have contributed to this (GCP; CIAT, 2013). In the Pacific Ocean region, El Niño events have brought extreme lack of rain and consequently, led to low flows in rivers in the area, particularly in the northeast of the Amazon. In relation to the Atlantic Ocean, according to the Brazilian Centre for Weather Forecast and Climate Studies (CPTEC/INPE) and the National Institute of Meteorology (INMET), the sea surface temperature went from 0.5 to 1.5 °C above the average for the North Atlantic for the period between September 2004 and September 2005. In other words, an abnormal and persistent warming phenomenon was registered (Freitas, 2006). This phenomenon, possibly responsible for the 2005 drought in the Amazon, ended up altering moist air mass in the Amazon region, particularly in important areas of the drainage basins of the Solimões, Negro, Madeira, Juruá and other rivers. In relation to hydroelectric potential, the trend of creating smaller capacity reservoirs will leave the region more vulnerable in terms of hydroelectric generation in years when water deficit is experienced.

6.6 Impacts on Water Resources

Among the consequences of climate change, one can expect a drop in water quality in general, as well as risks to the quality of drinking water, despite conventional treatment, due to interconnected factors like: temperature increase; increase in sediments, nutrients and pollutant loads from strong rainfall; increase in pollutant concentration during droughts; and interruption of treatment facilities during floods (IPCC, 2014a). According to Soito and Freitas (2011), climate change would probably not have a strong influence on water demand in cities and industry in general, but on consumption for irrigation, as a response to temperature increase and agricultural losses by evaporation (Freitas, 2004).

For the IPCC, there is sound evidence that climate change throughout the twenty-first Century will significantly reduce surface and underground hydro resources in most dry subtropical regions, increasing competition for water among sectors. Similarly, according to IPCC AR5 (IPCC, 2014a), currently dry regions would present a higher drought frequency until the end of the twenty-first Century in RCP 8.5. In contrast, there is strong evidence that water resources will increase in elevated latitudes. Globally, the prediction is that the negative impacts of climate change on water resources should outweigh possible benefits (IPCC, 2014a).

Hydrologists have developed an interest in simulating flow rates in several scales for several reasons, like: availability for irrigation, flood control, transport of sediments, hydroelectric generation, etc. (Xu, 1999). Therefore, climate change impacts on water availability may not be disassociated from the purpose the resource is allocated towards. In case of human needs, socio-economic development plays an essential role in assessing future water availability (Arnell, 2004; Dvorak, Hladny, & Kasparek, 1997). Climate change impacts on water resources are not just subject to alterations in the hydrological cycle. Climate change may also affect demand for water by increasing irrigation, alterations in land use and population displacement, among others (Frederick & Major, 1997; Lucena, 2010).

For the generation of electricity through hydraulic power, flow variation in rivers and levels of lakes related to global climate change depends on alterations in volume, intensity and rainfall time (Chiew, 2006). It also depends on evapotranspiration, which in turn is a function of temperature, insolation, atmospheric humidity and wind speed. Different water basins respond differently to changes in climate variation, depending on their physical and hydrogeological characteristics, as well as quantity of water stored on the surface and underground (Kundzewicz et al., 2008). Climate change will affect the function and operation of infrastructure elements like flood control, drainage and irrigation systems, in addition to water resources management. Finally, it can be said that climate change makes it more difficult for past hydrological experience to be used for predicting future conditions (Lucena, 2010).

Although there are several studies in the literature on impacts of climate change on water resources, papers that focus on Brazil are limited in number. Salati, Campanhol, and Nova (2007) calculated the future water balance of four Brazilian basins – Amazon, Paraguay, Northeast and Prata Basin – for the IPCC SRES A2 and B2 scenarios, based on the results of two GCMs (HadCM3 and GFDL) and the result average of five GCMs. The study is inconclusive in relation to aggregate impacts on the studied basin, as the results reached are contradictory, sometimes because of the GCMs and sometimes because of emission scenarios. This fact is an example of the great level of uncertainty of studies on hydrological impacts of climate change, including Brazil.

In Salati et al. (2007), the water balance for Brazil was calculated, using the HadRM3P model[2] for the periods 2011–2041, 2041–2070 and 2071–2100 for scenarios A2 and B2, compared with the reference period of 1961–1990. The water balance was calculated using georeference, in a geographic information system with spatial resolution of 0.5° by 0.5° of latitude/longitude (50 km by 50 km), utilising long term monthly climate averages. This study presented alarming results, with an estimated sharp drop in flows by 2100 in the Eastern Northeast and East Atlantic basins, reaching close to zero. For the South Atlantic and Uruguay basins, results point to a small tendency for flows to increase until 2100.

In addition, in the referred study the water surplus was analysed for eight basins, using the average reached for 15 alternative climate models with approximate scale of 2° by 2° of latitude/longitude. The results provide evidence of the discrepancies in prediction of several global climate models and their implications in terms of water surplus. It is particularly important to point out that less dramatic decreasing projections for the Northeast region - for example, the average of the other models predicts a surplus of 56% by 2100 for the Parnaíba, contrary to the 14% forecast by the Hadley Centre model (B2 scenario). In the case of the Paraguay and Parana basins, projections have gone against one another: surplus of 40% instead of 147% (Paraguay), and 47% instead of 110% (Paraná, both in scenario A2 in 2100).

Similarly, Marengo et al. (2010) used the HadRM3P of the Hadley Centre regional climate modelling system PRECIS. Initially, the regional model was

[2] HadRM3P is a model of the regional climate modelling system PRECIS (Providing Regional Climates for Impact Studies) of the Hadley Centre in the United Kingdom, which has a 50 km horizontal resolution. HadAM3P is a global climate model used for downscaling climate scenarios.

integrated to reach the climatology model for the present climate (1961–1990) and then, for future climate projections (2071–2100) for scenarios A2 and B2. In this study, the conclusion was that the areas considered the most vulnerable in Brazil are the Amazon and the Northeast. Average warming may reach 5 °C in 2100 in scenario A2 and 3 °C in B2, although gradual temperature increase in the Amazon could reach 7–8 °C or 4–6 °C in 2100, respectively. In the results presented, rainfall tends to drop during the twenty-first Century, with more intense reductions in the Northeast (up to 2–2.5 mm/day) and the Amazon (up to 1–1.5 mm/day). Situation that coincides with what has been presented by AR5 (IPCC, 2014a), where changes in rainfall patterns in the Amazon stand out, as well as degradation of rainforest, land and deforestation trends. For the whole of Brazil, projections indicate an increase in temperature and extreme heat, as well as reductions in the frequency of frost, due to a rise in the minimum temperature, particularly in the Southeast, South and Central-West states.

According to IPCC AR5 (IPCC, 2014a), Group II that addresses 'Impacts, Adaptation and Vulnerability', the following are the main adverse effects that may affect Brazil in the future due to global climate change: 1) The flow of water resources from the underground layer to the surface may diminish drastically in Brazil's Northeast region. 2) Very high probability of arid and semiarid areas in Brazil's Northeast region being particularly vulnerable to impacts of global climate change on water resources, reducing water supply. This scenario is even more relevant if the increase expected in the demand of water is considered, due to population growth in Brazil's Northeast region, temperature increase and reduction in quantity of water in the soil. 3) Increase of extreme rainfall events, aggravating impacts caused by erosion. Brazil's Northeast is vulnerable, as erosion in this region has already caused the sedimentation of reservoirs and consequently, has reduced water storage and supply capacity (MCTI, 2010).

In Nóbrega et al. (2011), uncertainties in climate projections associated to green house gas emission scenarios (A1b, A2, B1, B2) were considered, as well as the global average temperature from 1 to 6 °C for HadCM3. In addition, a comparison was done using 6 GCMs (CCCMA CGCM31, CSIRO Mk30, IPSL CM4, MPI ECHAM5, NCAR CCSM30, UKMO HadGEM1), considering HadCM3 as the baseline for an average temperature increase of 2 °C. The study was applied to the Grande River basin, located in the country's Southeast. In general, the results showed a tendency for the flow in the Grande River to increase with temperature increase. However, there is also evidence that the choice of a GCM is the most quantifiable uncertainty in studies on predicted impacts of climate change in the flow rate of rivers.

According to Soito and Freitas (2011), the most relevant characteristic of climate change in relation to vulnerability and adaptation of water resources is related solely on average trends, but also on noticeable variability alterations in hydrological systems or extreme events. The study points out that in the projections developed so far, results for South America have not reached an agreement in relation to flow predictions. First of all, because of rainfall prediction differences and secondly, as a consequence of differing expected evaporation values. In the same way, in countries exposed to water stress, a negative effect on the flow of rivers, and the refilling of underground water reservoirs and aquifers is expected from climate change.

As mentioned in other papers (Lucena, Szklo, Schaeffer, Souza, et al., 2009; Nóbrega et al., 2011; Salati 2010), Soito and Freitas (2011) point out that regional climate change projections in water resource terms, in other words, predictions related to rainfall anomalies in water basins in the Brazilian territory, vary according to the model used. According to the authors, a 1% increase in CO_2 concentration resulted in positive average flow variations in the Paraná River basin using the model HadCM3 (+50 to +150 mm/year in the margin of the basin) and negative using HadCM2 (−50 to −150 mm/year in the margin of the basin). Furthermore, the study indicates that the models are more ambiguous for research done in the southern hemisphere, as its hydro-meteorological observation network is smaller and was established more recently than in the northern hemisphere.

Soito and Freitas (2011) also present results of flow projections for 2050, using an average of 12 models used by the IPCC in its 2007 report for scenario A1B. According to these projections, water basins in the Amazon and Tocantins-Araguaia showed to drop, which would be worrying, particularly for new hydroelectric generation investments focused on making the most of these basins' high potential. However, in terms of the Paraná River basin, according to the authors, the increasing trend could be kept, which above anything else, would favour the already installed hydroelectric capacity in the region.

Intensifying peak rainfall and temperature increase could have serious effects on erosive processes. Lelis et al. (2011) assess impacts on sediment production and surface outflow, brought about by possible climate change in the São Bartolomeu River, based on the SWAT model.[3] The results show that the basin is extremely sensitive to climate alterations. In the scenarios with more frequent peak rainfall, there is a great rise in sediment production and surface outflow, with rates higher that the ones found today (up to three times more), which leads to higher erosion rates, having a impact on the use of water resources and sustainable development in the area.

Generally speaking, the scientific literature presented shows different methodologies to estimate water resources in the CO_2 emission scenarios projected by the IPCC. When comparing the results of these studies, great uncertainties in quantifying the impacts on the country's hydrological regime are seen. The main source of these uncertainties is the choice of climate model. However, a tendency to converge may be seen for the following projections: increasing frequency and intensity of extreme events, higher water stress in the Northeast, big fall in rainfall in the Amazon region and small flow increases in the South's basins.

6.7 Impacts on Hydro Power Generation

Studies that look into the impacts of climate change on hydroelectric power production or potential, tend to stick to the following schematic: start with a set of global climate change predictions and downscale to the regional level. These climate

[3] SWAT is a public domain tool developed for studying and predicting outflow, sediment production, pollutant load and water quality in water basins.

projections, which consist in rainfall and temperature data are then converted into hydrological impacts (i.e. on flow), which later feed the hydroelectric power generation model (Schaeffer, Szklo, Frossard Pereira De Lucena, Soria, & Chavez-Rodriguez, 2013). Some of the first studies to consider the possible impacts that climate change could have on hydroelectric power generation approached the issue too simply. Indeed, they pointed out the fact that historical hydrology may not be enough to assess the hydroelectric projects to be implemented (Gundry & Whittington, 1998).

Different methodological approaches were used in the studies of climate change impacts on hydroelectricity, as the characteristics of hydroelectric systems have strong influence on deciding which methodology to use in hydrological, as well as energy modelling. Hydroelectric systems may differ in terms of size, plant number and type, geographic extension, hydraulic (cascade plants) and energy (transmission) interconnections, share of hydroelectricity in total generation, among others (Schaeffer et al., 2008). The magnitude and geographic dispersion of the hydroelectric system studied, ultimately played an essential role in the choice of approach for the hydrological modelling. It is important to point out that different from the hydrological impact studies, where different hydrological cycle components are the object of the final study, researching the effects on hydroelectric systems focuses on energy. Therefore, the energy model itself, its characteristics and data requirements are crucial factors when choosing the hydrological model (Lucena, 2010).

The complexity of the hydroelectric simulation may vary from the modelling of one plant to an intricate plant system interconnected hydraulically and/or by energy. Without taking the characteristics of the plants into consideration, basically, two factors may influence the complexity of a hydroelectric system:

- Its geographic dispersion and integration level. Integration may be in terms of hydraulic - with more than one plant along the same river or its affluents - and energy connections, through the transmission of electricity. Operating cascade systems will maximise the amount of power produced not only in an isolated plant, but also through all the flow rate (D'Araujo, 2009), which should be taken into consideration in energy models.
- If the h assessments should use more conservative measures, like firm energy. For example, this is the case in Lucena, Szklo, Schaeffer, Souza, et al. (2009) and Schaeffer et al. (2013). Hydroelectric system is complementary or complemented by other sources of electricity generation. This distinction will in turn, determine the type of result that ought to be sought after based on the energy modelling. In systems where hydroelectricity represents a small share, complementing other sources, supplying for the energy demand reliably does not depend in great extent on the hydrological scenario. Therefore, variation in the average amount of energy produced per year or month by the hydroelectric system is a sufficient measure of possible climate change impacts, as is the case of the studies by Hamlet et al. (2010) and Vicuna et al. (2007). On the other hand, in a system where hydroelectricity predominates, being complemented by other sources, variation in the average amount of energy generated is not a sufficiently good measure to assess climate change impacts. In such a system, reliable elec-

tricity supply depends mainly on the hydrological scenario and thus, climate impact assessments should use more conservative measures, like firm energy.[4]

The system's reliability may be measured by the probabilistic concept of secured energy. While firm energy indicates the maximum amount of energy the system may supply at any moment, considering the worst hydrological scenario, secured energy is the most the system may offer in percentage terms at a given time (Kelman, 1987). However, in order to reach the secured energy, a set of future simulations have to be run for the operation (D'Araujo, 2009), through which a probability distribution may be determined, allowing for the calculation of a certain percentile of extreme flows (D'Araujo, 2009). Therefore, applying this concept to the analysis of climate change impacts implies on a series of methodological challenges related to the downscaling of GCM results, as well as hydrological and energy assessments.

One of the possible impacts of climate change on hydroelectric generation comes from alterations to flow variation or changes to the seasonality regime. Therefore, the vulnerability of a hydroelectric plant or system, depends on the water storage capacity of reservoirs to a great extent (Schaeffer et al., 2012). For example, Vicuna et al. (2007) conducted a sensitivity analysis of water storage capacity in reservoirs in the plants studied and found out that the impacts of climate change on electricity generation dropped as the reservoirs increased. Lucena, Szklo, Schaeffer, Souza, et al. (2009) found evidence that the impacts on electricity generation are not proportionate to the impacts on flow, because of the water storage capacity of reservoirs in Brazilian plants.

However, reservoirs may be put to other uses, which could affect the generation of electricity. The non-energy uses reservoirs include: flood control, other water uses like human and animal consumption, irrigation and leisure, among others. Inclusion of these variables in the modelling may increase its complexity. In fact, climate change may affect other uses of the water like irrigation, which adds another indirect impact on the generation of electricity. Notwithstanding, this analysis may be incorporated to the energy model, as done by Hamlet et al. (2010).

Different methodologies allow for the considerations mentioned above to be included or not. Three types of models were used in the studies of climate impacts on hydroelectric systems: simulation models (Harrison & Whittington, 2002), optimisation models (Costa, 2007; Vicuna et al., 2007) and econometric models (IIMI, 2007). Again, it is important to point out that the complexity of the hydroelectric system will define the approach to be used in assessing climate change impacts.

In the case of extremely complex systems like Brazil's, just the one model may not be enough to describe the hydroelectric operation. For instance, planning the operation of the Brazilian hydrothermal system is done based on three models,

[4] Firm energy may be defined for a hydroelectric system as the biggest amount of energy obtained, considering the hydrological scenario, generally based on the historical experience. Alternatively, it may be defined as the biggest amount of energy produced in a critical period, when the system's storage capacity goes from the maximum to the minimum, without intermediate refilling (CEPEL, 2007). In other words, it is the period when the energy accumulated in reservoirs is depleted without being fully replaced.

which approach the issues with different level of details and differing planning horizons (Costa, 2007; Pereira & Pinto, 1984). Despite climate change in principle being assessed in the long term, the possibility of developing methodologies that approach the issue from different perspectives should be considered.

Actually, Brazil's hydroelectric system is really 'hydrothermal', in the extent that such plants are used to complement generation through water sources, increasing the system's firm capacity (D'araujo, 2009). Therefore, Brazil has the SIN (National Interconnected System), which includes different generation sources in a transmission system the size of a continent, where hydroelectricity plays a major role. Today, around 60% of the electrical installed capacity in the country is hydro based (ANEEL, 2016a).

Therefore, as mentioned earlier, the fact that renewable sources correspond to a high share of Brazil's energy matrix, analysing climate change impacts in the sector becomes extremely important. Schaeffer et al. (2008) used rainfall and temperature projections in Brazil for the 2071–2100 period, with a 50 km by 50 km resolution, obtained from the CPTEC/INPE team, using the PRECIS model for scenarios A2 and B2, to estimate impacts of climate change on the Brazilian hydroelectric system, which are the result of alterations to the average flow of power producing rivers.

Flow variations were used in the Suishi-O[5] model to assess electricity production. The basins analysed in the study by Schaeffer et al. (2008) showed a drop in average flow, particularly in the São Francisco (23.4% in scenario A2, 26.4% in scenario B2), Tocantins-Araguaia (14.7% in A2, 15.8% in B2) and Parnaíba River basins (10% in both scenarios). In the Tocantins-Araguaia basin, seasonal impacts of global climate change would emerge more in the rainy, instead of the dry season. Generally speaking, the trend projected for the annual average flow in the system is negative (−10%). The results show predictions of falling electricity production throughout the analysed period, except for the Paraná River and the Grande River basin, which could see a small rise in generation in scenario A2 (~1%). The most significant falls were found in the São Francisco basin (4.3% in A2, 7.7% in B2).

Lucena, Szklo, Schaeffer, Souza, et al. (2009) simulate a series of natural inflow for 148 reservoirs of SIN hydroelectric power plants, using the statistical framework of time series. Based on the simulated series, an operation simulation model (SUISHI-O) was used to quantify variation in average and firm energy. The results of the study point to strong negative regional impacts, particularly in basins located in the Northeast and Central-West (Parnaíba, São Francisco and Tocantins-Araguaia). Energy related results also show a drop in generation due to lower flow rates in basin, particularly the São Francisco, which would experience an over 7% decrease in production in scenario B2, despite the impacts simulated for the aggregate system not being very relevant. The system showed a 1.58% and 3.15% drop in firm energy in scenarios A2 and B2 respectively. According to the study, impacts in the system are not relevant due to availability of reservoirs in most hydroelectric power plants.

---

[5] *Modelo de Simulação a Usinas Individualizadas de Subsistemas Hidrotérmicos Interligados* (Model for Simulation of individuas Power Plants in Hydrothermal Interconnected Systems), developed by CEPEL (2007).

Schaeffer et al. (2010) analysed the impact of climate change on firm energy and average power based on projections of water resources by Salati (2010). This study also uses the Suishi-O model to assess impacts in scenarios A2 and B2. Like Lucena, Szklo, Schaeffer, Souza, et al. (2009), the main impact identified was a fall in the system's reliability and strong regional effects on hydroelectric generation in the North and Northeast. However, in this study, there is a sharp drop in firm energy, around 30% in both scenarios (A2 and B2). When results are divided by basin, they show that the most affected are in the Northeast (mainly) and in the North, in terms of average, as well as firm energy. In fact, aggregate average energy remains regular because of a positive variation in the South and Southeast basins, particularly in the Paraná River that plays a major role in electricity generation. Moreover, transmission capacity, which acts as a shock absorber for the impacts, allows the increase in generation in some areas to compensate for losses elsewhere. In the Parnaíba and East Atlantic basins, water surplus drops up to 80% in some points of the projections, with a strong fall in energy production.

It is worth mentioning that SIN's storage capacity, through the use of reservoirs, has reduced the vulnerability of the hydroelectric system in relation to climate conditions. The more water storage capacity, the more suited the system is to deal with climate variability. However, due to growing environmental restrictions in relation to the building of plants with large reservoirs, it is expected that making the most of the remaining hydroelectric potential will be based on run-of-the-river plants, with small reservoirs. Therefore, the system's ability to compensate for weather variations may diminish, making it more vulnerable to climate change.

In contrast, results presented by MCTI (2010), based on the GCM HadCM3, downscaling using Eta-CPTEC for scenario A1B, indicate that in the short and medium term (2011–2040), the impact on electricity generation in Brazil will not be negative, as hydroelectricity tends to benefit from the climate scenarios produced. This study used the MSUI for the energy modelling and the results indicate an average power increase in the system of 12 to 16%, and 14 to 20% rise in firm energy in the short term.

In Scianni (2014), secured energy[6] is used as a measurement to assess the impacts of climate change on hydroelectricity. The studies are based in scenario A1B in the horizon from 2000 to 2100. In the paper, four members of the global HadCM3 model were built based on variations on scenario A1B referred to as 'high, medium, low and temperature control'. These variations on this scenario point to different temperature horizons at the end of 2100, with global warming varying from 2 °C to 6 °C. The downscaling process is done through the Eta model, reaching a horizontal scale of 20 km. Secured energy is calculated through the NEWAVE programme.[7] Two generation installed capacity scenarios were used in the calculation: one with the set of plants in place, referred to as existing infrastructure and another with plants projected by PNE (National Energy Plan) 2030, referred to as future infra-

[6]The secured energy (EASS) represents the amount that a generation park can produce with 5 per cent of energy risk deficit.

[7]Model developed by CEPEL and adopted by the Brazilian electricity sector to represent the medium and long term hydrothermal planning problem.

structure. The average secured energy loss for the Eta model, considering the existing park reaches 15%, while for the future park hits 25% from 2040 onwards.

Previous percentages are related to the amount of secured energy reached in 1990, based on climate data of the Eta model for the 1961–1990 period. For the basin simulation, the distributed hydrological model MGB-IPH (Hydrological Model for Big Basins) was selected, developed to represent the transformation process of rain into flow in large scale basins (Collischonn, 2001; Collischonn et al., 2007; Paiva, Collischonn, & Buarque, 2013). The sharpest fall in energy in the future generation infrastructure would be seen in plants in the Amazon, as the Eta model is predicting a reduction in rain. Another factor that contributes to this fall is a reduction in the wet periods and an increase in dry periods, even if big changes to the average annual rainfall are not seen. As the new plants have low regulation power, this has a negative effect on the resulting energy.

A study that assesses the impacts of climate change on the electricity production of a run-of-the-river power plant (Teles Pires in the Tapajós basin) is presented by Mohor et al. (2015). In the study, the hydrological model MHD-INPE is used (regular grid model to resolves water balance equations), considering four global climate models (MIROC-5, CSIRO-Mk3.6.0, IPSL-CM5A-LR and HadGEM2-ES) with projections based on scenario RCP 4.5. Additionally, dynamically downscaled data are used for atmospheric model HadC-M3, in this case utilising four members of scenario A1B. The impact assessment consisted rainfall and discharge statistical variations, where energy production was evaluated developing annual power curves. The paper considered a minimum flow, under which the plant cannot operate, as well as the maximum installed capacity. Considering the minimum operational flow of the permanent plant during the period studied (2041–2070), the most critical prediction indicates that the power station would not operate 59% of the time, as minimum flow would not be reached. Taking into account that the result does not include water consumption scenarios, it could be aggravated. For scenario A1B, the result of this study shows that energy production could be 82% lower than the historical production, which is based on historical data. However, for the IPSL model in scenario RCP 4.5, the reduction in annual energy production is only 3%, with 16 days/year of non-operation time. This result shows how vulnerable run-of-the-river hydroelectricity generation is in dry seasons, which presents a challenge to the country's energy security. Indeed, the country has been investing strongly in hydroelectric projects in the Amazon (GCP; CIAT, 2013).

Andrade et al. (2012) use the impacts predicted for hydroelectricity generation in order to map out environmental, economic, financial and operational variables that may be affected by climate change, in an attempt to determine risks and vulnerabilities for the sector.

Finally, Schaeffer et al. (2015) assessed impacts on the Brazilian hydroelectric generation system, based on two different paths for the sector's evolution: one where concern with mitigation is not explicit and another, where there are such efforts. Employing a flow series based on RCPs 4.5 and 8.5 for all the 188 existing hydroelectric facilities provided for in the SIN, the study used the SDDP model, which uses its stochastic approach to assess probabilistic elements in relation to the system, like deficit risk.

According to the hydrological scenarios, the average flow rate drops drastically in Brazilian water basins, particularly in RCP8.5 in the HadCM3 model. In energy terms, the results show that on average, although generation does not drop in proportion to water loss (as other sources manage to secure electricity production in part), the deficit risk in supply may reach over 90%. This suggests that there is no adaptation option other than huge investments in new electricity generation capacity by other sources.

Table 6.2 presents a summary of impacts of climate change in Brazilian hydroelectricity, analysed in this report.

6.8 Vulnerability Assessment of the Brazilian Energy System in a Scenario of Extreme Climate Change

As discussed earlier, several studies assessed the possible impacts of climate change on electricity generation - particularly hydroelectricity - in Brazil. No study looked specifically at climate impacts that corresponded to temperature increase over 4 °C, for even more extreme scenarios than A2 or RCP 8.5 for example. However, studies show that climate change may have a significant impact on the Brazilian electricity system, due to its predominant reliance on hydroelectricity, even in less extreme scenarios.

The vulnerability to extreme scenarios will depend on how the Brazilian energy system develops in the future. Lucena et al. (2015) made a comparison of different scenarios for the expansion of the Brazilian energy system. The scenarios were designed by six modelling groups, taking into consideration four scenarios with mitigation efforts, in addition to a baseline. Results show that demand for electricity may more than double by 2050. In order to supply for this additional demand, when there are no mitigation efforts (baseline), the Brazilian energy system could become more carbon intensive, using coal and natural gas to generate electricity. In mitigation scenarios, renewable sources like wind and solar become more important, in addition to fossil fuels coupled with carbon capture and storage (CCS). Regardless of the scenario, despite corresponding to a lower share, hydroelectricity remains relevant until 2050.

Therefore, it could be said that in qualitative terms and based on a quantified impact for RCP 8.5, which has already showed to be extreme in previous studies (Schaeffer et al., 2015), that an electricity supply deficit would be inevitable in an extreme climate scenario. Like Schaeffer et al. (2008) and Sathaye et al. (2013) point out higher temperatures may also put a stress on the electricity system in relation to demand, due to more air conditioning being used. Therefore, a drop in supply

Table 6.2 Studies on climate change impacts on hydroelectricity in Brazil

Studies	Emission scenario – Range of temperature[a]	Climate Model	Observations	Impacts (final remarks of study)
Schaeffer et al. (2008)	SRES A2 (2 °C to 5.4 °C) and B2 (1.4 °C to 3.8 °C)	HadCM3 and PRECIS for downscaling	The SUISHI-O model was used to assess electricity production.	The results showed a drop in average flow in the São Francisco (23.4% in scenario A2, 26.4 percent in scenario A2-BR, 26.4% in B2-BR), Tocantins-Araguaia (14.7% -A2-BR, 15.8% -B2-BR) and Parnaíba River basins (10% in both scenarios). Generally speaking, the trend projected for the annual average flow in the Brazilian system is negative (−10%). The results show predictions of falling electricity production throughout the analysed period, except for the Paraná River and the Grande River basin, which could see a small rise in generation in scenario A2 (~1%). The most significant falls were found in the São Francisco basin (−4.3%-A2, 7.7%-B2).
Lucena, Szklo, Schaeffer, Souza, et al. (2009)	SRES A2 (2 °C to 5.4 °C) and B2 (1.4 °C to 3.8 °C)	HadCM3 and PRECIS for downscaling	A series of natural inflow was simulated for 148 reservoirs of SIN hydroelectric power plants, using the statistical framework of time series. Based on the simulated series, an operation simulation model (SUISHI-O) was used to quantify variation in average and firm energy.	The results of this study point to strong negative regional impacts, particularly in basins located in the northeast and central-west (Parnaíba, São Francisco and Tocantins-Araguaia). Energy related results also show a drop in generation due to lower flow rates in basins, particularly the São Francisco, which would experience an over 7% decrease in production in scenario B2. The system showed a 1.58% and 3.15% drop in firm energy in scenarios A2 and B2 respectively.

Schaeffer et al. (2010)	SRES A2 (2 °C to 5.4 °C) and B2 (1.4 °C to 3.8 °C)	HadCM3	New flow series were projected to feed the SUISHI-O model.	The main impact identified was a fall in the hydroelectric system's reliability and strong regional effects on generation in the north and northeast. In this study, there is a sharp drop in firm energy, around 30% in both scenarios (A2-BR and B2-BR). When results are divided by basin, they show that the most affected are in Brazil's northeast (mainly) and in the north, in terms of average, as well as firm energy. In fact, aggregate average energy remains regular because of a positive variation in the south and southeast basins, particularly in the Paraná River that plays a major role in electricity generation; and in transmission capacity, which acts as a shock absorber for the impacts. On the other hand, in the Parnaíba and East Atlantic basins, water surplus drops up to 80% in some points of the projections, with a strong fall in energy production.
Lucena (2010)	SRES A2 (2 °C to 5.4 °C) and B2 (1.4 °C to 3.8 °C)	HadCM3 and PRECIS for downscaling	Energy modelling MAED-MESSAGE	New capacity factors in scenarios A2 and B2 show a relevant fall in firm capacity in small and medium sized plants in the country's north and northeast regions. Indeed, a drop of up to 48% will be seen in relation to the historical capacity factor.
MCTI (2010)	SRES A1B (1.7 °C to 4.4 °C)	GCM HadCM3 and downscaling using Eta-CPTEC	This study used MSUI for energy modelling	For scenario A1B, the study indicates that in the short and medium term (2011–2040), the impact on electricity generation in Brazil will not be negative. On the contrary, the results indicate an increase in the system's average energy from 12 to 16%, as well as a rise in firm energy from 14 to 20%.
Scianni (2014)	SRES A1B (1.7 °C to 4.4 °C)	HadCM3 and Eta for downscaling	The distributed hydrological model MGB-IPH (Hydrological Model for Big Basins) was used. The NEWAVE programme was used to calculate secured energy.	Two generation parks were used to calculate secured energy: one with the set of plants in place, referred to as existing park and another with plants provided for in PNE (National Energy Plan) 2030, referred to as future park. The average secured energy loss for the existing park reaches 15%, while for the future park hits 25% from 2040 onwards.

(continued)

Table 6.2 (continued)

Studies	Emission scenario – Range of temperature[a]	Climate Model	Observations	Impacts (final remarks of study)
Mohor, Rodriguez, Tomasella, and Júnior (2015)	SRES A1B (1.7 °C to 4.4 °C)	MIROC-5, CSIRO-Mk3.6.0, IPSL-CM5A-LR and HadGEM2-ES, HadCM3		Considering that the minimum operational flow of the plant remains constant for the period studied (2041–2070), the most critical prediction (the four members in scenario A1B) indicates that the power station would not operate for 59% of the time looked at, as minimum flow would not be reached. For scenario A1B, the result of this study shows that energy production could be 82% lower than the historical production, which is based on historical data. However, for the IPSL model in scenario RCP 4.5, the reduction in annual energy production is only 3%, with 16 days/year of non-operation time.
Andrade et al. (2012)	SRES A2 (2 °C to 5.4 °C) and B2 (1.4 °C to 3.8 °C)	-	This study uses the impacts predicted for hydroelectricity generation in order to map out environmental, economic, financial and operational variables that may be affected by climate change, in an attempt to determine risks and vulnerabilities for the sector.	In the specific case of CHESF, the conclusion was that proactive investments need to be made in order for actions to adapt to climate change impacts to be implemented. For example, using wind, solar or wave power, which are energy sources found in CHESF's area of operation.

Schaeffer et al. (2015)	RCP 4.5 (1.7 to 3.2 °C and RCP 8.5 (3.2 and 5.4 °C)	MIROC-5 and HadCM3	This study used the SDDP model, which uses its stochastic approach to assess probabilistic elements in relation to the system, like deficit risk.	The results show that on average, although generation does not drop in proportion to water loss (as other sources manage to secure part of the electricity production), the deficit risk in supply may reach over 90%. This suggests that there is no adaptation option other than huge investments in new electricity generation capacity by other sources.

[a]In the case of SRES scenarios, the average temperature rises shown refer to the 2090–2099 period in relation to 1980–1999. However, in the case of the RCP scenarios, the average temperature rises shown refer to the 2081–2100 period in relation to the reference period (pre-industrial era)

would be even more aggravated due to a rise in demand given the use of air conditioning, resulting from extreme temperatures.

In relation to other energy sources, as mentioned earlier, they would all be vulnerable to climate change to a higher or lower degree. Biomass power would also suffer the impacts of extreme climate, reducing Brazilian renewable resources even more. This impact may be the result of losing areas suited to growing energy related crops, as well as a higher probability of their failure, increasing the system's exposure to the climate.

In the case of wind power, even if some studies indicate that it would not suffer with negative impacts, its intermittent nature would not allow it to secure the system, weakened by the loss of hydroelectricity and increase in demand. The same occurs with solar power, which is also variable. Despite this variability being reduced in heliothermic energy generation (through CSP technology with thermal storage or hybridisation for example), this source would not be capable of securing the Brazilian electricity system.

Thus, one may infer that in an extreme climate change scenario, with high temperature rises, renewable sources would be hardly capable of supplying for the demand for electricity. Using fossil fuels presents itself as a possibility to avoid energy supply deficit in extreme scenarios, like the ones discussed herein. Despite being subject to climate impacts, as seen in Chap. 2, fossil fuels are less vulnerable than renewable sources (Schaeffer et al., 2012). However, it is important to point out that using fossil fuels would aggravate the climate change problem even more, due to CO_2 emissions associated to their use. Therefore, in extreme scenarios a vicious cycle may be expected, where the impacts generated would lead to the use of sources that would aggravate climate change even more.

6.9 Adaptation Options

Energy systems are subject to impacts from changes in climate in the production of power in its several formats, as well as its consumption (Lucena, Szklo, Schaeffer, Souza, et al., 2009; Schaeffer et al., 2012), as presented in previous Chapters. Renewable sources are particularly relevant in this sense, as their renewability depends on the climate.

The Brazilian energy system is mainly based on renewable sources, particularly for the generation of electricity through hydro power plants. Therefore, impacts caused by global climate change, particularly in extreme temperature scenarios are very relevant to this system. Thus, adaptation plans are also important, which have been proposed by some and are presented below.

According to Lucena et al. (2010) and Margulis, Marcovitch, and Dubeaux (2010), in the case of the Brazilian energy sector, in order to face climate change and adapt the system, an extra installed capacity to generate around 162 TWh and153 TWh per year would be needed from 2035, as provided for by IPCC's A2

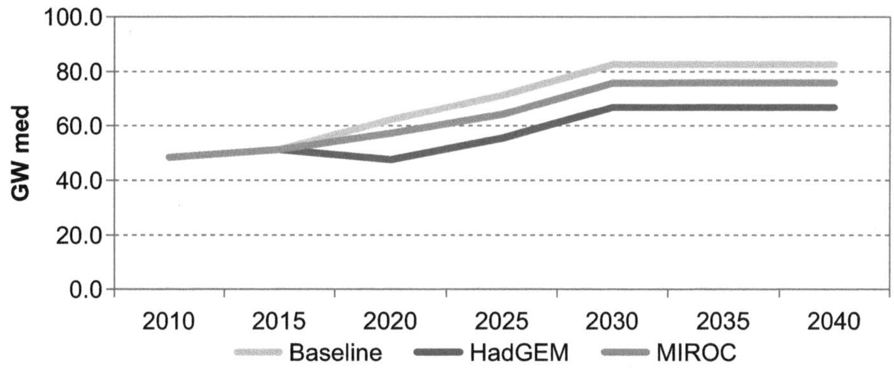

Fig. 6.1 Total hydroelectricity generation in scenarios RCP8.5 with and without impacts (Source: Schaeffer et al., 2015)

and B2 scenarios respectively. In addition, both analyses indicated that this extra capacity would be achieved mainly with natural gas.

In the study *'Adaptação às mudanças climáticas no Brasil: cenários e alternativas'* (Schaeffer et al., 2015), conducted by a cooperation endeavour between COPPE and PSR, an attempt was made to assess the impacts of different climate scenarios in relation to the Brazilian energy system, discussing alternative adaptation strategies. The study used hydrological projections provided by SAE (2015).

In that study two scenarios were developed: RCP 8.5, which presupposes an energy system that does not show to be concerned with mitigating greenhouse gas (GHG) emissions, which would lead to a radiative forcing of 8.5 W/m2 by 2100; RCP 4.5, where measures aimed at reducing emissions are adopted, based on the principle that Brazil would be part of the global mitigation effort compatible with a radiative forcing of 4.5 W/m2 by 2100. Therefore, in the extent that energy systems compatible with the different RCPs differ from one another, impacts predicted on these systems may not be the same.

Applying the exogenous shocks resulting from climate change impacts to the reference scenarios, an assessment on the effects on the system's operation may be drawn. In addition, a new expansion configuration pointing to lower cost adaptation options may be provided.

In the case of the more extreme climate scenario RCP 8.5, an expansion optimisation model – MESSAGE-Brasil – was ran to assess lower cost adaptation options to the predicted impacts. The MESSAGE-Brasil results presented by the study considered lower water availability in SIN's operation, reflected in a smaller capacity factor for hydroelectric power plants. Figure 6.1 shows the generation considered for each scenario.

Based on the falling capacity factor of hydro power plants, electricity generation ends up being 20% lower in the HadGEM scenario and 8% less in MIROC. Because of this loss in the system, the MESSAGE-Brasil model chose the generation portfo-

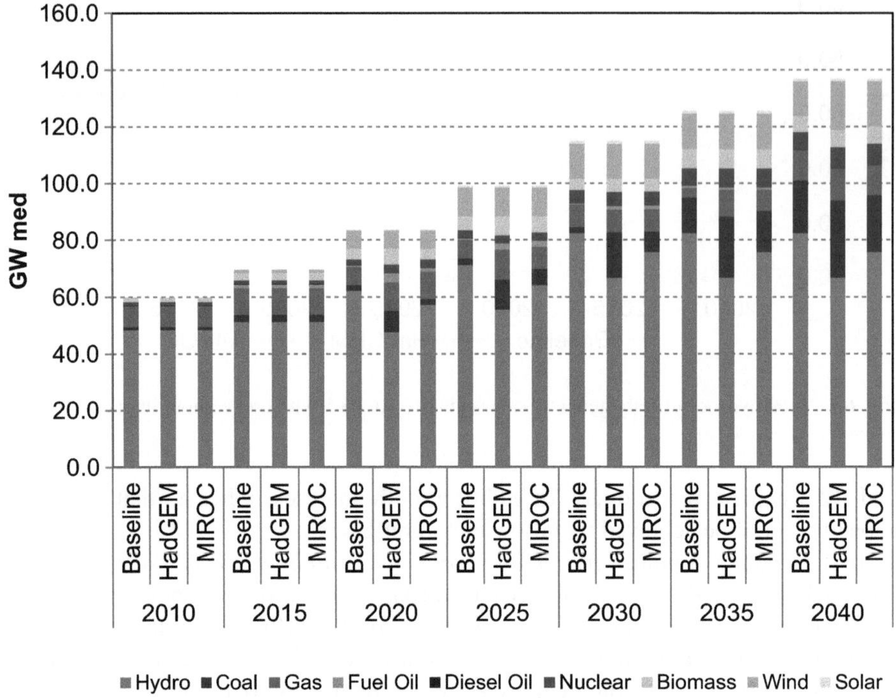

Fig. 6.2 Comparison of RCP 8.5 scenarios – Electricity adaptation alternatives (Source: Schaeffer et al., 2015)

lio presented in Fig. 6.2, increasing generation capacity of other sources to compensate for lower water availability.

In the RCP 8.5, as there are no explicit mitigation policies, the model chooses lower cost options for generating electricity, thus using fossil fuels, particularly coal and natural gas. As it may be seen, because it has smaller impacts on hydroelectricity, the MIROC 8.5 scenario needs less additional generation as adaptation, resulting in the lower coal based power.

Consequently, there is an increase in consumption of these primary energy sources and in CO2 emissions from electricity generation, when taking impacts of climate change into consideration (Fig. 6.3 and Fig. 6.4).

In terms of cost, investments necessary to build this additional generation capacity, in order to adapt to scenarios HadGEM 8.5 and MIROC 8.5 stand at 280 and 79 billion dollars until 2040 respectively.

In terms of operation, the results show that according to the hydrological scenarios of Schaeffer et al. (2015), impacts on SIN's operation are huge. This will lead the system's structure to crumble and to unacceptable deficit levels, much above the 5% currently considered in SIN's operation. In a scenario of extreme impact (HadGEM 8.5), deficit rates are likely to reach over 90%. In addition to high deficit

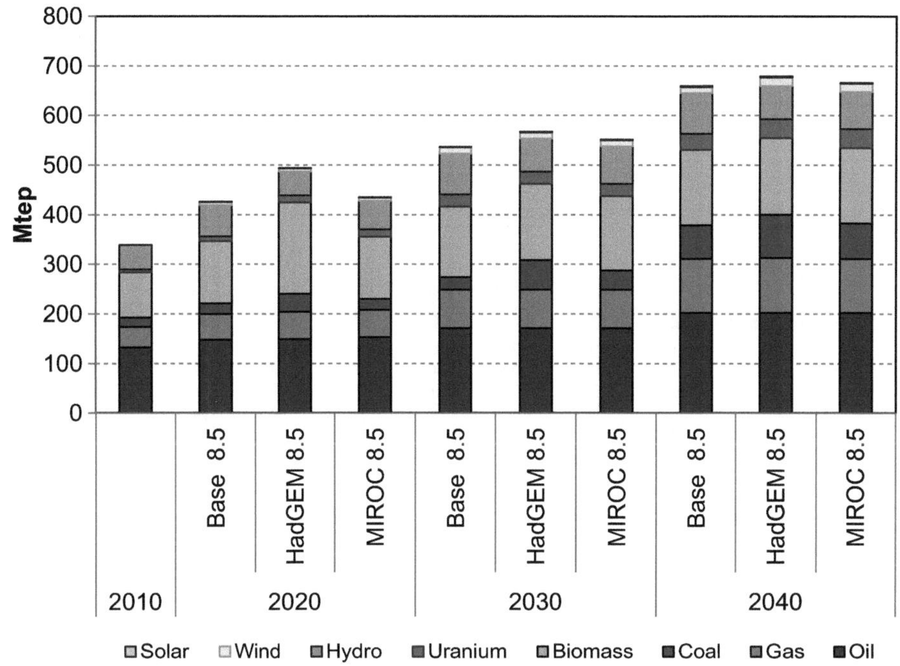

Fig. 6.3 Comparison of RCP8.5 Scenarios – Primary energy (Source: Schaeffer et al., 2015)

related costs, this would lead to very expensive operations (varying between a 3.5 and 16.7 fold increase in relation to the baseline, in smaller/bigger impact scenarios respectively), as existing plants would have to work at full capacity for the period, increasing fuel price. Currently, SIN works in a hydrothermal perspective, where thermal plants are used variably, only compensating for dry periods.[8] As a result of the role they play, thermal plants tend to have lower capital cost. However, they have higher operational cost and are less efficient. For example, when they operate uninterruptedly due to a climate impact scenario, the system's operational costs increases greatly.

Therefore, adapting by expanding would be inevitable, as SIN cannot work at such a high risk level. Furthermore, expansion investment costs may reduce the system's operational costs significantly, as lower cost alternatives are adopted, as seen in the integrated analysis.

When adapting by expanding SIN's generation capacity, Schaeffer et al. (2015) verified that the adaptation-mitigation interaction may very much affect the best alternatives to address a drop in availability in hydroelectric plants. The results of the adaptation-mitigation integrated analysis show that the cost to adapt to the severe impacts predicted by the HadGEM and MIROC models are significantly

[8] This, however, has not been the case over the last few years, when thermal power generation capacity has been operating on a more constant basis in Brazil.

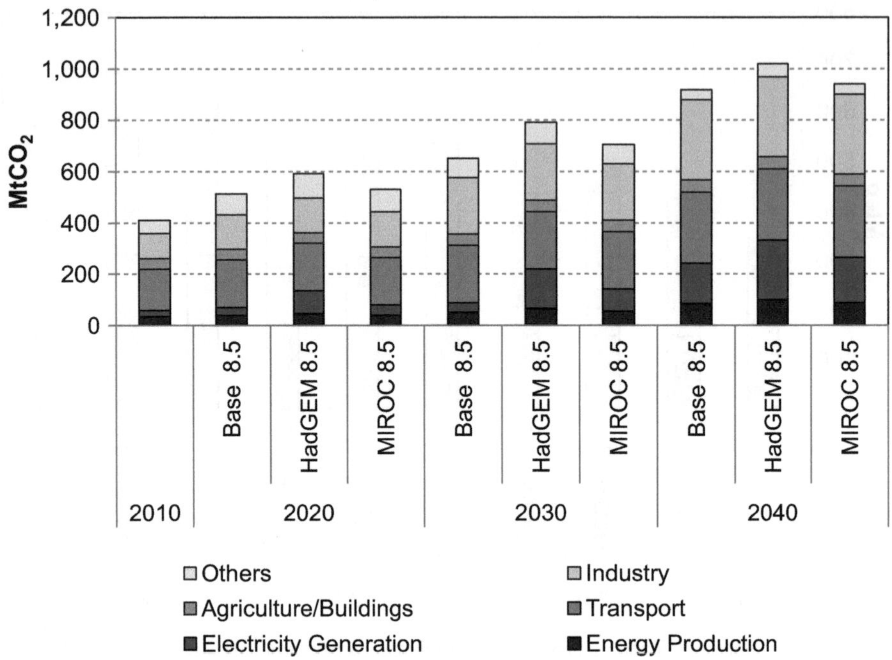

Fig. 6.4 Comparison of RCP8.5 Scenarios – Primary energy (Source: Schaeffer et al., 2015)

lower than in RCP4.5. In this scenario, it was supposed that mitigation measures would be taken in the extent that Brazil would be involved in the international effort aimed at stabilising the radiative forcing in 4.5 W/m2 by 2100.

On one hand, this is due to climate change impacts resulting from a scenario with lower radiative forcing (opposite to RCP 8.5) in the two climate models used (HadGEM and MIROC). On the other hand, this is more due to the presuppositions made in relation to the energy system compatible with a RCP 4.5 world. The supposed demand reduction through energy efficiency, made the RCP 4.5 scenarios arrive at adaptation costs significantly lower for operation and expansion. The risk of deficit predicted for this scenario is also lower for the same reason.

A high carbon cost was also supposed for RCP4.5 scenarios after 2030. It was observed that internalising the cost of carbon has a very important effect on lower cost adaptation alternatives through the system's expansion. By internalising the cost of carbon, coal is no longer a low price alternative to address the fall in hydroelectric generation, making room for sources like (in order of importance) wind, biomass and solar. In these scenarios, natural gas still plays a relevant, despite relatively small role, because it offers flexible generation in order to deal with the variabilities inherent to these sources. However, it is worth checking if a high use of these sources would lead to electrical problem in the network operation like control, voltage, frequency, harmonic distortion and flickering.

Nevertheless, internalising the cost of carbon did not make adaptation costs in RCP 4.5 scenarios higher than seen in RCP8.5. Again, the role played by energy efficiency should be highlighted. In other words, due to high energy efficiency, decarbonisation via incorporating the cost of carbon in the energy system does not increase adaptation prices, when compared to a scenario with high demand and no concern with mitigation.

However, although this result holds significant importance, by highlighting the role of energy efficiency in mitigating GHG emissions, as well as adapting to climate change impacts on hydroelectricity in Brazil, some uncertainties should be pointed out. Brazil's total electricity efficiency potential in scenarios with policies aimed at the electricity sector, or focusing on climate change, is still uncertain. In addition to economic and population growth, the potential of more efficient technologies being used will depend on buildings owning and utilising equipment. Changes in equipment and processes in industrial sectors will also play a role in this, as well as electricity becoming a bigger and more relevant player in the transport sector as a fuel source. Without a detailed assessment on consumption sectors, it is not possible to affirm the effective potential of SIN's energy efficiency. But this is not covered by this paper's scope.

Actual energy efficiency costs are also difficult to estimate. Although it is common for a series of negative cost measures to be identified - which means that the energy efficiency measure would be paid throughout time with at a certain discount rate - there are a number of market barriers that stop efficiency measure from gaining ground. Such barriers go from hidden costs (transaction costs) to information barriers, financing to behavioural barriers. Therefore, there is a need for policies that encourage energy efficiency measures. Generally speaking, such policies are a combination of market measures, and command and control measures, whose costs and potential are difficult to quantify. In any case, there are costs associated to their implementation. Despite the adaptation costs presented here being lower than in scenarios RCP 4.5 due to energy efficiency measures gaining ground, costs related to possible policies that could lead to such efficiency levels have not been calculated. To keep it simple, the study supposed that these costs would be small, when compared to efficiency costs.

Few studies assessed adaptation alternatives to climate change for the Brazilian energy system systematically, like the ones presented in this paper. The results of these studies indicate that impacts on hydro power generation may have significant adaptation costs. In extreme scenarios, this cost may not be that high, but could also lead to a high probability in power supply deficit, as there is a certain inertia in relation to investments in electricity generation. In this case, if climate change takes place rapidly and in an extreme format, the system's capacity to adapt is put in question.

6.10 Final Remarks

Assessments by the Intergovernmental Panel on Climate Change (IPCC) indicate that developing countries are among the most vulnerable to climate change. The more difficult it is for a country to address climate's natural variability and its extreme events, the more effort it will need to adapt to these alterations (IPCC, 2014b; Soito & Freitas, 2011).

This study presented a scientific literature review available in Brazil, on possible climate change impact in the electricity sector, as well as adaptation measures in course to deal with these effects. The extreme climate scenario assessment was done based on these studies' results, which despite not using very extreme scenarios, showed significant impacts to the Brazilian energy system.

Therefore, in extreme climate scenarios, a vicious cycle would emerge. An increase in demand for electricity would be seen due to global warming, while energy production would fall short in its supply, as temperature rises would affect renewable sources directly. These results do not depend on future settings of the Brazilian energy system, as renewable sources remain predominant in the medium-long term.

Although some research institutions in Brazil have started to study these issues better, for operation and expansion planning of the National Interconnected System (SIN), climate variables are still seen as following historical patterns (stationary), rather than incorporating the evidence of already felt impacts from climate change (i.e. it is supposed that its statistical properties remain constant throughout time). Therefore, the Brazilian electricity system is poorly prepared to deal with these issues, as the impacts climate change may have on the energy system are not taken into account in conventional planning at the moment. Hence, analysing the energy system's vulnerabilities and creating appropriate policies is crucial.

References

Acker, T. L., et al. (2012). Integration of wind and hydropower systems: Results of IEA wind task 24. *Wind Engineering, 36*, 1–18.

Andrade, E. M., et al. (2012). The vulnerability of hydroelectric generation in the northeast of Brazil: The environmental and business risks for CHESF. *Renewable and Sustainable Energy Reviews, 16*(8), 5760–5769 1364-0321.

ANEEL. (2016a). BIG - Banco de Informações de Geração. Available at: <http://www.aneel.gov.br/aplicacoes/capacidadebrasil/capacidadebrasil.cfm>. Accessed 20 Jan 2016.

ANEEL. (2016b). Capacidade instalada por fonte. Available at: <http://www.aneel.gov.br/aplicacoes/capacidadebrasil/OperacaoCapacidadeBrasil.cfm>. Accessed 21 Jan 2016.

Arnell, N. W. (2004). Climate change and global water resources: SRES emissions and socio-economic scenarios. *Global Environmental Change, 14*(1), 31–52.

Brown, R. A., Rosenberg, N. J., Hays, C. J., Easterling, W. E., & Mearns, L. O. (2000). Potential production and environmental effects of switchgrass and traditional crops under current and greenhouse-altered climate in the Central United States: A simulation study. *Agriculture, Ecosystems & Environment, 78*(1), 31–47.

Bull, S. R., et al. (2007). Effects of climate change on energy production and distribution in the United States. A report by the U. S. Climate change science program and the subcommittee on global change research. Washington.

Burkett, V. (2011). Global climate change implications for coastal and offshore oil and gas development. *Energy Policy, 39*(12), 7719–7725.

CCPS. In: Wilbanks Thomas, J., Vatsal, B., Bilello Daniel, E., Bull Stanley, R., James, E., Horak William, C., Joe Huang, Y., Levine Mark, D., Sale Michael, J., Schmalzer David, K., & Scott Michael, J., (Eds.), (2007). Effects of climate change on energy production and use in the United States. A report by the U.S. climate change science program and the subcommittee on global change research. Washington, DC., USA: Department of Energy, Office of Biological, Environmental Research. p. 160. http://www.climatescience.gov/Library/sap/sap4–5/finalreport/sap4–5-final-all.pdf

CEPEL. (2001). Atlas do Potencial Eólico Brasileiro . Brasilia: Centro de Pesquisa em Energia Elétrica. Available at: <http://www.cresesb.cepel.br/publicacoes/download/atlas_eolico/Atlas do Potencial Eolico Brasileiro.pdf>.

CEPEL. (2007). Manual de Referência do Modelo SUISHI-O . Rio de Janeiro: Centro de Pesquisa em Energia Elétrica.

Chiew, F. H. S. (2006). Estimation of rainfall elasticity of streamflow in Australia. *Hydrological Sciences Journal, 51*(4), 613–625.

Collischonn, W. (2001). Simulação hidrológica de grandes bacias. Universidade Federal do Rio Grande do Sul.

Collischonn, W., et al. (2007). The MGB-IPH model for large-scale rainfall—Runoff modelling. *Hydrological Sciences Journal, 52*(5), 878–895.

Costa, J. P. (2007). Árvore reduzida de cenários para o planejamento da operação de médio prazo. Universidade Federal do Rio de Janeiro.

Crook, J. A., Jones, L. A., Forster, P. M., & Crook, R. (2011). Climate change impacts on future photovoltaic and concentrated solar power energy output. *Energy & Environmental Science, 4*(9), 3101–3109. https://doi.org/10.1039/c1ee01495a

Cutforth, H. W., & Judiesch, D. (2007). Long-term changes to incoming solar energy on the Canadian Prairie. *Agricultural and Forest Meteorology, 145*(3), 167–175.

D'araujo, R. P. (2009). Setor Elétrico Brasileiro - Uma aventura mercantil. 1°. ed. Brasilia: Coronário Editora Gráfica Ltda.

Dutra, R., & szklo, A. (2008). Assessing long-term incentive programs for implementing wind power in Brazil using GIS rule-based methods. *Renewable Energy, 33*(12), 2507–2515.

Dvorak, V., Hladny, J., & Kasparek, L. (1997). Climate change hydrology and water resources impact and adaptation for selected river basins in the Czech Republic. *Climatic Change, 36*(1–2), 93–106.

Feeley, T. J., Skone, T. J., Stiegel, G. J., McNemar, A., Nemeth, M., Schimmoller, B., et al. (2008). Water: A critical resource in the thermoelectric power industry. *Energy, 33*(1), 1–11.

Fenger, J. (2007). *Impacts of climate change on renewable energy sources their role in the Nordic energy system.* Copenhagen, Denmark: Nordic Council of Ministers. 978-92-893-1465-7.

Feres, J., et al., (2010). Impactos da Mudança do clima no uso da terra . São Paulo: Economia da Mudança do Clima no Brasil: Custos e Oportunidades.

Fidje, A., & Martinson, T. (2007). Solar energy. In J. Fenger (Ed.), *Impacts of climate change on renewable energy sources: Their role in the Nordic energy system.* Copenhagen, Denmark: Nordic Council of Ministers.

Frederick, K. D., & Major, D. C. (1997). Climate change and water resources. *Climatic Change, 37*(1), 7–23.

Freitas, M. A. V. (2004). *Vulnerabilidade, Impactos e Adaptação à Mundaça do Clima.* Brasilia, Brazil: CGEE - Centro de Gestão e Estudos Estratégicos.

Freitas, M. A. V. (2006). *Vulnerabilidade climática e Antrópica dos Recursos Hídrico da Bacia Amazônica.* Brasilia, Brazil: Projeto GEF Amazonas - OTCA/PNUMA/OEA.

Garreaud, R. D., & Falvey, M. (2009). The coastal winds off western subtropical South America in future climate scenarios. *International Journal of Climatology, 29*(4), 543–554.

GCP; CIAT. (2013). Amazonia security agenda. Brasil: Global canopy Programme - Centro Internacional de Agricultura tropical.

Gundry, S. W., & Whittington, H. W. (1998). Global climate change and hydroelectric resources. *Engineering Science & Education Journal, 7*(1), 29–34. http://digital-library.theiet.org/content/journals/10.1049/esej_19980107

Hamlet, A. F., Lee, S., Mickelson, K. E. B., & Elsner, M. M. (2010). Effects of projected climate change on energy supply and demand in the Pacific Northwest and Washington State. In *The Washington climate change impacts assessment: evaluating Washington's future in a changing climate. Chapter 4: Energy. JISAO climate impacts group* (pp. 165–190). Seattle, WA: University of Washington.

Harrison, G. P., & Wallace, A. R. (2005). Climate sensitivity of marine energy. *Renewable Energy, 30*(12), 1801–1817.

Harrison, G. P., & Whittington, H. B. W. (2002). Susceptibility of the Batoka gorge hydroelectric scheme to climate change. *Journal of Hydrology, 264*(1), 230–241.

Harsem, Ø., Eide, A., & Heen, K. (2011). Factors influencing future oil and gas prospects in the Arctic. *Energy Policy, 39*(12), 8037–8045.

IIMI, A. (2007). Estimating global climate change impacts on hydropower projects: Applications in India, Sri Lanka and Vietnam. The World Bank: Sustainable development network. Finance, economics, and urban development department.

INPE. (2007). *Caracterização do clima atual e definição das alterações climáticas para o território brasileiro ao longo do Século XXI, Relatório No. 6: Mudanças Climáticas e Possíveis Alterações nos Biomas da América do Sul CPTEC*. São Paulo, Brazil: INPE; 2007.

IPCC - Intergovernmental Panel on Climate Change. Climate Change. (2014a). Impacts, adaptation and vulnerability. Summary for policy makers. Contributions of the working group II to the fifth assessment report.[Field, C.B., V.R. Barros, D.J. Dokken, K.J. Mach, M.D. Mastrandrea, T.E. Bilir, M. Chatterjee, K.L. Ebi, Y.O. Estrada, R.C. Genova, B. Girma, E.S. Kissel, A.N. Levy, S. MacCracken,: Cambridge University Press, Cambridge, UK, and New York, pp. 1–32. , 2014.

IPCC - Intergovernmental Panel on Climate Change. Climate Change. (2014b). *Mitigation of climate change. Contribution of working group III to the fifth assessment report of the intergovernmental panel on climate change*. Cambridge, UK, and New York: Cambridge University Press.

IPCC- Intergovernmental Panel on Climate Change. (2014). In R. K. Pachauri & L. A. Meyer (Eds.), *Climate Change 2014: Synthesis report. Contribution of working groups I, II and III to the fifth assessment report of the intergovernmental panel on climate change* (p. 151). Geneva, Switzerland: IPCC.

Kehlhofer, R., Hannemann, F., Rukes, B., & Stirnimann, F. (2009). *Combined-cycle gas & steam turbine power plants*. Tulsa, OK: Pennwell Books.

Kelman, J. (1987). Modelos estocásticos no gerenciamento de recursos hídricos. ______. *Modelos para Gerenciamento de Recursos Hídricos I. São Paulo: Nobel/ABRH*, 378.

Kopytko, N., & Perkins, J. (2011). Climate change, nuclear power, and the adaptation–mitigation dilemma. *Energy Policy, 39*(1), 318–333.

Kundzewicz, Z. W., et al. (2008). The implications of projected climate change for freshwater resources and their management. *Hydrological Sciences Journal, 53*(Fevereiro), 3–10.

Lehner, B., Czisch, G., & Vassolo, S. (2005). The impact of global change on the hydropower potential of Europe: A model-based analysis. *Energy Policy, 33*(7), 839–855.

Lelis, T. A., Calijuri, M. L., da Fonseca, A. S., & de Lima, D. C. (2011). Impactos causados pelas mudanças climáticas nos processos erosivos de uma bacia hidrográfica: Simulação de cenários/Impacts of climate change on erosion of a watershed: Simulation of scenarios. *Revista Ambiente & Água, 6*(2), 282.

Lucena, A. F. P. (2010). Proposta metodológica Para avaliação da vulnerabilidade às mudanças climaticas globais no setor hidroelétrico. Universidade Federal do Rio de Janeiro, 208 p.
Lucena, A. F. P., Szklo, A. S., & Schaeffer, R. (2009a). Renewable energy in an unpredictable and changing climate. *Modern Energy Review, 1*(Copenhagen Edition).
Lucena, A. F. P., Szklo, A. S., & Schaeffer, R. (2009b). Renewable energy in an unpredictable and changing climate. *Modern Energy Review, 1*, 22–25.
Lucena, A. F. P., Szklo, A. S., Schaeffer, R., & Dutra, R. M. (2010). The vulnerability of wind power to climate change in Brazil. *Renewable Energy, 35*, 904–912.
Lucena, A. F. P., Szklo, A. S., Schaeffer, R., Souza, R. R., Borba, B. S. M. C., Costa, I. V. L., et al. (2009). The vulnerability of renewable energy to climate change in Brazil. *Energy Policy, 37*, 879–889.
Lucena, A. F. P., et al. (2015). Climate policy scenarios in Brazil: A multi-model comparison for energy. *Energy Economics, 56*.
Marengo, J. A., et al. (2004). Vulnerabilidade, impactos e adaptação à mudança do clima no semi-árido do Brasil. Cachoeira Paulista v. CPTEC/INPE, p. 16.
Marengo, J. A., et al. (2010). *Cenários futuros sobre o clima no Brasil. Economia da Mudança do Clima no Brasil: Custos e Oportunidades*. São Paulo, Brazil: CPTEC/INPE.
Margulis, S., Marcovitch, J., & Dubeaux, C. B. S. (Org.), (2010). Economia das Mudanças do Clima no Brasil: custos e oportunidades. Available at http://www.economiadoclima.org.br
Martins, F. R., & Pereira, E. B. (2011). Enhancing information for solar and wind energy technology deployment in Brazil. *Energy Policy, 39*(7), 4378–4390.
MCTI - Ministério da Ciência, Tecnologia, Inovações e Comunicações. (2010). *Segunda Comunicação Nacional do Brasil à Convenção-Quadro das Nações Unidas sobre mudança do Clima*. Brasilia, Brazil: Ministério de Ciência, Tecnologia e Inovação.
MME – Ministério de Minas e Energia. (2014). Energia Eólica no Brasil e Mundo. Brasília.
Mohor, G. S., Rodriguez, D. A., Tomasella, J., & Júnior, J. L. S. (2015). Exploratory analyses for the assessment of climate change impacts on the energy production in an Amazon run-of-river hydropower plant. *Journal of Hydrology: Regional Studies, 4*, 41–59.
Nobre, C. (2007). *Caracterização do clima atual e definição das alterações climáticas para o território brasileiro ao longo do século XXI*. São Paulo, Brazil: CPTEC, INPE.
Nóbrega, M. T., et al. (2011). Uncertainty in climate change impacts on water resources in the Rio Grande Basin, Brazil. *Hydrology and Earth System Sciences, 15*(2), 585–595.
ONS - Operador Nacional do Sistema Elétrico. (2015a). Dados Relevantes - Operação do SIN. Rio de Janeiro: Operador Nacional do Sistema Elétrico. Available at: <http://www.ons.org.br/biblioteca_virtual/publicacoes_operacao_sin.aspx>.
ONS - Operador Nacional do Sistema Elétrico. (2015b). Geração de Energia. Available at: <http://www.ons.org.br/historico/geracao_energia.aspx>. Accessed on: 21 Nov 2015.
Paiva, R. C., Collischonn, W., & Buarque, D. C. (2013). Validation of a full hydrodynamic model for large-scale hydrologic modelling in the Amazon. *Hydrological Processes, 27*(3), 333–346.
Pereira, E. B., Martins, F. R., Pes, M. P., da Cruz Segundo, E. I., & Lyra, A. D. A. (2013). The impacts of global climate changes on the wind power density in Brazil. *Renewable Energy, 49*, 107–110.
Pereira, M. V. F., & Pinto, L. M. V. G. (1984). Operation planning of large-scale hydroelectric systems. Proc. 8th Power System Computation Cont, 2, 3.
Perez, P. (2009). Potential impacts of climate change on California's energy infrastructure and identification of adaptation measures. California Energy Commission v. CEC-150-20, n. January, p. 1–19.
Pinto, H., & Assad, E. (2008). *Aquecimento global e cenários futuro da agricultura brasileira*. Campinas, Brazil: UNICAMP.
Pryor, S. C., & Barthelmie, R. J. (2010). Climate change impacts on wind energy: A review. *Renewable and Sustainable Energy Reviews, 14*(1), 430–437.
SAE. (2015). Brasil 2040, Resumo Executivo. Available at http://www.mma.gov.br/images/arquivo/80182/BRASIL-2040-Resumo-Executivo.pdf. Acessed 01 Feb 2018.

Sailor, D. J., Hu, T., Li, X., & Rosen, J. N. (2000). A neural network approach to local downscaling of GCM output for assessing wind power implications of climate change. *Renewable Energy, 19*(3), 359–378.
Sailor, D. J., Smith, M., & Hart, M. (2008). Climate change implications for wind power resources in the Northwest United States. *Renewable Energy, 33*(11), 2393–2406.
Salati, E. (2010.) Estimativas da oferta de recursos hídricos no Brasil em cenários futuros de clima (2015–2100). São Paulo: Economia da Mudança do Clima no Brasil: Custos e Oportunidades.
Salati, E., Campanhol, T., & Nova, N. (2007). Tendências de Variações Climáticas para o Brasil no Século XX e Balanços Hídricos para Cenários Climáticos para o Século XXI. São Paulo: Relatório N°4: Mudanças Climáticas e Possíveis Alterações nos Biomas da América do Sul, CPTEC, INPE.
Sathaye, J. A., Dale, L. L., Larsen, P. H., Fitts, G. A., Koy, K., Lewis, S. M., et al. (2013). Estimating impacts of warming temperatures on California's electricity system. *Global Environmental Change, 23*(2), 499–511.
Schaeffer, R., Szklo, A., Frossard Pereira De Lucena, A., Soria, R., & Chavez-Rodriguez, M. (2013). The vulnerable Amazon: The impact of climate change on the untapped potential of hydropower systems. *Power and Energy Magazine, IEEE, 11*(3), 22–31.
Schaeffer, R., Szklo, A., & Lucena, A. (2008). *Climate change and energy security*. Rio de Janeiro, Brazil: UFRJ/COPPE/PPE.
Schaeffer, R., Szklo, A. S., de Lucena, A. F. P., Borba, B. S. M. C., Nogueira, L. P. P., Fleming, F. P., et al. (2012). Energy sector vulnerability to climate change: A review. *Energy, 38*(1), 1–12.
Schaeffer, R., et al. (2008). Mudanças Climaticas e Segurança Energética no Brasil. Rio de Janeiro, Brazil: PPE/COPPE/UFRJ.
Schaeffer, R., et al., (2010). Impactos na demanda e na confiabilidade no setor de energia. São Paulo: Economia da Mudança do Clima no Brasil: Custos e Oportunidades.
Schaeffer, R., et al. (2015). "Adaptação às Mudanças Climáticas no Brasil: Cenários e Alternativas". Relatório Final. Brasil 2040: Cenários e Alternativas de Adaptação à Mudança do Clima, Rio de Janeiro: Secretaria de Assuntos Estratégicos.
Scianni, L. (2014). Avaliação Preliminar do Efeito das Mudanças Climáticas na Geração de Energia Elétrica. Universidade Federal de Itajubá, 91 p.
Siqueira, O. J. W., Steinmetz, S., Salles, L. A. B., & Fernandes, J. M. (2001). Efeitos Potenciais das Mudanças Climáticas na Agricultura Brasileira e Estratégias Adaptativas para Algumas Culturas. In: Mudanças Climáticas Globais e a Agropecuária Brasileira. Jaguariúna: Embrapa Meio Ambiente. p. 397 [Chapter 2], pp. 33–63.
Soito, J., & Freitas, M. (2011). Amazon and the expansion of hydropower in Brazil: Vulnerability, impacts and possibilities for adaptation to global climate change. *Renewable and Sustainable Energy Reviews, 15*(6), 3165–3177. http://linkinghub.elsevier.com/retrieve/pii/S1364032111001432
Streck, N. A., & Alberto, C. M. (2006). Simulação do impacto da mudança climática sobre a água disponível do solo em agroecossistemas de trigo, soja e milho em Santa Maria. *Ciência Rural, 36*(2), 424–433.
Tolmasquim, M., Szklo, A., & Soares, J. (2003). Mercado de Gás natural na Indústria Química e no Setor Hospitalar Brasileiro Edições CENERGIA, Rio de Janeiro, 2003.
United Nations. (2016). Compendium on methods and tools to evaluate impacts of, and vulnerability and adaptation to climate change. Available at: <http://unfccc.int/adaptation/nairobi_work_programme/knowledge_resources_and_publications/items/5449.php>. Accessed 21 Jan 2016.
VicuñA, S., Leonardson, R., Dracup, J. A., Hanemann, M., & Dale, L. (2005). Climate change impacts on high elevation hydropower generation in California's Sierra Nevada: a case study in the Upper American river". In: *California climate change Center. California Energy Commission,* December.

Vicuna, S., et al. (2007). Climate change impacts on high elevation hydropower generation in California's Sierra Nevada: A case study in the Upper American River. *Climatic Change, 87*(1), 123–137.

Whittington, H. W., & Gundry, S. (1996). Global climate change and hydroelectric power. In *Opportunities and Advances in International Electric Power Generation, International Conference on (Conf. Publ. No. 419)* (pp. 40–43). IET.

Wigley, T. M. L., Jones, P. D., Briffa, K. R., & Smith, G. (1990). Obtaining sub-grid-scale information from coarse-resolution general circulation model output. *Journal of Geophysical Research: Atmospheres, 95*(D2), 1943–1953.

Wild, M., Folini, D., Henschel, F., Fischer, N., & Müller, B. (2015). Projections of long-term changes in solar radiation based on CMIP5 climate models and their influence on energy yields of photovoltaic systems. *Solar Energy, 116*, 12–24 https://doi.org/10.1016/j.solener.2015.03.039

Xu, C. Y. (1999). Climate change and hydrologic models: A review of existing gaps and recent research developments. *Water Resources Management, 13*(5), 369–382.

Chapter 7
Increase Risk of Drought in the Semiarid Lands of Northeast Brazil Due to Regional Warming above 4 °C

Jose A. Marengo, Ana Paula Cunha, Wagner R. Soares, Roger R. Torres, Lincoln M. Alves, Sheila S. de Barros Brito, Luz A. Cuartas, Karinne Leal, Germano Ribeiro Neto, Regina C. S. Alvalá, and Antonio R. Magalhaes

7.1 Introduction

Human-induced warming reached a global average of about 1 °C above pre-industrial levels in 2016, increasing at 0.1–0.25 °C per decade (IPCC, 2013; IPCC, 2014). Many regions have already experienced greater warming and significant changes in rainfall, as well as increases in the frequency and intensity of extreme events (e.g. droughts and floods) that result in resource depletion, conflict and forced migration that are affecting economic development worldwide. It is this rising risk that underpins the ambition of the Paris COP21 agreement, to pursue efforts to limit the rise in global temperatures to 2.0 °C above pre-industrial levels.

The World Bank Reports from 2012 and 2013 (World Bank, 2012; World Bank, 2013) outline that if the world warms by 2 °C—warming which may be reached in 20 to 30 years—it will cause widespread droughts and floods, unprecedented heatwaves, and more intense cyclones. In the near-term, climate change, which is already unfolding, could affect vulnerable areas and greatly harm the lives and the hopes of individuals and families who have had little hand in raising the Earth's temperature. Global mean temperature change of 4 °C is close to the difference between the temperatures of the present day and those of the last ice age, and the current change—human induced—is occurring at a much faster rate, over a century, not millennia. If the currently planned adaptation and mitigation actions are not fully implemented, a warming of 4 °C could occur as early as the 2060s. Such a

J. A. Marengo (✉) · A. P. Cunha · W. R. Soares · R. R. Torres · L. M. Alves
S. S. de Barros Brito · L. A. Cuartas · K. Leal · G. Ribeiro Neto · R. C. S. Alvalá
National Center for Monitoring and Early Warning of Natural Disasters (CEMADEN), São Paulo, Brazil
e-mail: jose.marengo@cemaden.gov.br

A. R. Magalhaes
Center for Strategic Studies, CGEE, 70308-200 Brasilia, Brazil

C. A. Nobre et al. (eds.), *Climate Change Risks in Brazil*,
https://doi.org/10.1007/978-3-319-92881-4_7

warming level by 2100 would not be the end point: a further warming to levels over 6 °C would likely occur over the following centuries.

A world in which warming reaches 4 °C above preindustrial levels would be one of unprecedented heat waves, severe droughts, and major floods in many regions around the world (IPCC, 2014), with serious impacts on natural and human systems, ecosystems, and associated services. One of those regions would be the semiarid region of Northeast Brazil (NEB), which is one of the regions most vulnerable to food, water and energy insecurity due to extremes of climate variability and possibly to climate change (IPCC, 2014). The region covers an area of 614,870 km^2, covered by mixed grasslands–croplands (Fig. 7.1) and includes 1069 municipalities (Brito et al., 2017; Marengo et al., 2017; Martins et al., 2015).

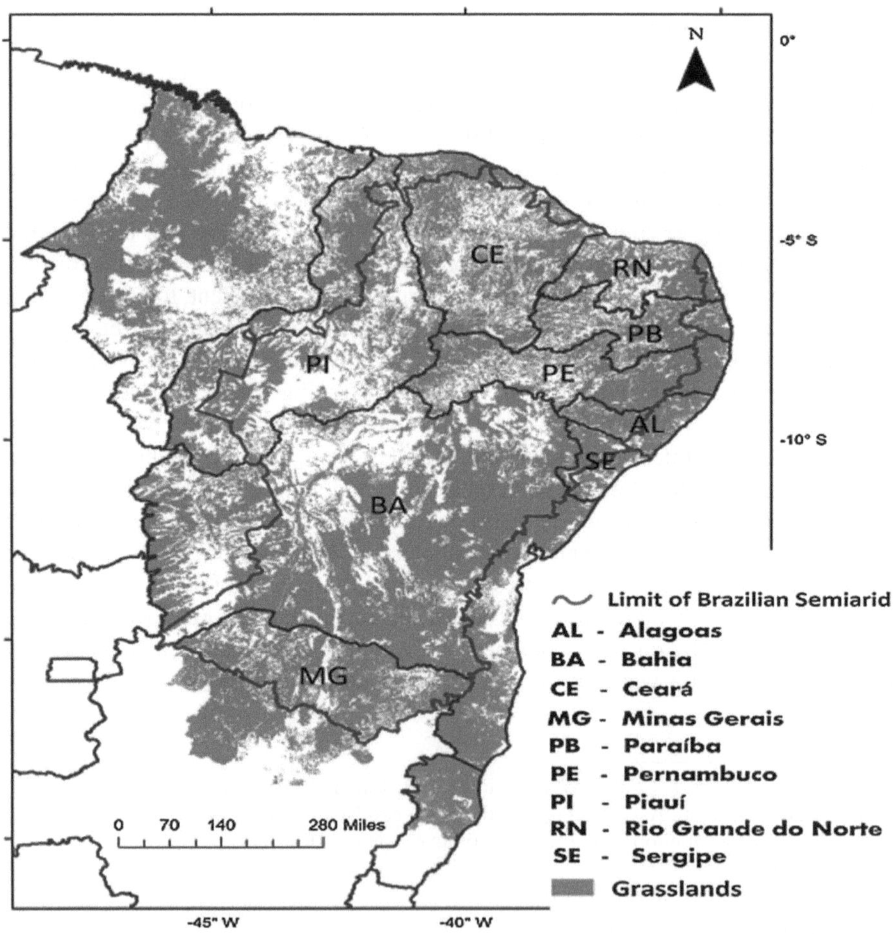

Fig. 7.1 Geographic limits of the semiarid region of Northeast Brazil (*pink line*) and distribution of mixed grasslands-croplands (Adapted from and Vieira et al., 2013)

In this region, droughts affect more people than any other natural hazard owing to their large scale and long-lasting nature. They are recurrent in the region and while El Nino has driven some of these events, others are more dependent on the tropical North Atlantic sea surface temperature fields. Some measures have been taken by the governments to mitigate their impacts on the population, but there is still a perception that residents, mainly in rural areas, are not yet adapted to this hazard. Brazil has made efforts to adapt to climate change and variability, for example, through the use of early warning systems and climate forecast for subsistence agriculture in NEB (Alvalá et al., 2017). However, during the recent drought situation that started in 2012 it has been noted that local traditional politics relied on patron-client relationships with drought-affected households to maintain power, suggesting that there was little incentive for adaptation policies that dramatically decreased their level of vulnerability. Integrating drought monitoring and seasonal forecasting provides an efficient means of assessing impacts of climate variability and change, identifying vulnerabilities and allowing for better adaptation measures, not only for medium- and long-term climate change, but also for extremes of the interannual climate variability, particularly droughts.

If regional warming exceeds the extreme limit of 4 °C in relation to the pre-industrial era, large areas of Brazil, including NEB, may suffer dangerous climate change, affecting population, human activities and natural ecosystems. In this chapter as well as in other chapters of this book we assess the impacts of regional warming above 4 °C, because the sectorial impacts may be higher under a regional intense warming. In fact, Fig. 7.2 shows that regional warming in NEB is larger than the mean global warming, and thus regional impacts may be stronger than global. For instance, for sea level rise, global warming would be more important over a region than regional warming, while for sectors such as health the regional warming would be more relevant and have more impacts than the global one.

This chapter presents an analysis of the increased drought risk in NEB under various climate change scenarios. Risk is usually defined in the context of drought

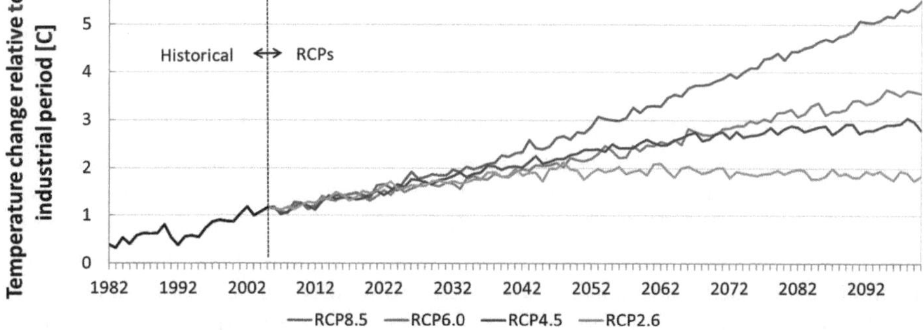

Fig. 7.2 Projected CMIP5 annual temperature changes (C) for RCP2.6 (*blue line*), RCP4.5 (*purple line*), RCP6.0 (*green line*) and RCP8.5 (*red line*)

and other natural hazards as a product of the exposure to the hazard and societal vulnerability. This chapter is largely focused on climatic features and risk as defined only by the exposure to the hazard, with occasional mention of how this change might affect vulnerability.

7.2 History of Droughts in the Region

In the NEB there are frequent occurrences of dry periods during the rainy season, which, depending on the intensity and duration, can cause significant damage to family-farmed crops with a farming system characterized by low productivity indices. However, rain-fed agriculture has large economic expression and high social importance due to the region is densely occupied, and contributes to the establishment of communities in the countryside. During periods of extreme drought, food security for the most vulnerable communities, such as those in the semiarid regions, suffers pressures with reduced subsistence production, reduced income and increased pricing of agricultural products. Food price inflation, associated with loss of income due to harvest shortfalls, loss of productivity and unemployment, results in low food security and, in extreme cases, in famine.

Drought episodes in NEB have been reported since the sixteenth century, coming from various sources (Araújo, 1982; Gutierrez, Engle, De Nys, Molejon, & Martins, 2014; Magalhães & Coauthors, 1988; Marengo, Torres, & Alves, 2016; Marengo et al., 2017; Martins et al., 2015; Wilhite, Sivakumar, & Pulwarty, 2014). A list of events updated to 2017 follows: 1583, 1603, 1624, 1692, 1711, 1720, 1723–1724, 1744–1746, 1754, 1760, 1772, 1766–1767, 1777–1780, 1784, 1790–1794, 1804, 1809, 1810, 1816–1817,1824–1825, 1827, 1830–1833, 1845, 1877–1879, 1888–1889, 1891, 1898, 1900, 1902–1903, 1907, 1915, 1919, 1932–1933, 1936, 1941–1944, 1951–53, 1958, 1966, 1970, 1976, 1979–1981, 1982–1983, 1986–87, 1992–1993, 1997–1998, 2001–2002, 2010 e 2012–2017. Intense droughts during strong El Nino years were reported in 1877–79, 1897, 1899, 1902–03, 1919, 1951, 1958, 1966, 1982–83, 1986–87, 1997–98, 2010, and 2015 (El Nino and La Nina years are listed in (https://www.esrl.noaa.gov/psd/enso/past_events). However, El Nino explains only a fraction of rainfall variability in NEB. In the recent droughts of 1992, 1998, 2002, and 2010 and now, 2012–2016, only those of 1998, 2002 and 2015–16 occurred during an El Nino event. On the other hand the drought in 2011–2012 occurred during a La Niña event (Rodrigues & MCphaden, 2014).

The recent long lasting drought still carries its effects since the end of 2011. The worst drought in terms of severity, frequency and duration of the past 36 years occurred during 2011–2016 (Brito et al., 2017). When compared with past decades, this drought affected a larger area with significant impacts for population, as well as economic activities. Only considering the hydrologic year (October to September) of 2015–2016, drought caused damage to agricultural production, with grain production in the Northeast region presenting a reduction of about 40% compared with the crop of 2014–2015. For sugarcane production, the reduction was in the order of

19% relative to the previous crop (CONAB, 2017). Regarding the impact on water supply, the water reserves of the equivalent reservoirs (storage capacity above 10 hm^3) in the Northeast have presented successive reductions since 2012, which resulted in a minimum stored volume of approximately 13.8% in March 2017 (ANA, 2017).

IPCC (2014), Marengo et al. (2016, 2017, and references quoted in) have shown that projections in future climate change scenarios suggest a decrease in water availability for irrigated agriculture and human use owing to reductions in precipitation and increases in evapotranspiration. Therefore, this chapter is dedicated to the assessment of the probability of occurrence of a regional warming above 4 °C as well as to the estimate potential changes in water availability that could lead to more droughts on this region. Previous chapters provide regional and sectorial impacts of various level of regional warming above 4 °C.

7.3 Climate Change Projections for NEB

Decreases in rainfall and increases in temperature can affect agricultural productivity in the short term, threatening the food security of the poorest population. In the long term, it could affect the water supply for the local and regional population. By the end of the twenty-first century meteorological droughts (less rainfall), agricultural droughts (drier soil) and hydrological droughts (less water availability) are projected to become longer, or more frequent and intense in some regions and some seasons, because of reduced rainfall or increased evaporation or both.

Figure 7.2 shows the projected future changes in annual-mean surface air temperature in relation to the pre-industrial period (1861–1890) in NEB for various radiative forcings. The IPCC AR5 analyzed the future radiative forcings through the new RCP scenarios (Representative Concentration Pathways, Moss et al., 2010), which adopt a more complete system and take into account emission impacts, i.e., how much the radiation balance will change in the Earth system. CMIP5 models were run under these RCPs. RCPs are identified by their total radiative forcing (expressed in W/m^2) to be reached during (or by the end of) the twenty-first century: RCP 2.6 (mitigation scenario, leading to a very low forcing level), RCP 4.5 and RCP 6.0 (two stabilisation scenarios), and RCP 8.5 (scenario with very high GHG emissions) (IPCC, 2014). Each RCP provides spatially distributed datasets on land-use changes and sectorial emissions of air pollutants, and specifies annual GHG concentrations and anthropogenic emissions up to 2100 (Burkett et al., 2014). More information on the RCPs can be found in Meinshausen, Smith, Calvin, et al. (2011), Van Vuuren et al. (2011) and IPCC (2013). For this analysis we use RCPs 2.6, 4.5, 6.0 and 8.5 with 32, 42, 25 and 29 simulations from different climate models, respectively.

From the RCP2.6 (Fig. 7.2) the Temperature Change Stabilizes in 2 °C above the Average around 2040. On the Other Hand, the RCP8.5 Scenario Drives End-of-Century Temperature Increases in Excess of 5 °C. The 3 °C Warming Level Is

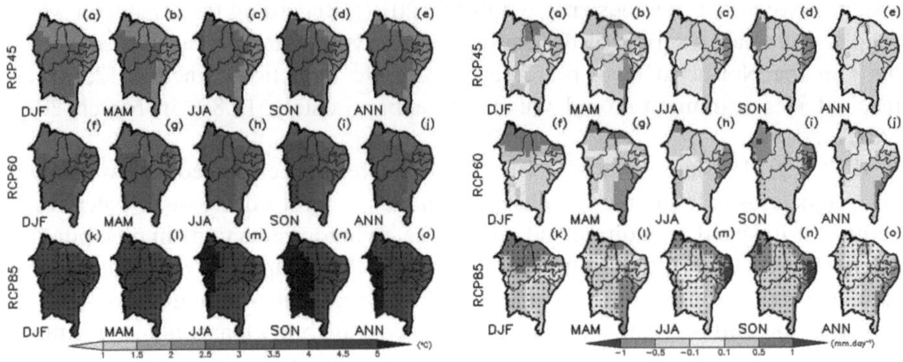

Fig. 7.3 Left-Change in seasonal and annual surface air temperature (°C); Right-Change in seasonal and annual precipitation-evaporation (mm/day). Projections are for 2081–2100 in the forcing scenarios RCP 4.5 (**a–e**), RCP 6.0 (**f–j**) and RCP 8.5 (**k–o**) in both the left and right panels. The changes are defined relative to 1861–1890. Dotted areas indicate regions whose regional temperature changes exceed 4 °C

Reached in 2050 for the RCP 8.5 and in the Middle 2070's and End of the 2090's for the RCP6.0 and 4.5 Respectively. The RCP8.5 Shows Warming of 4 °C Starting in 2070 and 5 °C Starting in 2090.

In future climate projections (2081–2100) from CMIP5 models, extremes drought events are projected under the RCP8.5 for NEB (Fig. 7.3), due mainly to the projected warming that surpasses the 4 °C throughout the year, but are more pronounced in the western part of the region. Water availability expressed as precipitation minus evaporation (Fig. 7.3) shows large decreases in the entire region particularly after March and in regions where warming is above 4 °C. An index of annual maximum number of consecutive dry days (CDD; IPCC, 2013) shown in Fig. 7.4 indicates an increase in the number of dry days in the entire region for the RCP8.5, particularly in the semiarid region of NEB, where projected warming is above 4 °C. Changes in amount and timing of precipitation (e.g., daily rainfall amount and the number of consecutive dry days) strongly influence the extent of the area with severe or extreme drought. One explanation for the relationship is, presumably, that low water availability (precipitation-evaporation) decreases the soil moisture content and thus increases the potential for drought. All these conditions lead to a decrease in discharge into the reservoirs and lakes and to an increase in evaporation, affecting irrigation and agriculture as well as key water uses including hydropower, industry and drinking water, and thus, the welfare of the residents. In sum, the climate model projections suggest large impacts due to intense drought on population due a warming above 4 °C, and this could affect the adaptive capacity of the population in the region.

The situation in 2012–2016 at Northeast Brazil highlighted difficulties that many stakeholders face in managing drought risks. To show additional details of the drought situation, the 12-month Standardized Precipitation Index (SPI) (McKee,

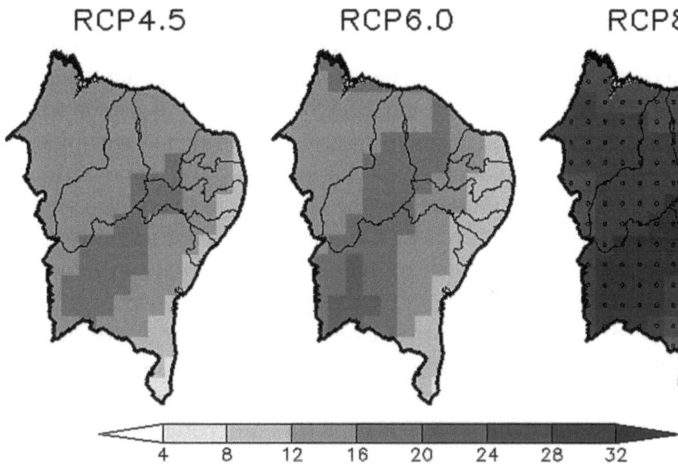

Fig. 7.4 Change in annual consecutive dry days (CDD) projected for 2081–2100 in the forcing scenarios RCP 4.5 (**a**), RCP 6.0 (**b**) and RCP 8.5 (**c**). The changes are relative to 1861–1890. Dotted areas indicate regions whose regional temperature changes exceed 4 °C. Units are in days

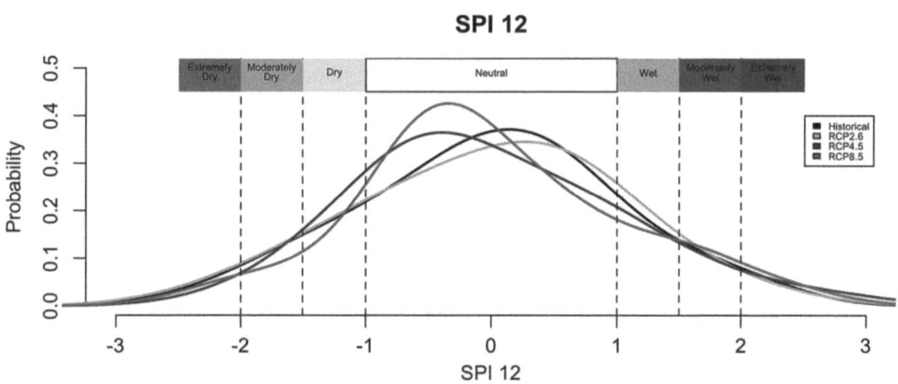

Fig. 7.5 Probability Density Functions for 12-month SPI over the end of twenty-first century (2071–2100) relative to the reference period 1961–2005 under RCP2.6, RCP4.5 and RCP8.5 from the multimodel mean by CMIP5 global climate models for Northeast Brazil

Doesken., & Kleist, 1993) is used to assess extreme events and meteorological droughts based on CMIP5 projections (Fig. 7.5). While other indices such as the Standardized Precipitation-Evapotranspiration Index (SPEI) may be the best for assessment of drought severity, it has a limitation due to the data sets needed to calculate it (e.g. temperature and potential evapotranspiration). So, we decided to use SPI. Moreover, the SPI combined with Vegetation Health Index (VHI) results can describe well drought conditions and potentially damaging extent. Furthermore, the WMO endorse the SPI as the standard for determining the existence of meteorological drought (Hayes, Svoboda, Wall, & Widhalm, 2011).

Figure 7.5 shows the probability density function of 12-month SPI at the end of twenty-first century (2071–2100) compared with the current climate (1961–2005). Their shapes are rather different and the changes are more significant toward the dry conditions, especially under RCP4.5 and RCP8.5. This is consistent with larger surface temperature increases and changes in seasonal and annual precipitation over the tropics by CMIP5 climate models (Figs. 7.4 and 7.5). The shape of the RCP2.6 is symmetric relative to current climate and the change is not very large. In general, the results show that most projections of future extreme events lead to significant changes regarding variance and mean on regional basis mainly in frequency and severity of dry events due to warming.

For assessment of drought scenarios under climate change we selected the Vegetation Health Index (VHI) developed by Kogan (1995), Kogan (1997), Kogan (2002)), Bokusheva, Kogan, Vitkovskaya, Conradt, and Batyrbayeva (2016), Kogan and Guo (2017), which has been widely accepted as an operational tool for global drought monitoring. VHI is as additive combination of Vegetation Condition Index (VCI) and the Temperature Condition Index (TCI) of comparable magnitude. VCI is obtained by normalizing the Normalized Difference Vegetation Index (NDVI) values by their multi-year absolute minimum and maximum values in the analysed period. The VCI not only reflects the spatial and temporal vegetation variability but also allows quantifying the impact of weather and climate on vegetation (Kogan, 1994; Unganai & Kogan, 1998). TCI is similar to VCI, but relates to the brightness temperature (BT) estimated from the thermal band of Advanced Very High Resolution Radiometer (AVHRR, channel 4). The TCI provides an opportunity to identify subtle changes in vegetation health due to thermal effect as drought proliferates when moisture shortage is accompanied by high temperature (Kogan, 2000a; Kogan, 2000b; Kogan, 2002; Kogan et al., 2005). VHI is expressed as:

$$VHI_i = \alpha VCI_i + (1 - \alpha) TCI_i \tag{7.1}$$

where α and (1-α) is a coefficient to determine the contribution of each index. Due to the lack of more accurate information, α has been usually assigned a value of 0.5, assuming equal contribution of both variables to the VHI (Kogan, 2000a, Kogan, 2000b). The VHI change from zero (quantifying extreme vegetation stress) to 100 (indicating optimal condition). VHI values less than 40 represents presence of vegetation stress and greater than 40 favours good condition for vegetation (Kogan, 2001).

In overall, the combined utilization of NDVI (VCI) and BT (TCI) is based on the strong negative correlation between those two variables, due to increase in evaporation with a decrease in soil moisture, caused by higher temperatures, which results in a decline of vegetation cover, where water is the main limiting factor for vegetation growth (Karnieli, Bayasgalan, Bayarjargal, Khudulmur, & Tucker, 2017 in press; Lambin & Ehrlich, 1996; Nemani & Running, 1989). Thus, the VHI has been widely considered in different application, such as drought detection, assessment of drought severity and duration, drought-related losses of crop and pasture production, wildfire risk and early drought warning (Kogan & Guo, 2017; Seiler, Kogan, &

Table 7.1 Drought severity classes for VHI (Kogan, 2001)

Severity Class	VHI
Normal	100 > VHI > 40
Mild drought	30 < VHI ≤ 40
Moderate drought	20 < VHI ≤ 30
Severe drought	10 < VHI ≤ 20
Extreme drought	0 < VHI ≤ 10

Sullivan, 1998). VHI product is provided weekly by the National Oceanic and Atmospheric Administration (http://www.star.nesdis.noaa.gov/smcd/emb/vci/VH/vh_ftp.php.) For the drought scenarios under climate change, we used the VHI at 4 km spatial and 7-day composite temporal resolution, from 1982 to 2016.

The VHI index has been used to study drought in the recent times, including the current drought affecting the region since end of 2011 (Alvalá et al., 2017; Brito et al., 2017; Marengo et al., 2017). In particular, Brito et al. (2017) adopted VHI and SPI indices to monitor and quantify the NEB drought impacts, once VHI is associated with impacts in vegetation and SPI in hydro-meteorological aspects. These indices demonstrated to be suitable as methods and procedures for drought identification in this region. To evaluate the impact of climate change on drought events we used monthly temperature and precipitation data from the CMIP5 models data for the past and projected climate during the period from 1862 up to 2100. Many studies have reported that seasonal average of the VCI is directly correlated with cumulative seasonal rainfall (Bhuiyan, 2008; Davenport & Nicholson, 1993; Herrmann, Anyamba, & Tucker, 2005; Ibrahim, Balzter, Kaduk, & Tucker, 2015; Liu, Wu, Wu, & Liu, 2013), and seasonal average of the TCI shows strong correlation with average seasonal aboveground temperature (Bhuiyan, 2008). Correlation and multiple regression analyses were used to build a predictive model using the VHI (modelled as the dependent variable) and monthly climate variables (independent variables) from the CMIP5 (Table 7.1).

From Fig. 7.2 and considering only the changes in temperature, many studies have pointed out that high temperatures might have more intense and direct effect on vegetation and may cause permanent damage to plant growth and development (Hasanuzzaman, Nahar, Alam, Roychowdhury, & Fujita, 2013). Extremely high temperatures generate thermal stress in vegetation, lower moisture content and thereby adversely affect the vegetation health (Bhuiyan, Saha, Bandyopadhyay, & Kogan, 2017). Since rainfall is the principal supplier of moisture to plants (mainly in semiarid regions), it is also a key driver of VHI. Thus, intense thermal stress accompanied by moisture-stress can intensify the vegetative drought (Gusso, Ducati, Veronez, Sommer, & Da Silveira Jr, 2014). These points can be verified from the VHI projections under various RCP scenarios (Fig. 7.6) over grasslands in semiarid for the present and future.

The results indicate that the semiarid NEB region can experience more intense levels of drought (severe and extreme drought) by the middle of the twenty-first century, while moderate levels can be detected since 2040. Recently, moderate

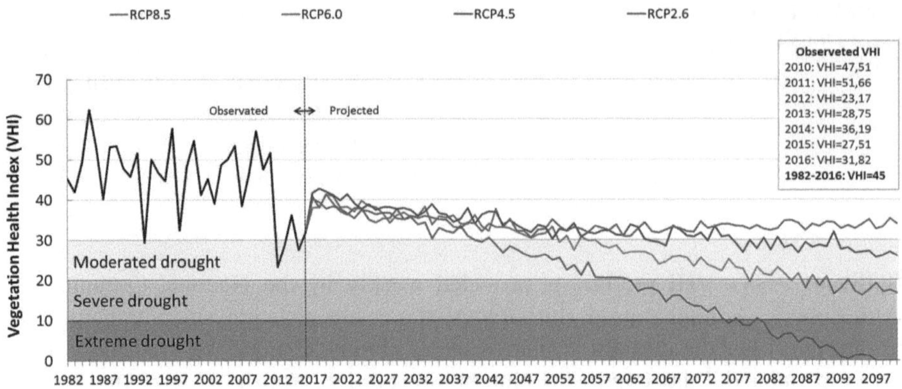

Fig. 7.6 Time series of VHI for the period of 1982–2016 (*black line*) and future projection (2017–2100) obtained from CMIP5 models over grassland in Brazilian semiarid. RCP2.6 (*blue line*), RCP4.5 (*purple line*), RCP6.0 (*green line*) and RCP8.5 (*red line*)

drought conditions were observed from 2012 to 2016. From the VHI-RCP4.5, the moderate level is reached in the middle of the century, while for the RCP6.0 moderate levels are reached in 2050 evolving to severe drought by 2086. For the RCP8.5 moderate drought was detected in 2040, evolving to severe drought in 2080 to extreme drought by 2076. In the last decade of the twenty-first century, the VHI reached almost zero values when warming was above 4 °C, suggesting that the region that is now covered with grassland may not show any photosynthetic activity (becoming bare soil) due to higher temperatures and dryness. This suggests irreversible degradation leading to arid conditions at that time and with drought becoming permanent with warming above 4 °C.

7.4 A Review of Possible Impacts to be Caused by Regional Warming Exceeding 4 °C in NEB

According to Gemenne (2011), warming above 4 °C would bring unprecedented changes to the environment, which may affect human mobility in different ways. The relationship between environmental changes and migrations is highly complex and depends on many specific contexts and variables. It cannot be reduced to a direct causal relationship. Thus, the effects associated with a 4 °C warming may affect not only the magnitude of population movements induced by climate change, but could also particularly affect their nature. However, studies like this have not been performed for NEB.

In other chapters of this book, impacts on agriculture, human health, biodiversity and energy have documented regional impacts of warming above 4 °C in NEB (Assad, Ribeiro, & Nakai, 2018, Hacon, de Oliveira, & Silveira, 2018, Scarano, 2018, Schaeffer et al., 2018):

The chapter about impacts on agriculture for a regional warming above 4 °C (Assad et al., 2018) shows a need to include switching to new crop types, installing irrigation systems, agricultural intensification, shifting agricultural and to new areas and/or the use of genetically modified crops resistant to future warmer climate. NEB is the most affected region for rice production, going to 79% risk in 2070. For beans, with a 4 °C increase, there will be an increase in the low risk areas for the crop and the most significant loss will take place in NEB, and the loss risk analysis for sunflower shows that the Northeast region goes from a current risk of 24% to 63% in 2070. For maize loss risk analysis indicates that the Northeast would go from a current risk of 31% to 64% by 2070. With larger reductions in crop yields projected for NEB by Assad et al. (2018) this raises the potential for substantial human migrations having impacts such as demand for land and water in the regions into which they move.

- If the regional temperature rises above 4 °C in the RCP8.5, climatic conditions in municipalities of the states of Maranhão and Piauí in NEB, may pose a high mortality risk, with average temperatures equal to or greater than 30 °C. In the period of 2071 to 2099, thermal conditions in the municipalities of NEB will be even more favourable for the *Aedes aegypti*, vector of diseases like dengue, chikungunya and zika (average temperatures between 22 °C and 32 °C). Health conditions in approximately 80% of the municipalities in NEB are very highly vulnerable, representing close to 40% of the region's child and elderly population, which may be exposed to an average increase of 5 °C in the region (Hacon et al., 2018).
- A regional increase equal to or greater than 4 °C in the average temperature will result in a 15.7% global increase in species extinction. Extinction will be highest in South America, including NEB, with an expected rate of 23% (Scarano, 2018).
- An average growth trend from 15 to 30% for onshore wind power density for most of NEB is detected in the RCP8.5 for NEB. Energy related results also show a drop in generation due to lower flow rates in basins, particularly the São Francisco River Basin in NEB, which would experience more than a 7% decrease. The main impact identified was a fall in the system's reliability and strong regional effects on hydroelectric generation in NEB (Schaeffer et al., 2018).

7.5 Impacts of Warming above 4 °C on the Risk of Drought in NEB

In this chapter, we assess the impact of warming above 4 °C on the risk of having severe and extreme drought in NEB by estimating changes in the area covered with crop and grasslands at the municipality level. Then, drought hazard (projected VHI from precipitation and temperature trends) and vulnerability maps are integrated to derive a drought risk map at the municipality level (Kussul et al., 2010; Skakun, Kussul, Shelestov, et al., 2016). This analysis is performed for the present (the

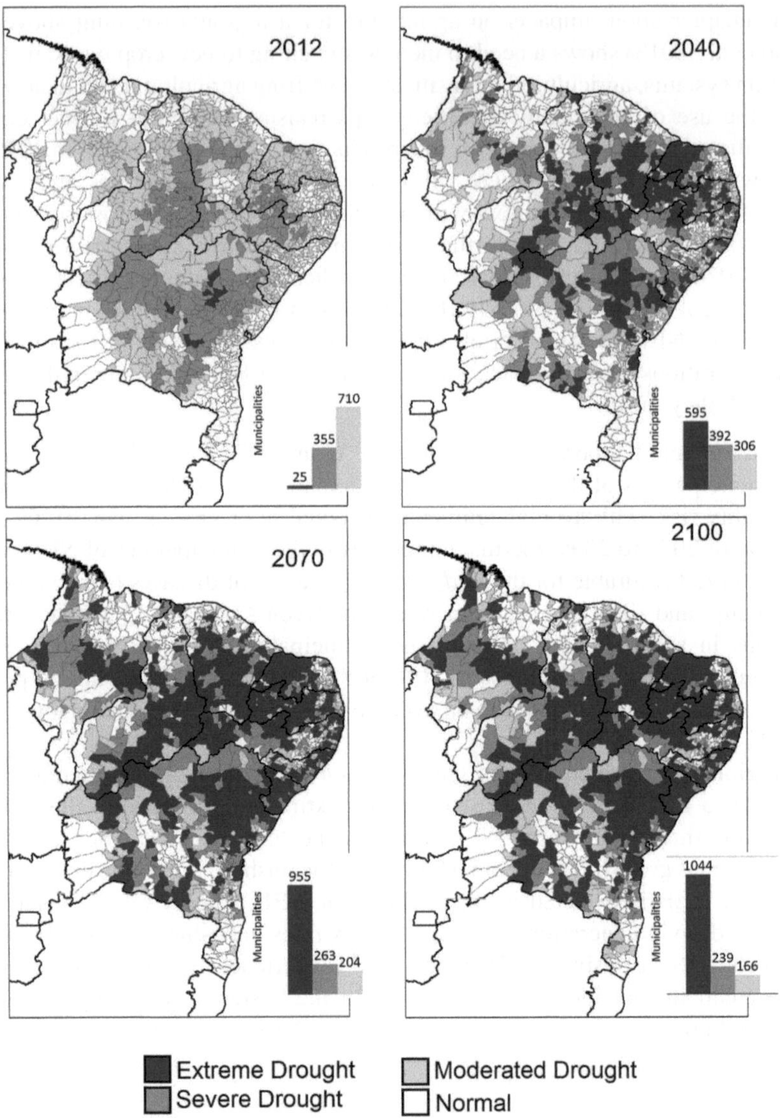

Fig. 7.7 Municipalities affected by the drought, according to observed VHI for 2012 (**a**), future projections for 2040 (**b**), 2070 (**c**) and 2100 (**d**)

observed 2012–2013 drought) and for projections from the CMPI5 models with the high scenario, RCP8.5, which is the projection that shows warming above 4 °C (Fig. 7.7). Drought risk is defined by the intensity of the VHI and the percentage of the area of the municipality affected by drought. To define extreme drought, the VHI index varies from 0 to 10 and at least 75% of the area of a municipality is affected by drought. For severe drought, the VHI varies between 10 and 20, and the area affected by drought should be at last 50% of the municipality.

When a municipality shows high risk of having intense or extreme drought they receive financial emergency funds from the federal government to cope with the impacts of droughts. For instance, as a consequence of 2012–2013 droughts, 1118 municipalities (about 1 million smallholders) had a loss of their harvest. Only considering the *Garantia-Safra* insurance payment, a loss of approximately R$ 980 mi (around U$ 280 mi) were spent. Considering the entire period from 2012 to 2016, this amount exceeded R$ 3 bi (around U$ 860 mi) in agricultural insurance payments (SAF/MDA, 2017). The *Garantia-Safra* Program is operated by the Brazilian Federal Government and aims to guarantee minimum conditions of survival of the rural family farmers affected systematically by drought or excess rainfall (Alvalá et al., 2017).

To assess the drought risks under climate change scenarios at the municipality level, multiple regression analyses were performed for each grid point in the region. For this in addition to the monthly VHI data from 1982 to 2015, the Climatic Research Unit (CRU) TS4 dataset (available online at https://crudata.uea.ac.uk/cru/data/hrg/) were used. This is the most complete and updated dataset of gridded monthly precipitation and temperature at the global scale, with a spatial resolution of 0.5° (Harris, Jones, Osborn, & Lister, 2014).

Compared to 2012, which was the most affected year of the last drought event in Northeast Brazil (Marengo et al., 2017), show projections of a number of municipalities with potential conditions of extreme drought for 2040, 2070 and 2100, with 595, 955 and 1044 municipalities under the risk of extreme drought, respectively (Fig. 7.7). In 2012–13, the number of the municipalities in Northeast Brazil that were classified in severe and extreme drought conditions was 21 (about 20%). In the future projections, these values may reach 55%, 68% and 71%, respectively (Table 7.2). In addition, in 2012, municipalities that experienced extreme and severe drought conditions were identified only in the semiarid region, while in the future projections, it is noted that municipalities outside of the semiarid boundary

Table 7.2 Summary of drought risk assessment in NEB

		Years			
		2012–13	2040[a]	2070[a]	2100[a]
Number of municipalities affected by various levels of drought (and projections)	Extreme drought	25	595	955	1044
	Severe drought	355	392	263	239
	Moderated drought	710	306	204	166
Percentage of NEB municipalities affected by extreme and severe drought (from a total of 1793)		21%	55%	68%	71%
Mean annual temperature increase at NEB region		+1 °C	+2 °C	+4 °C	+5.5 °C

[a]years 2040, 2070 and 2100 are future projections

also may present extreme and severe drought conditions, including those in the *zona da mata,* where annual precipitation between 800–2000 mm.

It is highlighted that in 2012, the average temperature in the Northeast Brazil was 1°C above the average, and in the future projections for the decades centred on 2040, 2070 and 2100, an increase of 2°C, 4°C and 5.5°C above the average, respectively.

In sum, regional warming above 4 °C is likely to increase the drought risk in Northeast Brazil, with increase temperature and decrease precipitation resulting in lower vegetal productivity and more unpredictable harvests. In municipalities, where smallholder livelihoods are not very diversified and are dominated by subsistence agriculture, even a moderate drought (as in 2012) can cause a decline in harvests; and, with an increased drought risk (as the future projections), the harvest scenario can still be worse and devastating for regional and national food security and economy.

7.6 Probability of Warming Exceeding 4 °C

According to the World Bank (2012), global warming above 4 °C will shift temperatures in the tropics more than 6 standard deviations for all months in the year. In particular, countries in tropical South America, Central Africa, and all tropical islands in the Pacific will see unprecedented extreme temperatures become the new norm throughout the year. According to IPCC (2014) intense regional warming will take place on various regions of the planet, varying from 4 °C to 10 °C, showing a dramatic increase in the intensity and frequency of high temperature extremes and drought periods in South America, for instance.

Figure 7.8 shows the probabilities of annual temperature projected for the RCP8.5 for warming exceeding 4, 5, 6, 7 °C for the NEB region. These probabilities were calculated as described in Chap. 2 of this book. Regional warming rates higher than 4 °C and beyond 2100 are assessed for RCP 8.5 only. Figure 7.8 shows probabilities higher than 50% from mid-2070, reaching 100% by 2120. Probabilities higher than 50% for warming above 6 °C start occurring in 2120 and reach 100% by 2160.There is a 50% probability of warming above 7 °C by 2050 and it exceeds 80% by 2180.

7.7 Final Remarks

A number of studies have shown that although semiarid vegetation is usually resistant and highly resilient to water deficits, vegetation activity in semiarid regions is highly controlled by interannual variations in water availability and decrease in water availability may trigger land degradation (Vicente-Serrano et al., 2013).

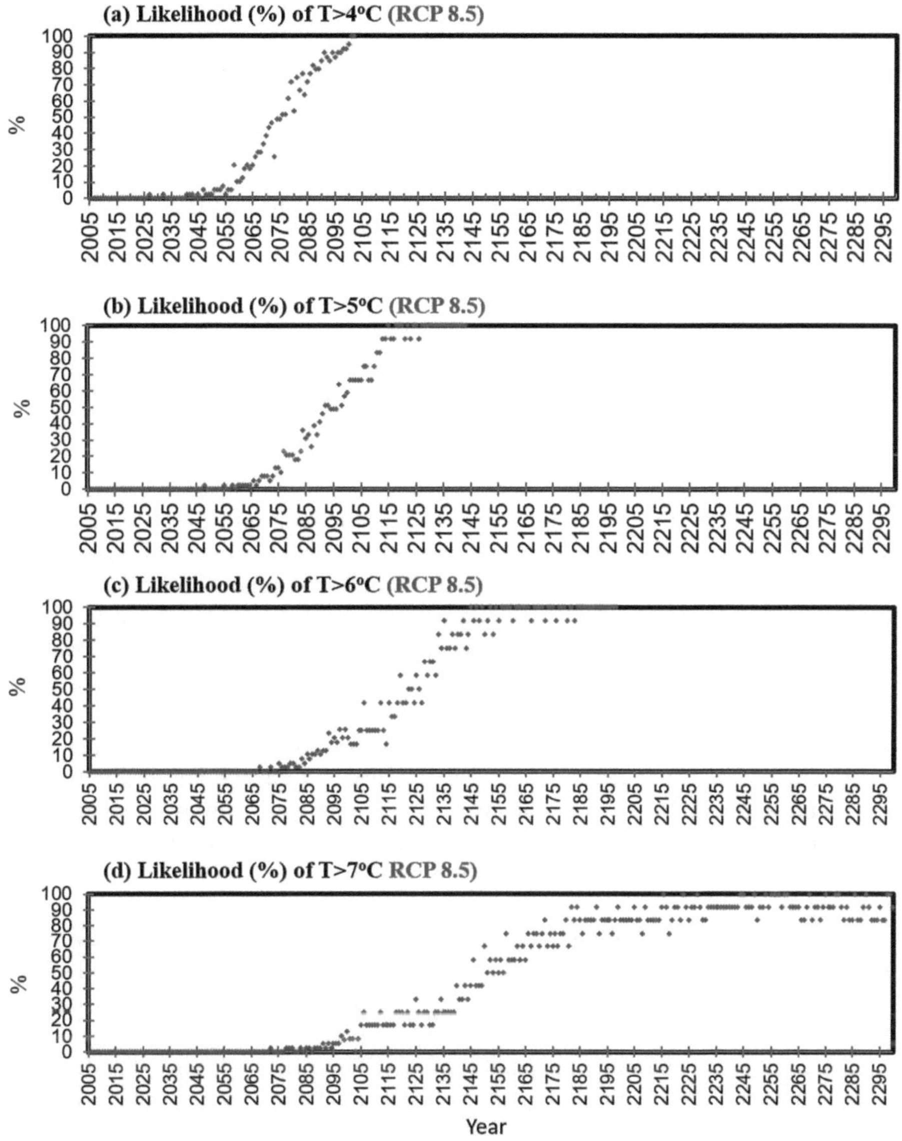

Fig. 7.8 Probabilities of annual average temperature projected by RCP 8.5 exceeding: (**a**) 4 °C, (**b**) 5 °C, (**c**) 6 °C, and (**d**) 7 °C. The warming probabilities were obtained from the values of temperature anomalies in CMIP5 simulations. Probabilities of warming above 4 °C were not found in RCPs 2.6, 4.5 and 6.0. The unit is %

Drought is a recurrent phenomenon and NEB and this region is vulnerable to climate extremes and droughts. Recurrent droughts conditions in semiarid regions can produce a progressive loss of resilience that affects negatively the ability of recovering the initial state, rendering plants to be more vulnerable to a recurring disturbance (Cunha, Alvalá, Nobre, & Carvalho, 2015; Martiny, Richard, & Camberlin, 2005; Philippon, Mougin, Jarlan, & Frison, 2005). Thus, the vegetation can be durably affected by a drought if it is preceded by another dry year, often leading to vegetation degradation. The drought affecting this region continuously during the last 6 years shows an intensity and impact not seen in several decades in the regional economy and society. In fact, this is the most prolonged below-average precipitation period since rainfall records exist in the region beginning in the second half of the nineteenth Century (Alvalá et al., 2017; Marengo et al., 2017; Martins et al., 2015; Martins et al., 2016; Martins, Vieira, Biazeto, & Molejon Quintana, 2016). Droughts also affect the social, economic and political structures in the region.

For the RCP8.5 scenario, the probability of having a warming of 4 °C varies from 50% from mid-2070 to 100% by 2120. There is a 50% probability of warming above 7 °C by 2050 and it exceeds 80% by 2180. While droughts are projected to increase by 2100 and the increase in the likelihood of warming above 4 °C by the middle of this century, it is still uncertain what these rainfall and soil moisture deficits might mean for prolonged reductions of stream flow and groundwater levels in NEB (IPCC, 2014). Even with the possibility of longer and more extensive drought in NEB as projected by climate models it seems that adaptation processes to cope the drought are often poorly implemented and losses in the regional economy and ecosystem services induced by climate variability and change are generally high and yet unexplored. Thus, there is an urgent need for proactive drought management and preparedness strategies as well as integrated assessments considering the synergy of impacts and limits to adaptation in multiple sectors and regions in a 4 °C warming for NEB.

As suggested by D. Wilhite (personal communication), with regard to drought, the push now is through the United Nations agencies, development banks and others is to approach drought management through what is commonly referred to as the 'three pillar approach' of drought monitoring and early warning, vulnerability and impact assessment, and mitigation and response. This process leads to the development of a proactive drought policy aimed at risk reduction, both under current climates and under a changing climate.

The challenges involved in keeping the mean global warming below 1.5–2 °C have if anything increased since the 2015 Paris agreement. With decisive action, a 4 °C world can be avoided and warming can likely be held below 2 °C. Numerous studies show that there are technically and economically feasible emissions pathways to achieve this. Thus, the level of impact that developing countries and the rest of the world experience will be a result of government, private sector, and civil society decisions and choices about climate change which includes, unfortunately, inaction.

References

Agência Nacional de Águas ANA. (2017). Boletins Mensais de Monitoramento. Available at: http://www2.ana.gov.br/Paginas/servicos/saladesituacao/boletinsmonitoramento.aspx

Alvalá, R. C., Cunha, A. P. M. A., Brito, S. B., Seluchi, M. E., Marengo, J. A., Moraes, O. L. L., et al. (2017). *Drought monitoring in the Brazilian semiarid region*, In Press, Annais da Academia Brasilieira de Ciencias do Brasil.

Araújo, J. (1982). Barragens no Nordeste do Brasil: Experiência do DNOCS em Barragens na Região Semi-Arida. Departamento Nacional de Obras contra as Secas, Ministerio do Interior, pp 15.

Assad, E. D., Ribeiro, R. R., & Nakai, A. M. (2018). Agricultural sector. Assessments and how an increase in temperature may have an impact on agriculture in Brazil and mapping of the current and future situation, Chapter 3. In C. A. Nobre, J. A. Marengo, & W. R. Soares (Eds.), *Climate change risks in Brazil* (pp. xx–xx). New York: Springer.

Bhuiyan, C. (2008). Desert vegetation during droughts: Response and sensitivity, the international archives of the photogrammetry. *Remote Sensing And Spatial Information Sciences, 37*, 907–912.

Bhuiyan, C., Saha, A. K., Bandyopadhyay, N., & Kogan, F. N. (2017). Analyzing the impact of thermal stress on vegetation health and agricultural drought–a case study from Gujarat, India. *GIScience & Remote Sensing*, 1–22.

Bokusheva, R., Kogan, F., Vitkovskaya, I., Conradt, S., & Batyrbayeva, M. (2016). Satellite-based vegetation health indices as a criteria for insuring against drought-related yield losses. *Agricultural and Forest Meteorology, 220*, 200–206.

Brito, S. S. B., Cunha, A. P. M. A., Cunningham, C. C., Alvalá, R. C., Marengo, J. A., & MAraujo. (2017). Frequency, duration and severity of drought in the Brazilian semiarid. *International Journal of Climatology, 38*, 517. https://doi.org/10.1002/joc.5225

Burkett, V. R., et al. (2014). Point of departure. In *Climate change 2014: Impacts, adaptation, and vulnerability. Part A: Global and sectoral aspects. Contribution of working group II to the fifth assessment report of the IPCC* (pp. 169–194). Cambridge, UK: Cambridge University Press.

CONAB. (2017). Companhia Nacional de Abastecimento. Levantamento de Safras. Available at: http://www.conab.gov.br/conteudos.php?a=1253&t=2. Acessado em março de 2017.

Cunha, A. P., Alvalá, R. S., Nobre, C. A., & Carvalho, M. A. (2015). Monitoring vegetative drought dynamics in the Brazilian semiarid region. *Agricultural and Forest Meteorology, 214–215*, 494–505.

Davenport, M. L., & Nicholson, S. E. (1993). On the relation between rainfall and the normalized difference vegetation index for diverse vegetation types in East Africa. *International Journal of Remote Sensing, 14*, 2369–2389.

Gemenne, F. (2011). Climate-induced population displacements in a 4 ° C+ world. *Philosophical Transactions of the Royal Society of London A: Mathematical, Physical and Engineering Sciences, 369*(1934), 182–195.

Gusso, A., Ducati, J. R., Veronez, M. R., Sommer, V., & Da Silveira Jr., L. G. (2014). Monitoring heat waves and their impacts on summer crop development in southern Brazil. *Agricultural Sciences, 5*, 353–364. https://doi.org/10.4236/as.2014.54037

Gutierrez, A. P. A., Engle, N. L., De Nys, E., Molejon, C., & Martins. (2014). Drought preparedness in Brazil. *Weather and Climate Extremes, 95*(3), 95–106.

Hacon, S. S., de Oliveira, B. F. A., & Silveira. (2018). Health sector: Review of literature on the health sector and impacts of 4 °C or mopre warming, Chapter 4. In C. A. Nobre, J. A. Marengo, & W. R. Spares (Eds.), *Climate change risks in Brazil* (pp. xx–xx). New York: Springer.

Harris, I., Jones, P. D., Osborn, T. J., & Lister, D. H. (2014). Updated high-resolution grids of monthly climatic observations – The CRU TS3.10 data set. *International Journal of Climatology, 34*, 623–712. https://doi.org/10.1002/joc.3711

Hasanuzzaman, M., Nahar, K., Alam, M. M., Roychowdhury, R., & Fujita, M. (2013). Physiological, biochemical, and molecular mechanisms of heat stress tolerance in plants. *International Journal of Molecular Sciences, 15*, 9643–9684.

Hayes, M. J., Svoboda, M. D., Wall, N., & Widhalm, M. (2011). The Lincoln declaration on drought indices — Universal meteorological drought index recommended. *Bulletin of the American Meteorological Society, 92*, 485–488. https://doi.org/10.1175/2010BAMS3103.1

Herrmann, S. M., Anyamba, A., & Tucker, C. J. (2005). Recent trends in vegetation dynamics in the African Sahel and their relationship to climate. *Global Environmental Change, 15*, 394–404.

http://www.star.nesdis.noaa.gov/smcd/emb/vci/VH/vh_ftp.php

https://crudata.uea.ac.uk/cru/data/hrg/

https://www.esrl.noaa.gov/psd/enso/past_events

Ibrahim, Y. Z., Balzter, H., Kaduk, J., & Tucker, C. J. (2015). Land degradation assessment using residual trend analysis of GIMMS NDVI3g soil moisture and rainfall in sub-Saharan West Africa from 1982 to 2012. *Remote Sensing, 7*, 5471–5494.

IPCC. (2013). Climate change 2013: The physical science basis. In T. F. Stocker, D. Qin, G.-K. Plattner, M. Tignor, S. K. Allen, J. Boschung, A. Nauels, Y. Xia, V. Bex, & P. M. Midgley (Eds.), *Contribution of working group I to the fifth assessment report of the Intergovernmental panel on climate change* (p. 1535). Cambridge, UK and New York: Cambridge University Press.

IPCC. (2014). Central and South America. In V. R. Barros, C. B. Field, D. J. Dokken, M. D. Mastrandrea, K. J. Mach, T. E. Bilir, M. Chatterjee, K. L. Ebi, Y. O. Estrada, R. C. Genova, B. Girma, E. S. Kissel, A. N. Levy, S. Mac Cracken, P. R. Mastrandrea, & L. L. White (Eds.), *Climate change 2014: Impacts, adaptation, and vulnerability. Part B: Regional Aspects. Contribution of working group II to the fifth assessment report of the Intergovernmental panel on climate change* (pp. 1499–1566). Cambridge,UK and New York: Cambridge University Press.

Karnieli, A., Bayasgalan, M., Bayarjargal, Y., Khudulmur, S. & Tucker C. J. (2017). Comments on the use of the vegetation health index over Mongolia. International Journal of Remote Sensing, in press.

Kogan, F. N. (1994). Application of vegetation index and brightness temperature for drought detection. *Advances in Space Research., 15*, 91–100. https://doi.org/10.1016/0273-1177(95)00079-T

Kogan, F. N. (1995). Application of vegetation index and brightness temperature for drought detection. *Advances in Space Research, 15*, 91–100.

Kogan, F. N. (1997). Global drought watch from space. *Bulletin of the American Meteorological Society, 78*, 621–636.

Kogan, F. N. (2000a). Global drought watch from space. *Bulletin of the American Meteorological Society, 78*, 621–636.

Kogan, F. N. (2000b). Satellite-observed sensitivity of world land ecosystems to El Nino/La Nina. *Remote Sensing of Environment, 74*, 445–462.

Kogan, F. N. (2001). Operational space technology for global vegetation assessment. *Bulletin of the American Meteorological Society, 82*, 19491964.

Kogan, F. N. (2002). World droughts in the new millennium from AVHRR-based vegetation health indices. Eos, transactions. *American Geophysical Union, 83*(48), 562–563.

Kogan, F. N., & Guo, W. (2017). Strong 2015-2016 El Niño and implication to global ecosystems from space data. *International Journal of Remote Sensing, 38*(1), 161–178.

Kogan, F. N., et al. (2005). Modelling corn production in China using AVHRR-based vegetation health indices. *International Journal of Remote Sensing., 26*, 2325–2336.

Kussul, N. N., Sokolov, B. V., Zyelyk, Y. I., Zelentsov, V. A., Skakun, S. V., & Shelestov, A. Y. (2010). Disaster risk assessment based on heterogeneous geospatial information. *Journal of Automation and Information Sciences, 42*(12), 32–45.

Lambin, E. F., & Ehrlich, D. (1996). The surface temperature-vegetation index space for land cover and land-cover change analysis. *International Journal of Remote Sensing, 17*(3), 163–487.

Liu, J., Wu, J., Wu, Z., & Liu, M. (2013). Response of NDVI dynamics to precipitation in the Beijing–Tianjin sandstorm source region. *International Journal of Remote Sensing, 34*, 5331–5350.

Magalhães, A., & Coauthors. (1988). The effects of climate variations on agriculture in Northeast Brazil. In M. Parry, T. Carter, & N. Konijn (Eds.), *The impact of climate variations on agriculture. vol 2. assessments in semiarid regions* (pp. 277–304). Amsterdam: Kluwer Academic Publishers.

Marengo, J. A., Alves, L. M., Alvalá, R. C., Cunha, A.P., Brito, S.S., & Moraes, O. L. L. (2017). *Climatic characteristics of the 2010–2016 drought in the semiarid Northeast Brazil region*, Anais da Academia Brasileira de Ciências (2017) (Annals of the Brazilian Academy of Sciences) Printed version ISSN 0001–3765 / Online version ISSN 1678–2690 http://dx.doi.org/DOI.

Marengo, J. A., Torres, R. R., & Alves, L. M. (2016). Drougth in northeast Brasil – Past, presente and future. *Theoretical and Applied Climatology, 20*, 1–12.

Martins, E. S., De Nys, E., Molejon Quintana, C., Biazeto, B., Vieira, R. F., & Engle, N. (2015). Monitor de Secas do Nordeste, em busca de um novo paradigma para a gestão de secas. Série Água, 10.

Martins, E. S., Molejon Quintana, C., Dias, S., Vieira RF, M. A. F., Biazeto, B., Foratini, G. D., et al. (2016). The technical and institutional case: The northeast drought monitor as the anchor and facilitator of collaboration, Chapter 3. In N. Engle, E. De Nys, & A. R. Magalhaes (Eds.), *Drought in Brazil: Proactive management and policy, Drought and Water Crises* (pp. 37–48). Boca Raton, FL: Taylor & Francis., 2016.

Martins, E. S., Vieira, R. F., Biazeto, B., & Molejon Quintana, C. (2016). Northeast drought monitor: The process, Chapter 11. In N. Engle, E. De Nys, & A. R. Magalhaes (Eds.), *Drought in Brazil: proactive management and policy, Drought and Water Crises* (pp. 143–166). Boca Raton, FL: Taylor & Francis., 2016.

Martiny, N., Richard, Y., & Camberlin, P. (2005). Interannual persistence effects in vegetation dynamics of semi-arid Africa. *Geophysical Research Letters, 32*, L24403. https://doi.org/10.1029/2005GL024634

McKee, T. B., Doesken., N. J., & Kleist, J. (1993). *The relationship of drought frequency and duration of time scales. Eighth conference on applied climatology* (pp. 179–186). Anaheim, CA: American Meteorological Society., Jan17–23, 1993.

Meinshausen, M., Smith, S. J., Calvin, K., et al. (2011). *Climatic Change, 109*, 213. https://doi.org/10.1007/s10584-011-0156-z

Moss, R. H., Edmonds, J. A., Hibbard, K. A., Manning, M. R., Rose, S. K., Van Vuuren, D. P., et al. (2010). The next generation of scenarios for climate change research and assessment. *Nature, 463*(7282), 747–756.

Nemani, R. R., & Running, S. W. (1989). Estimation of regional surface resistance to evapotranspiration from NDVI and thermal-IR AVHRR data. *Journal of Applied Meteorology, 28*, 276–284.

Philippon, N., Mougin, E., Jarlan, L., & Frison, P.-L. (2005). Analysis of the linkages between rainfall and land surface conditions in the west African monsoon trough CMAP, ERS-WSC, and NOAA-AVHRR data. *Journal of Geophysical Research, 110*, D24115. https://doi.org/10.1029/2005JD006394

Rodrigues, R. R., & MCphaden, M. J. (2014). Why did the 2011–2012 La Niña cause a severe drought in the Brazilian Northeast? *Geophysical Research Letters, 4*, 1012–1018.

SAF/MDA. (2017). Secretaria Especial de Agricultura Familiar e do Desenvolvimento Agrário. Disponível em: http:// www.mda.gov.br. Acesso em 07 de março de 2017.

Scarano, F. (2018). Biodiversity sector: Assessment of the literature of temperature increase in the biodiversity sector and ecosystems, Chapter 5. In C. A. Nobre, J. A. Marengo, & W. R. Soares (Eds.), *Climate change risks in Brazil* (pp. xx–xx). New York: Springer.

Schaeffer, R., Lucena, A. F. P., Costa, I., Vazquez, E., Viviescas, C., & Hubak, V. (2018). Energy sector: Climate change and the energy sector in Brazil, Chapter 6. In C. A. Nobre, J. A. Marengo, & W. R. Soares (Eds.), *Climate change risks in Brazil* (pp. xx–xx). New York: Springer.

Schlaepfer, D. R., et al. (2017). Climate change reduces extent of temperate drylands and intensifies drought in deep soils. *Nature Communications, 8*, 14196. https://doi.org/10.1038/ncomms14196

Seiler, R. A., Kogan, F., & Sullivan, J. (1998). AVHRR-based vegetation and temperature condition indices for drought detection in Argentina. *Advances in Space Research, 21*, 481–484.

Skakun, S., Kussul, N., Shelestov, A., et al. (2016). The use of satellite data for agriculture drought risk quantification in Ukraine. *Geomatics, Natural Hazards and Risk, 7*, 901–917.

Soares, W. R., Marengo, J. A., & Nobre, C. A. (2018). Assessment of warming projections and probabilities for Brazil, Chapter 2. In C. A. Nobre, J. A. Marengo, & W. R. Soares (Eds.), *Climate change risks in Brazil* (pp. xx–xx). New York: Springer.

Unganai, L. S., & Kogan, F. N. (1998). Drought monitoring and corn yield estimation in southern Africa from AVHRR data. *Remote Sensing of Environment, 63*, 219–232.

Van Vuuren, D. P. P., den Elzen, M., Lucas, P., Eickhout, B., Strengers, B., van Ruijven, B., et al. (2007). Stabilizing greenhouse gas concentrations at low levels: An assessment of reduction strategies and costs. *Climatic Change*. https://doi.org/10.1007/s/10584-006-9172-9

Van Vurren, D. P., Edmons, J., Kainuma, M., Riahi, K., Thomson, A., Hibbard, K., et al. (2011). The representative concentration pathways: An overview. *Climatic Change, 109*, 5–31.

Vicente-Serrano, S. M., Gouveia, C., Camarero, J. J., Beguería, S., Trigo, R., López-Moreno, J. I., et al. (2013). The response of vegetation to drought time-scales across global land biomes. *Proceedings of the National Academy of Sciences of the United States of America, 2013*(110), 52–57.

Vieira, R. M. S. P., Cunha, A. P. M. A., Alvalá, R. C. S., Carvalho, V. C., Ferraz Neto, S., & Sestini, M. F. (2013). Land use and land cover map of a semiarid region of Brazil for meteorological and climatic models. *Revista Brasileira de Meteorologia, 28*, 129–138.

Wilhite, D. A., Sivakumar, M. V. K., & Pulwarty, R. (2014). Managing drought risk in a changing climate: The role of national drought policy. *Weather and Climate Extremes, 3*, 4–13.

World Bank. (2012). *4C: Turn down the heat. A report for the World Bank by the Potsdam institute for climate impact research and climate analytics*, November 2012, Washington, DC 20433.

World Bank. (2013) Turn down the heat: Climate extremes, regional impacts, and the case for resilience. *A report for the World Bank by the Potsdam institute for climate impact research and climate analytics*. Washington, DC 20433.

Chapter 8
Assessing the Possible Impacts of a 4 °C or Higher Warming in Amazonia

Gilvan Sampaio, Laura S. Borma, Manoel Cardoso, Lincoln Muniz Alves, Celso von Randow, Daniel Andrés Rodriguez, Carlos A. Nobre, and Felipe Ferreira Alexandre

8.1 Introduction

The Amazon region occupies the largest contiguous forest area on the planet and is responsible for numerous environmental services, among which are: habitat of about 10 to 15% of land biodiversity (Hubbell et al., 2008; Lewinshohn and Prado, 2002; Nobre et al., 2016) climate regulation (Li et al., 2015) and freshwater production, contributing with around 15% - 20% of all freshwater discharged into oceans annually (Barthem et al., 2004; Molinier, Guyot, De Oliveira, & Guimarães, 1996).

From the point of view of climate regulation, the Amazon operates on several scales. At the global scale, it is responsible for the storage of about 150–200 Pg of carbon in living biomass and soils (Feldpausch et al., 2012), which could otherwise be released into the atmosphere with strong impacts on the global climate (e.g., Ballantyne, Alden, Miller, Tans, & White, 2012; Le Quéré et al., 2013; Pan et al., 2011; Phillips et al., 2009). On the local and regional scales, Amazon forest exerts control on rainfall and temperature through evapotranspiration (ET), in a process known as 'moisture recycling'. A number of studies attempted to quantify the water

G. Sampaio (✉)
Center for Weather Forecasting and Climate Studies/National Institute for Space Research (CPTEC/INPE), São Paulo, Brazil
e-mail: gilvan.sampaio@cptec.inpe.br

L. S. Borma · M. Cardoso · C. von Randow · F. F. Alexandre
Earth System Science Center/National Institute for Space Research (CCST/INPE), São Paulo, Brazil

L. M. Alves
National Institute for Space Research (INPE), São Paulo, Brazil

D. A. Rodriguez
Federal University of Rio de Janeiro (UFRJ), Rio de Janeiro, Brazil

C. A. Nobre
Institute for Advanced Studies, University of São Paulo, São Paulo, Brazil

C. A. Nobre et al. (eds.), *Climate Change Risks in Brazil*,
https://doi.org/10.1007/978-3-319-92881-4_8

balance in the Amazon basin. Despite uncertainties, estimates of evapotranspiration (ET) widely range from about 35% to over 80% of the precipitation (Costa & Foley, 1999; Eltahir & Bras, 1994; Marengo, 2004; Marengo, 2005; Marengo, Miller, Russell, Rosenzweig, & Abramopoulos, 1994; Nobre, Sellers, & Shukla, 1991; Roads, Kanamitsu, & Stewart, 2002; Salati, Dall'Olio, Matsu, & Gat, 1979; Salati & Vose, 1984; Vorosmarty et al., 1996; Zeng, 1999). Basin-wide and long-term water balances based on observations of precipitation and Amazon River discharge constrain the E/P ratio to about 0.45 to 0.60 (e.g., Nobre et al., 1991; Salati & Vose, 1984, among others). This aspect is even more relevant considering that in some places of the region ET in the dry season tends to be equal to or higher than the ET of the wet season (e.g. Shuttleworth, 1988). So, the high flow of water vapour flux generated by the forest's evapotranspiration during dry season would play an important role in the onset of the rainy season (Fu & Li, 2004).

However, the Amazonian forest's ability to provide environmental services is threatened by anthropogenic forcing at various scales, such as deforestation, fire, global and regional climate change, and extreme events (Borma, Nobre, & Cardoso, 2013; Nobre & Borma, 2009). In addition to the impacts resulting from each one of these drivers, the synergistic effects potentially increase the risks (Betts, Malhi, & Roberts, 2008; Malhi et al., 2008; Nobre et al., 2016). For example, large-scale deforestation and the prospects of global climate changes can intensify the perspectives of a drier and warmer Amazon climate (da Silva, Werth, & Avissar, 2008; Nobre et al., 1991). The extreme events of drought increase the incidence of fires (Nepstad et al., 2004), which, in turn, are more pronounced at the edges between vegetated and non-vegetated areas, indicating a relation between deforestation and fire (Aragão et al., 2008). All these pressures pose Amazon forest closer to its bioclimatic limit than other tropical forests (Good et al., 2011).

In general, the prospects resulting from these anthropogenic drivers point to a potential replacement of the Amazon forest by a type of vegetation more resilient to seasonal droughts, such as tropical savanna (Huntingford et al., 2013, Malhi et al., 2009; Nobre et al., 1991; Oyama & Nobre, 2003; Shukla, Nobre, & Sellers, 1990). This process, known as 'savanization' (Nobre et al., 1991) is part of studies related to the vulnerability of some regions to a biome shift caused by future climate change (e.g., Lenton et al., 2008; Scholze, Knorr, Arnell, & Prentice, 2006). In the Amazon, a number of model projections had pointed to the risk of the replacement of the evergreen vegetation to a more seasonal one (Lapola, Oyama, & Nobre, 2009; Oyama & Nobre, 2003; Salazar, Nobre, & Oyama, 2007). Model calculations also indicate that there may be a 'tipping point' related to the total area deforested before most of the tropical forest would be converted to savannas and that is 40% of deforested area (Sampaio et al., 2007). It should be mentioned that the "savannization" of the Amazon does not detract from the savannah as a biome, given the richness of species and potential environmental services provided by this biome. However, it may imply changes in the environmental services provided today by the tropical forest, in particular as regards its role as a key climate regulator and freshwater producer for the planet. Recent studies have shown a decrement in the net increase of aboveground biomass during the past decade, suggesting a decline of the carbon sink function of Amazon forests (Brienen et al., 2015). That very important carbon sink role is further weakened by forest clearing processes, especially fires. Additionally, considering the severity of the extreme droughts of 2005, 2010 and 2015–2016, drought-related fires have become an even more important source of carbon emission (Aragao, Anderson, Fonseca, et al., 2017; Anderson et al., 2015).

The Amazonian response to anthropogenic drivers will depend on the resilience of the forest. The ability of vegetation to access water from the deeper layers of the soil (Jipp, Nepstad, Cassel, & Reis De Carvalho, 1998) and the so-called CO_2 fertilization effect have been raised as potential aspects related to forest resilience to drought (Borma et al., 2013; Nobre & Borma, 2009). The effect of CO_2 fertilization is due to the increased concentration of this gas in the atmosphere, which potentially could favour the increase of forest productivity, counteracting emissions (Curtis & Wang, 1998; Norby et al., 2005; Prentice, Farquhar, & Fasham, 2001). On the other hand, it also reduces the evaporation back to the atmosphere and possibly reduces down-wind rainfall further. Whether or not this represents a net benefit, is not fully understood yet. Despite a number of uncertainties on the net benefit of this process (DeLucia, Moore, & Norby, 2005; Lapola et al., 2009; Rammig et al., 2010), some model projections consider the favourable effect of the CO_2 enrichment in the studies of biome-climate stability (Lapola et al., 2009). This is the case of the projections presented in the next sections of this chapter.

In the light of the above, this chapter aims to evaluate the future prospects for the Amazon in a scenario of 4 °C or higher warming resulting from anthropogenic climate change and the related hydrological cycle changes. For that, we analysed i) the prospect of increasing temperature in the Amazon basin according to the CMIP5 results; ii) potential occurrence of vegetation shift in the Amazon, based on the potential vegetation model (CPTEC-PVM2), and iii) impacts of climate change in river discharge. The results of these predictions are analysed in light of the other anthropogenic drivers that operate in the basin, in particular deforestation and increase of forest fires. In the light of this book's rationale, we also focus our attention on those extreme scenarios likely to cause catastrophic impacts to contribute elements to a concerted risk analysis for Amazonia, that is, events of uncertain likelihood, but of severe impacts.

Some model projections (e.g., Nobre et al., 2016) have shown that over the next several decades there is a risk of an abrupt and irreversible change over a part or perhaps the entirety of Amazonia, with forest being replaced by savanna-type vegetation with large-scale loss of biodiversity and loss of livelihoods for people in the region, and with impacts of climate in adjacent regions. It is not a trivial scientific question to find out at which point the current stable state could switch (perhaps abruptly) to the savanna-type vegetation (Jones, Lowe, Liddicoat, & Betts, 2009), given the combined forcing of deforestation, forest fragmentation, increased forest fire, global warming with a likely consequence of more intense droughts such as the severe droughts observed in 2005 and 2010. All these factors may accelerate the "savannization" of Amazonia and if the current pace of change is kept unaltered we may well only find out that the climate-vegetation equilibrium has been reached after we have passed the threshold for its establishment. In this chapter, we describe some of the impacts of climate change under the possibility of a 4° C and higher increase in global warming.

8.2 Climate Change and Climate Modelling in the Amazon

Projections of climate change in Amazonia from the state-of-the-art CMIP5 (Coupled Model Intercomparison Project phase 5, Taylor et al., 2012) ensemble show a consistent signal for warming in the models over the twenty-first century, with greater warming under the higher Representative Concentration Pathways (RCPs). Maximum warming in the South America region occurs over the interior of the continent, including much of Amazonia. Greater temperature increases are also seen during the dry season than during the wet.

The mean seasonal change of temperature projected by CMIP5 multi-model ensemble shows an increasing trend in the surface air temperature in all future time slices for RCP8.5 scenario (Fig. 8.1). This is physically expected for this scenario, which is the most extreme in the family of RCP scenarios in terms of higher CO_{2e} concentrations. Clearly, this trend has a larger magnitude of increment of temperature changes exceeding 4 °C from the 2060s onwards and mainly in the months of August to November.

By the end of the century, the results of the models show that the projection of an increase in temperatures can reach even higher levels and reach approximately 7 °C in southern Amazon. The first decades of the twenty-first century also follow the tendency of increase according to the projections, but with a smaller magnitude, with an increase of up to 1 ° C already from the 2020 decade. The significantly

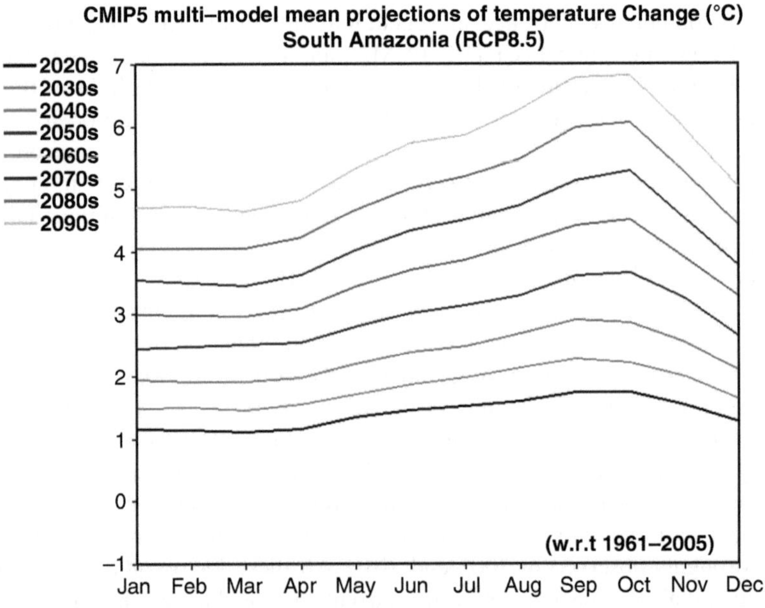

Fig. 8.1 CMIP5 multi-model ensemble projected change (°C) in annual cycle of surface air temperature in southern Amazonia (70°W - 50°W, 12.5°S - 5°S) under the RCP8.5 scenario for decadal time slices for the period 2020–2100 relative to the 1961–2005 baseline

increasing trend in mean temperature, considered without other changes, has a negative effect on the vegetation in the Amazon (Huntingford et al., 2013), and should be considered a potentially important stress factor on the forest.

For each decadal time slice, the percentage change in precipitation (Fig. 8.2) shows a reduction in precipitation for all months. However, it is more pronounced during dry and transition season of July to November (JASON) in southern Amazonia, which modifies the annual cycle and the length of the seasons. During the transition from dry season to wet season (ASO), the projections show the sharpest proportional decline in precipitation, implying a progressive prolongation of the dry season, resulting in a reduction in the length of the rainy season. In SH summer (December–March) the projections show very small magnitudes of changes of about 5%.

Changes in seasonal distribution, magnitude, and duration of precipitation may have significant impacts on Amazonia hydrology and other sectors, since rainfall reductions predominantly in dry and transition seasons, from July to November, increases of air temperature, such as observed during the 2005 and 2010 extreme droughts, may cause a decrease in the water levels to extremely low levels and an elevated fire-induced tree mortality (Marengo et al., 2008; Phillips et al., 2009). In this scenario, the role of fire in these potentially drier regions becomes a critical factor (Aragão et al., 2007).

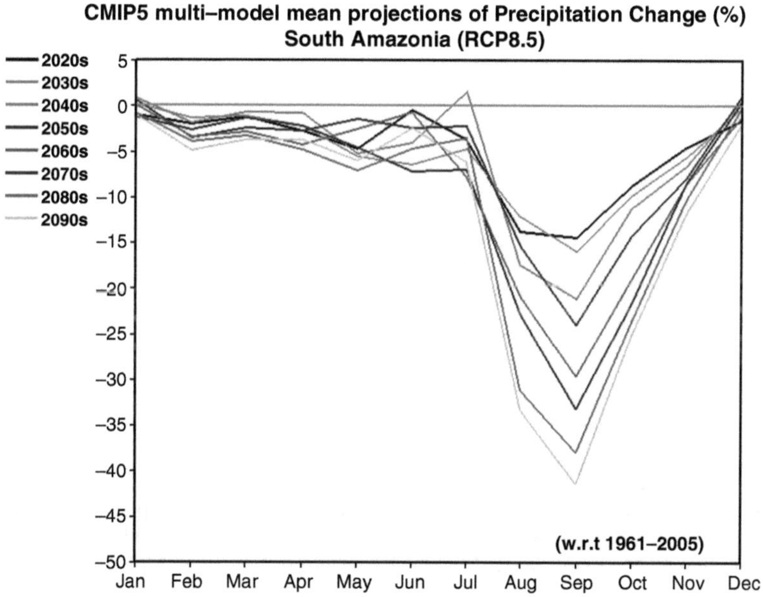

Fig. 8.2 CMIP5 multi-model ensemble projected change (%) in annual cycle of precipitation in southern Amazonia (70°W - 50°W, 12.5°S - 5°S) under the RCP8.5 scenario for decadal time slices for the period 2020–2100 relative to the 1961–2005 baseline

In general, projections of temperature and precipitation changes in southern Amazonia during SH summer (December–February), generated by CMIP5 multimodel ensemble under two different forcing scenarios (Fig. 8.3) show that this region under extreme conditions of temperature increase is subject to substantial changes in rainfall mainly in the RCP8.5 scenario. In such scenario, extreme droughts will become frequent such as the one observed in 2010, which caused drastic reduction in river water levels, as well as fires and damage to navigation and the local population (Borma et al., 2013). While deforestation is the most visible threat to the Amazon ecosystem, climate change is emerging as the most insidious threat to the region's future. Lengthening of the dry season and changes in the frequency and intensity of extreme drought episodes are probably the most critical factors for the Amazon. If the currently planned mitigation actions called for by the Paris Accord are not fully implemented, a warming of 4°C could occur as early as the 2070s, with a possible earlier date of approximately 2060 associated with the higher emissions scenarios and stronger carbon-cycle feedbacks (Betts et al., 2011). Such a warming level by 2100 would not be the end point: a further warming to levels over 6 °C would likely occur over the following century.

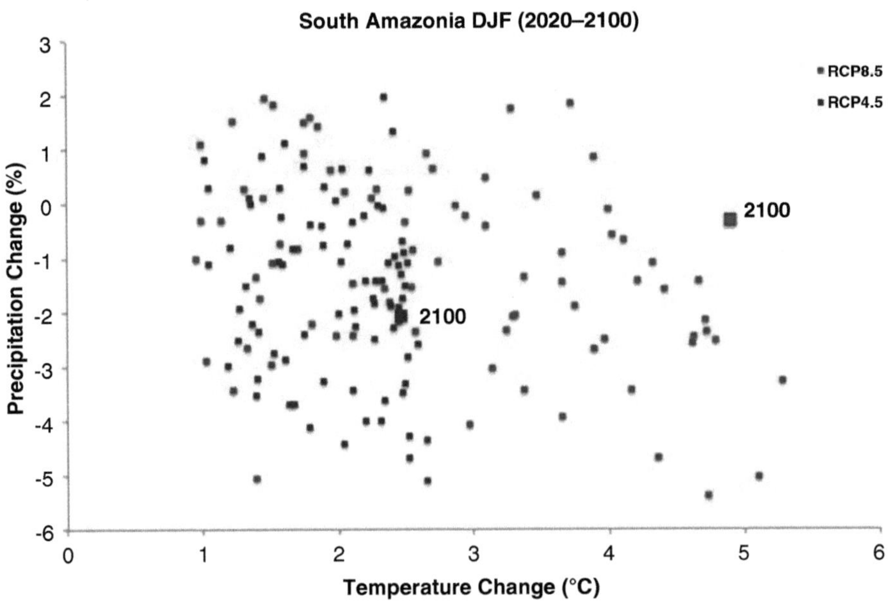

Fig. 8.3 Scatter diagram of interannual variability in change of mean temperature and precipitation for southern Amazonia region projected by CMIP5 multimodel ensemble under the RCP4.5 (*blue dots*) and RCP8.5 (*red dots*). Multimodel mean position by 2100 for each scenario is highlight

8.3 Impacts of Climate Change on the Biomes of South America

Field observations (e. g., Butt, de Oliveira, & Costa, 2011, Costa et al., 2003, Gash et al., 1996) and numerical studies (e.g., Nobre et al., 1991; Sampaio et al., 2007) revealed that large-scale deforestation in Amazonia could alter the regional climate significantly. Evapotranspiration is reduced and surface temperature is increased for pastures. That effect might lead to a 'savannization' of portions of the tropical forest domain. Oyama and Nobre (2003) showed that second stable biome-climate equilibrium with savannas covering eastern Amazonia and semi-deserts in Northeast Brazil may exist. However, there have been fewer studies on the impact of global climate change on South America, particularly on its biomes such as Cox et al. (2000), Malhi et al. (2009), Salazar, Nobre and Oyama (2007), Rammig et al. (2010). This session addresses this question by extending one calculation (Nobre et al., 2016) on how natural biomes could change in response to various scenarios of climate change.

The CPTEC Potential Vegetation Model version 2 (CPTEC-PVM2; Lapola et al., 2009) was used to study the possible alterations of South American biomes in response to the projected climate change scenarios of Figs. 8.1 and 8.2 for the end of this century, where the input data corresponds to the monthly climatology for years 2070–2100 of the CMIP5 multi-model ensemble of surface air temperature and temperature, under the RCP8.5 scenario. PVM2 is a model that describes reasonably well the large-scale distribution of world's biomes as a function of five climate variables and preserves the particularly good performance of its predecessor CPTEC PVM (Oyama & Nobre, 2003) for South American biomes. The model considers seasonality in precipitation as a determinant for the delimitation of tropical forests and savannas, and is able to account for varying atmospheric CO_2 concentration on plants primary productivity.

Simulations of equilibrium biome changes using CPTEC-PVM2 potential vegetation model forced with future scenarios of climate variables presented in Sect. 8.1 are shown in Fig. 8.4. Here we concentrate our results only on maps of potential biomes distribution derived from multi-model ensemble projections, and did not extend our analyses to explore the role of dry season length or different levels of CO_2 fertilization on dynamic processes, such as forest resilience. In these calculations, the CO2 fertilization effect was taken into consideration by 25%. They project extensive changes in biomes distribution over Amazonia. Fundamentally, it is expected expansion of tropical savannah and *caatinga* (dry shrubland) over areas currently occupied by tropical forests. Some of the initial questions that may arise are related to the likelihood of these results. The same questioning was already made and discussed in Lapola, Oyama, Nobre, and Sampaio (2008), where the

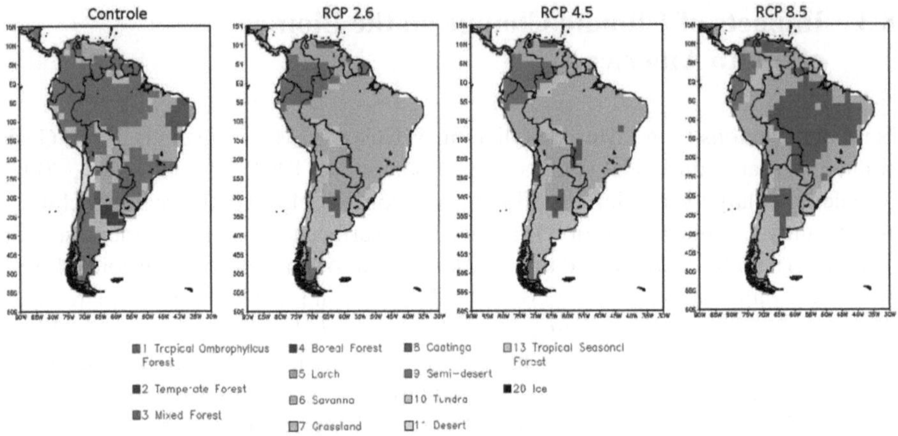

Fig. 8.4 Current and future distribution of equilibrium biomes in South America estimated with the potential vegetation model CPTEC-PVM2

authors examine the plausibility of this type of results in light of the CO_2 fertilization effect on stomata conductance, vegetation distribution and direct human disturbance.

As estimated by the potential vegetation maps simulated by Lapola et al. (2008), CO_2 fertilization has an overall effect of NPP enhancement in agreement with other studies. The estimated higher CO_2 concentration leading to changes in stomata conductance and an increase in water use efficiency also corroborates other studies. In the parameterizations used by the CPTEC-PVM2, the NPP seasonality thresholds for existence of tropical forests also agree with other studies showing forest retreat for dry season lengths longer than 4 months. As noted by the authors, the model does not account for direct human disturbance, which could enhance forest declining.

Taken together, the considerations above support the changes projected by the CPTEC-PVM2. In addition to that, the strength of the climate forcing that was considered is high. Four degrees Celsius in average for the region, above of the 1961–2005 baseline of observed temperature near the surface, represents a substantial change in the large-scale relations between climate and vegetation characteristics observed in present conditions. Given the size of this projected climate change, it makes sense that the results estimate such a large biome modification.

The hypothetical expansion of savanna and caatinga over areas of tropical forests would not directly eliminate all environmental services these areas can provide. For example, ecosystems in Brazilian savannas and caatinga are important for their biodiversity and other ecosystem services. However, as discussed in Nobre et al. (2016) the continued deforestation and the climate change will lead to high risks of irreversible change of biodiversity and ecosystems.

8.4 Exploring Likelihood of Impacts

Regardless of the possibility of a 4 °C increase in global warming, Amazonia is under threat also through the combined effects of unsustainable regional development and climate change. As summarized in the IPCC reports of 2007 and 2014, studies in the past 10 years have indicated that these changes can lead to deforestation, regional disturbance of temperatures and the water cycle, as well as loss of carbon stocks and biodiversity. In turn, these changes can lead to forest loss, droughts, low river levels, floods, loss of hydropower energy and many other ecosystem services and even enhanced risk of diseases and loss of agricultural productivity. We are only beginning to understand and quantify these threats individually.

By combining and developing integrated model frameworks and data-providing networks of different areas of knowledge, we should better understand how the system might respond to coupled global change drivers, and develop capacity to predict and identify the possible circumstances leading to loss of Amazonian ecosystem service provision. The AMAZALERT project (2011–2014, a consortium of 14 institutes in Europe and south-America, funded by European Union FP7 and national funds) applied multi-level integrated analysis to address the environmental and societal risks, to provide insights into the likelihood of future changes in vegetation, water and carbon cycles in the forests of Amazonia as a consequence of climate change and deforestation. The project has brought together a range of global climate predictions from the CMIP5 studies, improved several atmospheric and land surface models and combined them with new scenarios for regional land-use change to assess the likely impact on vegetation and water in the Amazon, in the twenty-first century.

The AMAZALERT project has found that when deforestation is kept to a minimum, projected changes in climate alone seem unlikely to bring about large- scale forest die-back. This conclusion was drawn from synthesizing different model results, including offline vegetation models and comparisons with coupled climate-vegetation simulations (e.g. Huntingford et al., 2013; Kay et al., 2014). The study reported by Kay et al., 2014 and later by Boulton, Booth, & Good, 2017, used model simulations of the forest changes until 2100 in response to three greenhouse gas concentration scenarios by an ensemble of the HadCM3C climate-carbon cycle model. Unlike the standard configuration of the model (Cox et al. 2000), the majority of the simulations showed resilient forest responses, but with increasing uncertainty in the forest state at 2100 with increasing severity of the emission scenarios (Boulton et al., 2017).

On average the forest is projected to remain largely unchanged in extent in the first two scenarios. In the third – and highest emission scenarios – scenario (RCP8.5), the mean average reduction in forest cover remains marginal but the risk of substantial loss is significant, whilst the probability of forest gain is minimal; in this last scenario, however there is much greater uncertainty in the extent of possible forest loss, ranging from almost no loss to 50% or more. However, as mentioned in Kay et al. (2014), "missing processes and biases (known and potential) in these climate-vegetation models are such that dieback is much harder to rule out than implied by these models alone. There are key uncertain processes, such as fire, CO_2 fertilization and regional rainfall dynamics, which could lead to substantial changes in model projections in the future. Further, the interactions between climate variability and change and land use change, particularly through fire, are likely to increase the probability of biome change, especially in regions such as the south and east of Amazonia that are already particularly vulnerable to these drivers of change."

Besides working with climate simulations, the AMAZALERT project also put a large effort to consult stakeholders and experts in a scenario participatory process, and held interviews and workshops to discuss specifically drivers and actions to decrease deforestation, in any of the giving scenarios. The main analyses concluded that in the worst case, with high deforestation, up to 50% of forests could disappear or degrade, which would promote drying effects on the regional climate. However, in the best case, forest degradation could be limited to the current 20% of Amazonia, with minimal effects on regional climate.

8.5 Hydrologic Change under Climate Change Scenarios

Brazilian basins holds 13% of the global availability for fresh water resources and 80% of these resources are located in Amazonia's basins (ANA, 2013). The large quantity of water resources also reflects in the hydroelectric power potential of the country. In this context, Brazil has one of the five largest potential for hydropower production and 40,5% of this potential is located in the Amazon's basin. Nowadays, only around 30% of this potential is exploited (ANEEL, 2008). However, the Amazon region plays an important role in affording the Brazilian future energy demand. The Growth Acceleration Program – PAC initiative of the Brazilian Government to accelerate economic growth has driven large investments for the construction of new hydropower plants in Amazonia (PAC, 2013), although the recent plans for construction of new hydropower plants in the Amazon has been downscaled and only a handful of new plants will be built (EPE, 2017).

Amazonia region is a hotspot for climate variability and change issues. Climate variability driven by surface temperature anomalies in Pacific and Atlantic Tropical oceans has been associated with severe drought and flood events in the Amazon basin (Borma et al., 2013; Marengo et al., 2008; Marengo, Tomasella, Soares, Alves, & Nobre, 2012; Ovando et al., 2016; Tomasella et al., 2010). Most of this large-scale droughts and floods have happened during the last 12 years, such as the 2005, 2010 and 2015–2016 droughts and the 2009, 2012 and 2014 floods, suggesting an intensification of the hydrological cycle in the region (Gloor et al., 2013). Although one cannot rule out that these events were due to natural climate variability, climate model projections suggest an intensification of frequency and intensity of these extreme events in the region, increasing vulnerability of natural and human systems (Cox et al., 2008; Marengo, Williams, Alves, Soares, & Rodriguez, 2016).

The Amazon River basin forest is being converted to either agricultural or pasture at fast rates (Trancoso et al., 2009). The effects of land use and land cover change locally interacts with the effects of climate variability and change modifying the hydrological response of basins. To assess the combined impacts of land use and land cover change and climate change on the basin hydrology, the hydrological models are integrated using potential future scenarios of land use and land cover (Aguiar et al., 2016; Soares-Filho et al., 2006) together with climate projections (Guimberteau et al., 2017; Jr Siqueira, Tomasella, & Rodriguez, 2015). Conversion from forest to either pasture or agriculture counterbalances the effects of climate change in the Madeira river basin (Jr Siqueira et al., 2015). Reduction in the evapotranspiration rate driven by land use and land cover change results in increasing discharges, which could overpass the effect of reducing precipitation, even resulting in increment of discharges.

Climate projections are translated into projections of hydrological change through the use of hydrological models. Hydrological models are implemented and calibrated for the basin and climate model simulations are used as input for the hydrological models after removing bias (Bárdossy & Pegram, 2011). In this approach, projections of future hydrology are compared to simulations of historical basin hydrology, in order to assess climate change impacts (Cloke, Wetterhall, He, Freer, & Pappenberger, 2013; Demaria, Maurer, Thrasher, Vicuña, & Meza, 2013; Guimberteau et al., 2017; Mohor, Rodriguez, Tomasella, & Júnior, 2015; Nobrega, Collischonn, & Tucci, 2011; Jr Siqueira et al., 2015).

Climate model projections are affected by irreducible uncertainties due mainly to model structure and lack of processes knowledge (Dessai & Hulme, 2004), which are propagated through the model chain (Jones, 2000). Such uncertainty drives a significant dispersion of hydrologic impact results (Nobrega et al., 2011; Demaria et al., 2013), which must be carefully assessed once probabilistic presentation of this information could misrepresent uncertainties and will lead to maladaptation (Hall, 2007). Besides the lack of agreement between projections, these results allow to carry out exploratory analyses of system's vulnerabilities (Dessai, Hulme, Lempert, & Pielke, 2009; Mohor et al., 2015) by analysing a strategy's robustness (Lempert & Schlesinger, 2000; Mohor et al., 2015). Following this approach, a better assessment of the impacts is carried out through an exploratory modelling approach, which is based in a number of model projections that offers a range of results (Bates, Kundzewicz, Wu, & Palutikof, 2008). The performance of the selected strategy is then analysed under the different scenarios identifying potential weakness (Mohor et al., 2015).

Projected hydrological changes for the Amazon basin have been assessed in literature using several climate change scenarios and different climate and hydrological models, and results show large dispersion. Projections for the Madeira basin (Jr Siqueira et al., 2015) using the Distributed Hydrological Model of National Institute for Space Research – MHD-INPE (Rodriguez & Tomasella, 2016) suggest that under climate change scenarios extreme discharges in the basin will be affected. In general there is a reduction in lower and higher discharges, however the magnitude of change differs between models (Jr Siqueira et al., 2015). The impacts of climate change in the Tapajós basin's discharges results in the decline of basin's discharges for most of the scenarios (Mohor et al., 2015). However, one of the projections suggests discharges during the wet season will be significantly increased. Also, for the Xingú basin the hydrological change under climate change conditions will reduce the discharges in the basin (da Cruz Junior, Rodrigues, & Lopes, 2017). Similar results with those of the southern tributaries of Amazon River were found in the Purus basin, in western Amazonia (Dalagnol, Borma, Mateus, & Rodriguez, 2017). Projections suggest a dry trend in the discharges of the basin, with gradual changes during the wet season and abrupt changes in the dry season.

A scenario at Madeira, Tapajós and Xingú rivers suggests climate change is likely to reduce discharges in these basins (Fig. 8.5). Such reduction is largely related to decreasing precipitation and increasing temperature, which overlapping effects enhance discharge reduction. In general, for the scenarios considered in these hydrological simulations, a larger decreasing precipitation scenario also has larger increments in temperature, which explains the rates of change in discharges (Mohor et al., 2015). Results suggest that for larger temperature increments, larger than 4 °C, discharges are more sensible to precipitation changes than for lower increments. However, climate sensibility largely varies between basins, affected by surface characteristics and basin's scale (Mohor et al., 2015).

However, the results showed in Fig. 8.5 do not consider changes in land use and vegetation cover to assess the water balance in the basins. Hydrologic projections considering conversion of the tropical forest to pasture and farming were carried out by Jr Siqueira et al. (2015) and Guimberteau et al. (2017), applying potential scenarios for land use and land cover change in Amazonian basins. As expected, the

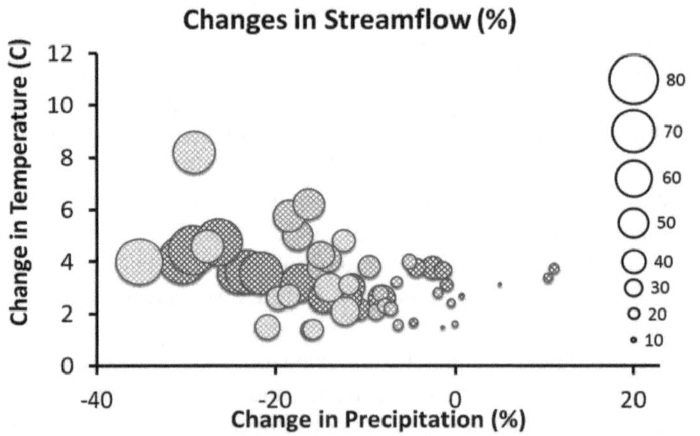

Fig. 8.5 Projected climate change impacts in discharges at streamflow gauge stations, relative to the historic period (1970–1990), in the Madeira (*Green hatched area*), Tapajós (*Black hatched area*) and Xingu (*Orange hatched area*) rivers as a function of precipitation and temperature changes. Blue/Red circle's borders indicate increasing/decreasing discharges. The size of the circle is proportional to absolute change in discharges

increasing deforestation in the basins results in lower rates of evapotranspiration and higher runoff generation, which drive the counterbalance of climate change effects in streamflow.

8.6 Conclusion Remarks

Future climate scenarios corresponding to the CMIP5 RCP8.5 high emission scenarios project progressively higher warming that may exceed 4 °C in Amazonia in the second half of the century, particularly during the dry season in the region. Associated with these scenarios, the amount and seasonal distribution of precipitation are also expected to change. In general, it is projected a reduction of precipitation year-round, being a substantial reduction predominantly in the dry and transition seasons, from July to November, and smaller reductions of the order of 5% for the SH summer, from December to March. Evaluating the consequences of such substantial climatic change, several negative effects in Amazonia can be anticipated, including short-term hydrological changes similar to the events associated to the extreme 2005 and 2010 droughts, and longer time-scale modifications of broad scale characteristics such as different biome distribution.

Based on hydrological models, it is generally expected a reduction in river discharges associated to precipitation decreases and temperature increases brought about by projected climate change, but with the magnitude of the changes differing between models.

Based on the CPTEC-PVM2, future climate change scenarios imply important changes in biomes distribution over Amazonia, with potential expansion of savannah and caatinga over large areas currently occupied by tropical forests. Considering factors like the effect of CO_2 on fertilization and stomata conductance, direct human disturbance on the ground, and the size of the expected climate forcing, these hypothetical large biome modifications make sense and can lead to high risks of irreversible change of biodiversity and ecosystems.

Numerous studies show that a world in which warming reaches 4°C above pre-industrial levels is so different from the current one that it comes with high uncertainty. In addition, they point to a new world where unprecedented extreme climate events might have serious impacts on human systems, ecosystems, and associated services. More specifically, this level of warming could bring increased frequency of flooding and droughts, risks for food production and irreversible loss of biodiversity. In terms of future biome distribution in Amazonia the synergistic combination of impacts due to both land cover and climate changes points out to 'savannization' of portions of its tropical forests. It is necessary a reduction to nearly zero in tropical deforestation and reducing land-cover emissions and mitigating climate change to avoid a dangerous interference with the ability of natural ecosystems to adapt to these possible changes.

References

AGÊNCIA NACIONAL DE ÁGUAS. (2013). Plano estratégico de recursos hídricos dos afluentes da margem direita do Rio Amazonas: diagnóstico. (http://margemdireita.ana.gov.br/autentica.asp?id_arquivo=128).

Aguiar, A. P. D., Vieira, I. C. G., Assis, T. O., Dalla-Nora, E. L., Toledo, P. M., Oliveira Santos-Junior, R. A., et al. (2016). Land use change emission scenarios: Anticipating a forest transition process in the Brazilian Amazon. *Global Change Biology, 22*(5), 1821–1840.

Anderson, L. O., Aragão, L. E., Gloor, M., Arai, E., Adami, M., Saatchi, S. S., et al. (2015). Disentangling the contribution of multiple land covers to fire-mediated carbon emissions in Amazonia during the 2010 drought. *Global Biogeochemical Cycles, 29*(10), 1739–1753.

ANEEL. (2008). Atlas de energia elétrica do Brasil / Agência Nacional de Energia Elétrica.3. ed. – Brasília : Aneel, 2008. 236 p. ISBN: 978-85-87491-10-7.

Aragao, L. E. O. C., Anderson, L. O., & Fonseca, M. G. et al. (2017). 21st century drought-related fires counteract the decline of Amazon deforestation carbon emissions. Nature Communications ISSN 2041-1723 (In Press).

Aragão, L. E. O. C., Malhi, Y., Barbier, N., Lima, A., Shimabukuro, Y., Anderson, L., et al. (2008). Interactions between rainfall, deforestation and fires during recent years in the Brazilian Amazonia. *Philosophical Transactions of the Royal Society of London. Series B, Biological Sciences, 363*(1498), 1779–1785. https://doi.org/10.1098/rstb.2007.0026

Aragão, L. E. O. C., Malhi, Y., Roman-Cuesta, R. M., Saatchin, S., Anderson, L. O., & Shimabukuro, Y. E. (2007). Spatial patterns and fire response of recent Amazonian droughts. *Geophysical Research Letters, 34*, L07701. https://doi.org/10.1029/2006GL028946

Ballantyne, A. P., Alden, C. B., Miller, J. B., Tans, P. P., & White, J. W. C. (2012). Increase in observed net carbon dioxide uptake by land and oceans during the past 50 years. *Nature, 488*(7409), 70–72. https://doi.org/10.1038/nature11299

Bárdossy, A., & Pegram, G. (2011). Downscaling precipitation using regional climate models and circulation patterns toward hydrology. *Water Resources Research, 47*, W04505.

Barthem, R. B., et al. (2004). Amazon Basin – GIWA regional assessment 40b. (Univ Kalmar, Kalmar, Sweden).

Bates, B. C., Kundzewicz, Z. W., Wu, S., & Palutikof, J. P. (2008). *Climate change and water. Technical paper of the intergovernmental panel on climate change* (p. 210). Geneva, Switzerland: IPCC Secretariat.

Betts, R. A., Collins, M., Hemming, D. L., Jones, C. D., Lowe, J. A., & Sanderson, M. G. (2011). *When could global warming reach 4°C? Philosophical Transactions of the Royal Society A - Mathematical Physical and Engineering Sciences, 369*, 67–84.

Betts, R. A., Malhi, Y., & Roberts, J. T. (2008). The future of the Amazon: New perspectives from climate, ecosystem and social sciences. *Philosophical Transactions of the Royal Society of London. Series B, Biological Sciences, 363*(1498), 1729–1735. https://doi.org/10.1098/rstb.2008.0011

Borma, L. S., Nobre, C. A., & Cardoso, M. F. (2013). Response of the Amazon tropical forests to deforestation, climate, and extremes, and the occurrence of drought and fire. In *Climate vulnerability: Understanding and addressing threats to essential resources* (Vol. 2, pp. 153–163). https://doi.org/10.1016/B978-0-12-384703-4.00228-8

Borma, L. S., et al. (2013). Response of the Amazon tropical forests to deforestation, climate, and extremes, and the occurrence of drought and fire. In R. Pielke (Ed.), *Vulnerability of food resources to climate. Climate vulnerability* (Vol. 2, pp. 153–163). New York: Elsevier.

Boulton, C. A., Booth, B. B. B., & Good, P. (2017). Exploring uncertainty of Amazon dieback in a perturbedparameter earth system ensemble. *Global Change Biology, 23*, 5032–5044. https://doi.org/10.1111/gcb.13733

Brienen, R. J. W., Phillips, O. L., Feldpausch, T. R., Gloor, E., Baker, T. R., Lloyd, J., et al. (2015). Long-term decline of the Amazon carbon sink. *Nature, 519*(7543), 344–348.

Butt, N., de Oliveira, P. A., & Costa, M. H. (2011). Evidence that deforestation affects the onset of the rainy season in Rondonia, Brazil. *Journal of Geophysical Research, 116*, D11120. https://doi.org/10.1029/2010JD015174

Cloke, H. L., Wetterhall, F., He, Y., Freer, J. E., & Pappenberger, F. (2013). Modelling climate impact on floods with ensemble climate projections. *Quarterly Journal of the Royal Meteorological Society, 139*(671), 282–297.

Costa, M. H., & Foley, J. A. (1999). Trends in the hydrologic cycle of the Amazon Basin. *Journal of Geophysical Research, 104*, 14189–14198.

Costa, M. H., et al. (2003). Effects of large-scale changes in land cover on the discharge of the Tocantins River, southeastern Amazonia. *Journal of Hydrology (Amsterdam), 283*(1–4), 206–217.

Cox, P. M., Betts, R. A., Jones, C. D., Spall, S. A., & Totterdell, I. J. (2000). Acceleration of global warming due to carbon-cycle feedbacks in a coupled climate model. *Nature 408*, 184–187.

Cox, P. M., Harris, P. P., Huntingford, C., Betts, R. A., Collins, M., Jones, C. D., et al. (2008). Increasing risk of Amazonian drought due to decreasing aerosol pollution. *Nature (London), 453*, 212–215.

Curtis, P. S., & Wang, X. (1998). A meta-analysis of elevated CO2 effects on woody plant mass, form, and physiology. *Oecologia, 113*, 299. https://doi.org/10.1007/s004420050381

da Cruz Junior, W. L. T., Rodrigues, D. A., & Lopes, L. G. (2017). Modelagem dos impactos das mudanças climáticas na operação dos reservatórios da bacia do rio Xingu. Relatório final PIBIC-IC - http://mtc-m21b.sid.inpe.br/col/sid.inpe.br/mtcm21b/2017/08.10.13.05/doc/Wellington%20Luis%20Teodoro%20da%20Cruz%20Junior.pdf

da Silva, R. R., Werth, D., & Avissar, R. (2008). Regional impacts of future land-cover changes on the Amazon basin wet-season climate. *Journal of Climate, 21*(6), 1153–1170. https://doi.org/10.1175/2007JCLI1304.1

Dalagnol, R., Borma, L. D. S., Mateus, P., & Rodriguez, D. A. (2017). Assessment of climate change impacts on water resources of the Purus Basin in the southwestern Amazon. *Acta Amazonica, 47*(3), 213–226.

DeLucia, E. H., Moore, D. J., & Norby, R. J. (2005). Contrasting responses of forest ecosystems to rising atmospheric CO2: Implications for the global C cycle. *Global Biogeochemical Cycles, 19*(3), 1–9. https://doi.org/10.1029/2004GB002346

Demaria, E. M. C., Maurer, E. P., Thrasher, B., Vicuña, S., & Meza, F. J. (2013). Climate change impacts on an alpine watershed in Chile: Do new model projections change the story? *Journal of Hydrology, 502*, 128–138.

Dessai, S., & Hulme, M. (2004). Does climate adaptation policy need probabilities? *Climate Policy, 4*(2), 107–128.

Dessai, S., Hulme, M., Lempert, R., & Pielke, R. (2009). Do we need better predictions to adapt to a changing climate? *Eos, Transactions American Geophysical Union, 90*(13), 111–112.

Eltahir, E. A. B., & Bras, R. L. (1994). Precipitation recycling in the Amazon Basin. *Quarterly Journal of the Royal Meteorological Society, 120*, 861–880.

EPE. (2017). Plano Decenal de Expansão de Energia 2026 / Ministério de Minas e Energia. Empresa de Pesquisa Energética. Brasília: MME/EPE, 2017 - http://www.mme.gov.br/documents/10584/0/PDE2026_versao_para_publicacao_08dez2017.pdf/f5d8f999-566d-4f5b-9167-d80b6ea8924e

Feldpausch, T. R., Lloyd, J., Lewis, S. L., Brienen, R. J. W., Gloor, M., Monteagudo Mendoza, A., et al. (2012). Tree height integrated into pantropical forest biomass estimates. *Biogeosciences, 9*(8), 3381–3403. https://doi.org/10.5194/bg-9-3381-2012

Fu, R., & Li, W. (2004). The influence of the land surface on the transition from dry to wet season in Amazonia. *Theoretical and Applied Climatology, 78*(1–3), 97–110. https://doi.org/10.1007/s00704-004-0046-7

Gash, J., et al. (1996). *Amazonian deforestation and climate*. New York: Wiley.

Gloor, M. R. J. W., Brienen, D., Galbraith, T. R., Feldpausch, J., Schöngart, W., Guyot, J. L., et al. (2013). Intensification of the Amazon hydrological cycle over the last two decades. *Geophysical Research Letters, 40*, 1729–1733. https://doi.org/10.1002/grl.50377

Good, P., Jones, C., Lowe, J., Betts, R., Booth, B., & Huntingford, C. (2011). Quantifying environmental drivers of future tropical forest extent. *Journal of Climate, 24*(5), 1337–1349. https://doi.org/10.1175/2010JCLI3865.1

Guimberteau, M., Ciais, P., Ducharne, A., Boisier, J. P., Aguiar, A. P. D., Biemans, H., et al. (2017). Impacts of future deforestation and climate change on the hydrology of the Amazon Basin: A multi-model analysis with a new set of land-cover change scenarios. *Hydrology and Earth System Sciences, 21*(3), 1455.

Hall, J. (2007). Probabilistic climate scenarios may misrepresent uncertainty and lead to bad adaptation decisions. *Hydrological Processes, 21*, 1127. https://doi.org/10.1002/hyp.6573

Hubbell, S. P., et al. (2008). Colloquium paper: How many tree species are there in the Amazon and how many of them will go extinct? *Proceedings of the National Academy of Sciences of the United States of America, 105*(Suppl 1), 11498–11504.

Huntingford, C., Zelazowski, P., Galbraith, D., Mercado, L. M., Sitch, S., Fisher, R., et al. (2013). Simulated resilience of tropical rainforests to CO2-induced climate change. *Nature Geoscience, 6*(4), 268–273.

Jipp, P. H., Nepstad, D. C., Cassel, D. K., & Reis De Carvalho, C. (1998). Deep soil moisture storage and transpiration in forests and pastures of seasonally-dry Amazonia. *Climatic Change, 39*, 395–412. https://doi.org/10.1023/A:1005308930871

Jones, C., Lowe, J., Liddicoat, S., & Betts, R. (2009). Committed terrestrial ecosystem changes due to climate change. *Nature Geoscience, 2*, 484–487. https://doi.org/10.1038/ngeo555

Jones, R. N. (2000). Managing uncertainty in climate change projections: Issues for impact assessment. *Climatic Change, 45*, 403–419.

Jr Siqueira, L., Tomasella, J., & Rodriguez, D. A. (2015). Impacts of future climatic and land cover changes on the hydrological regime of the Madeira river basin. *Climatic Change, 129*, 117–129. https://doi.org/10.1007/s10584-015-1338-x

Kay G., Alves L., Boisier J. P., Boorman P., Boulton C., Ciais P., et al. (2014). Report on estimated likelihood for irreversible collapse. Project "Raising the alert about critical feedbacks between climate and land use change in Amazonia (Amazalert)". European Commission's 7th Framework Programme, www.eu-amazalert.org

Lapola, D. M., Oyama, M. D., & Nobre, C. A. (2009). Exploring the range of climate biome projections for tropical South America: The role of CO2 fertilization and seasonality. *Global Biogeochemical Cycles, 23*(3), GB3003. https://doi.org/10.1029/2008GB003357

Lapola, D. M., Oyama, M. D., Nobre, C. A., & Sampaio, G. (2008). A new world natural vegetation map for global change studies. *Anais da Academia Brasileira de Ciências, 80*(2), 397–408.

Le Quéré, C., Andres, R. J., Boden, T., Conway, T., Houghton, R. A., House, J. I., et al. (2013). The global carbon budget 1959–2011. *Earth System Science Data, 5*(1), 165–185 https://doi.org/10.5194/essd-5-165-2013

Lempert, R. J., & Schlesinger, M. E. (2000). Robust strategies for abating climate change. *Climatic Change, 45*, 387–401.

Lenton, T. M., Held, H., Kriegler, E., Hall, J. W., Lucht, W., Rahmstorf, S., et al. (2008). Tipping elements in the Earth's climate system. *Proceedings of the National Academy of Sciences of the United States of America, 105*(6), 1786–1793. https://doi.org/10.1073/pnas.0705414105

Lewinshohn T. M., Prado P. I. (2002). Biodiversidade Brasileira. Síntese do estado atual do conhecimento (Contexto, São Paulo, Brazil).

Li, Y., Zhao, M., Motesharrei, S., Mu, Q., Kalnay, E., & Li, S. (2015). Local cooling and warming effects of forests based on satellite observations. *Nature Communications, 6*(6603), 8. https://doi.org/10.1038/ncomms7603

Malhi, Y., Roberts, J. T., Betts, R. A., Killeen, T. J., Li, W., & Nobre, C. A. (2008). Climate change, deforestation, and the fate of the Amazon. *Science (New York), 319*(5860), 169–172. https://doi.org/10.1126/science.1146961

Malhi, Y., et al. (2009). Exploring the likelihood and mechanism of a climate-change-induced dieback of the Amazon rainforest. *Proceedings of the National Academy of Sciences, 106*(49), 20610–20615

Marengo, J. A. (2004). Inter-decadal variability and trends in rainfall in the Amazon basin. *Theoretical and Applied Climatology, 78*, 79–96.

Marengo, J. A. (2005). The characteristics and variability of the atmospheric water balance in the Amazon basin: Spatial and temporal variability. *Climate Dynamics., 24*, 11–22.

Marengo, J. A., Miller, J. A., Russell, G., Rosenzweig, C., & Abramopoulos, F. (1994). Calculations of river-runoff in the GISS GCM: Impact of a new land surface parameterization and runoff routing on the hydrology of the Amazon River. *Climate Dynamics, 10*, 349–361.

Marengo, J. A., Nobre, C. A., Tomasella, J., Oyama, M. D., Oliveira, G. S., Oliveira, R., et al. (2008). The drought of Amazonia in 2005. *Journal of Climate, 21*, 495–516.

Marengo, J. A., Tomasella, J., Soares, W. R., Alves, L. M., & Nobre, C. A. (2012). Extreme climatic events in the Amazon basin. *Theoretical and Applied Climatology, 107*, 73–85.

Marengo, J. A., Williams, E. R., Alves, L. M., Soares, W. R., & Rodriguez, D. A. (2016). Extreme seasonal climate variations in the Amazon basin: droughts and floods. In *Interactions between biosphere, atmosphere and human land use in the Amazon Basin* (pp. 55–76). Berlin/Heidelberg, Germany: Springer.

Mohor, G. S., Rodriguez, D. A., Tomasella, J., & Júnior, J. L. S. (2015). Exploratory analyses for the assessment of climate change impacts on the energy production in an Amazon run-of-river hydropower plant. *Journal of Hydrology: Regional Studies, 4*, 41–59.

Molinier, M., Guyot, J. L., De Oliveira, E., & Guimarães, V. (1996). Les regimes hydrologiques de l'Amazone et de ses affluents. *IAHS Publication, 238*, 209–222.

Nepstad, D., Lefebvre, P., Da Silva, U. L., Tomasella, J., Schlesinger, P., Solorzano, L., et al. (2004). Amazon drought and its implications for forest flammability and tree growth: A basin-wide analysis. *Global Change Biology, 10*(5), 704–717.

Nobre, C. A., & Borma, L. S. (2009). Tipping points for the Amazon forest. *Current Opinion in Environmental Sustainability, 1*, 28–36. https://doi.org/10.1016/j.cosust.2009.07.003

Nobre, C. A., Sampaio, G., Borma, L., Castilla-Rubio, J. C., Silva, J. S., & Cardoso, M. (2016). The fate of the Amazon forests: Land-use and climate change risks and the need of a novel sustainable development paradigm. *Proceedings of the National Academy of Sciences*. www.pnas.org/cgi/doi/10/1073/pnas.1605516113, *113*, 10759.

Nobre, C. A., Sellers, P. J., & Shukla, J. (1991). Amazonian deforestation and regional climate change. *Journal of Climate, 4*(10), 957–988 https://doi.org/10.1175/1520-0442(1991)004<0957:ADARCC>2.0.CO;2

Nobrega, M. T., Collischonn, W., Tucci, C. E. M., & Paz, A. R. (2011). Uncertainty in climate change impacts on water resources in the Rio Grande Basin, Brazil. *Hydrology and Earth System Sciences, 15*, 585–595 www.hydrol-earth-syst-sci.net/15/585/2011/. https://doi.org/10.5194/hess-15-585-2011

Norby, R. J., DeLucia, E. H., Gielen, B., Calfapietra, C., Giardina, C. P., King, J. S., et al. (2005). Forest response to elevated CO2 is conserved across a broad range of productivity. *Proceedings of the National Academy of Sciences, 102*(50), 18052–18056. https://doi.org/10.1073/pnas.0509478102

Ovando, A., Tomasella, J., Rodriguez, D. A., Martinez, J. M., Siqueira-Junior, J. L., Pinto, G. L. N., et al. (2016). Extreme flood events in the Bolivian Amazon wetlands. *Journal of Hydrology: Regional Studies, 5*, 293–308.

Oyama, M. D., & Nobre, C. A. (2003). A new climate-vegetation equilibrium state for tropical South America. *Geophysical Research Letters, 30*(23), 2199. https://doi.org/10.1029/2003GL018600

Pan, Y., Birdsey, R. A., Fang, J., Houghton, R., Kauppi, P. E., Kurz, W. A., et al. (2011). A large and persistent carbon sink in the world's forests. *Science (New York), 333*(6045), 988–993. https://doi.org/10.1126/science.1201609

Phillips, O. L., Aragão, L. E. O. C., Lewis, S. L., Fisher, J. B., Lloyd, J., López-González, G., et al. (2009). Drought sensitivity of the Amazon rainforest. *Science (New York), 323*(5919), 1344–1347. https://doi.org/10.1126/science.1164033

Prentice, I., Farquhar, G., & Fasham, M. (2001). The carbon cycle and atmospheric carbon dioxide. *Climate Change 2001: The Scientific Basis, 57*, 183–237. https://doi.org/10.1256/004316502320517344
Programa de Aceleração do Crescimento—PAC. (2013). Ministério do Planejamento. Governo Federal. <http://www.pac.gov.br/pub/up/relatorio/ffaff442f57973143fbd0a6c39b8ae69.pdf>.
Rammig, A., et al. (2010). Estimating the risk of Amazonian forest dieback. *The New Phytologist, 187*, 694–706.
Roads, J., Kanamitsu, M., & Stewart, R. (2002). CSE water and energy budgets in the NCEP-DOE Reanalyses. *Journal of Hydrometeorology, 3*, 227–248.
Rodriguez, D. A., & Tomasella, J. (2016). On the ability of large-scale hydrological models to simulate land use and land cover change impacts in Amazonian basins. *Hydrological Sciences Journal, 61*(10), 1831–1846.
Salati, E., Dall'Olio, A., Matsu, E., & Gat, J. R. (1979). Recycling of water in the Amazon Basin : An isotopic study. *Water Resources Research, 15*(5), 1250–1258.
Salati, E., & Vose, P. B. (1984). Amazon basin: A system in equilibrium. *Science (New York), 225*(4658), 129–138. https://doi.org/10.1126/science.225.4658.129
Salazar, L. F., Nobre, C. A., & Oyama, M. D. (2007). Climate change consequences on the biome distribution in tropical South America. *Geophysical Research Letters, 34,* L09708.
Sampaio, G., et al. (2007). Regional climate change over eastern Amazonia caused by pasture and soybean cropland expansion. *Geophysical Research Letters, 34.*
Scholze, M., Knorr, W., Arnell, N. W., & Prentice, I. C. (2006). A climate-change risk analysis for world ecosystems. *Proceedings of the National Academy of Sciences of the United States of America, 103*(35), 13116–13120. https://doi.org/10.1073/pnas.0601816103
Shukla, J., Nobre, C., & Sellers, P. (1990). Amazonia deforestation and climate change. *Science, 247*, 1322–1325.
Shuttleworth, W. J. (1988). Evaporation from Amazonian rainforest. *Proceedings of the Royal Society B: Biological Sciences, 233*, 321–346.
Soares-Filho, B. S., Nepstad, D. C., Curran, L. M., Cerqueira, G. C., Garcia, R. A., et al. (2006). Modeling conservation in the Amazon Basin. *Nature, 440*, 520–523. https://doi.org/10.1038/nature04389
Taylor, K. E., et al. (2012). An overview of CMIP5 and the experiment design. *Bulletin of the American Meteorological Society, 93*(4), 485–498.
Tomasella, J., Borma, L. S., Marengo, J. A., Rodriguez, D. A., Cuartas, L. A., Nobre, C. A., et al. (2010). The droughts of 1996-1997 and 2004-2005 in Amazonia: Hydrological response in the river main-stem. *Hydrological Processes, 25*, 1228–1242.
Trancoso, R., Carneiro Filho, A., Tomasella, J., Schietti, J., Forsberg, B. R., & Miller, R. P. (2009). Deforestation and conservation in major watersheds of the Brazilian Amazon. *Environmental Conservation, 36*(4), 277–288.
Vorosmarty, C., Willmott, C., Choudhury, B., Schloss, A., Stearsns, T., Robertson, S., et al. (1996). Analysing the discharge regime of a large tropical river trough remote sensing, ground climatic data, and modeling. *Water Resources Research, 32*, 3137–3150.
Zeng, N. (1999). Seasonal cycle and interannual variability in the Amazon hydrologic cycle. *Journal of Geophysical Research, 104*, 9097–9106.

Chapter 9
Final Remarks and Recommendations

Jose A. Marengo, Carlos A. Nobre, Wagner R. Soares, and Ana P. Soares

9.1 Final Remarks

Mark Lynas laid out what would happen as the temperature rises in his award-winning book, *Six Degrees: Our Future on a Hotter Planet.* According to the best scientific estimate by IPCC AR5, if humans carry on with a "business as usual" approach using large amounts of fossil fuels, the Earth's average temperature will rise by between 2.6 and 4.8 °C above pre-industrial levels by 2100. 4 °C may not sound like much. But the world was only about 4°C to 7 °C cooler, on average, during the last ice age, when large parts of Europe and the United States were covered by glaciers. In 2014, the World Bank report on projected effects of 4 °C warming showed a growing number of studies that suggest that global food production could take a big hit under 3 °C or 4 °C of warming. Poorer countries like Bangladesh, Egypt, Vietnam, and parts of Africa and South America could see large tracts of farmland made unusable by rising seas. Perhaps most significantly, the World Bank report wasn't even sure if humanity could adapt to a 4 °C world. However, uncertainty remains about climate projections and the full nature and scale of impacts.

In the next decades, global warming and consequently, climate change will probably alter human and natural systems significantly. Variability and vulnerability limits will be pushed beyond historical values, leading to significant changes in what are considered to be normal conditions. Identifying places, periods and characteristics

J. A. Marengo (✉) · W. R. Soares
National Center for Monitoring and Early Warning of Natural Disasters (CEMADEN), São Paulo, Brazil
e-mail: jose.marengo@cemaden.gov.br

C. A. Nobre
Institute for Advanced Studies, University of São Paulo, São Paulo, Brazil

A. P. Soares
Center for Environmental Studies and Research, State University of Campinas (UNICAMP), São Paulo, Brazil

C. A. Nobre et al. (eds.), *Climate Change Risks in Brazil*,
https://doi.org/10.1007/978-3-319-92881-4_9

of these impacts to a certain level of global warming allows for appropriate adaptation strategies to be developed, or may motivate decisions aimed at mitigating climate change. However, these impacts are important, as they may alter effects, restrict response options, leading to indirect consequences in other regions, increasing adaptation challenges a lot.

IPCC-AR5 considers all possible result ranges, not just including high probability results, but also ones with low chance, but with greater consequences. In addition, it looks onto climate alterations as a risk management challenge, paving the way to a wide range of possible solutions. However, it is important to point out that climate change is a risk management challenge, which may be substantially reduced with mitigation. Due to the strong link between emission, energy use and economic activities, decarbonisation efforts will inevitably affect the global economy. Economic indicators relevant to mitigation challenges include carbon prices, which quantify the marginal cost of reducing emission in short and long term mitigation costs.

This book assessed sector risks for Brazil, based on o a regional warming of 4°C temperature increase predictions, as well as the impacts of scenarios linked to this rise in four key sectors for the country: agriculture, human health, biodiversity and energy and for the Northeast Brazil and Amazonia. Higher risks are usually seen in more pessimistic climate scenarios like RCP 8.5. In this scenario, Brazil has a high probability (over 70 per cent) of suffering temperature rises of over 4 °C before the end of the century. Nonetheless, it is important to point out that the scenario considered today as the most pessimistic, is nothing more than the current standard of greenhouse gas emissions. The 2014–2015 drought in Southeast Brazil was considered the fifth most costly disaster in the world in 2014 by Munich-Re, Germany's oldest reinsurance firm. Losses were estimated in 5 billion US dollars. The summer of 2014 was the warmer in 60 years and rainfall we below 50% normal. This shows that Brazil is vulnerable to the current extreme climate variability, which could worsen in the future due to global warming and lead to different impacts regionally. For the 4 °C or above warming provided by the RCP8.5 scenario, the semiarid Northeast Brazil could become arid and these dry conditions could lead to an irreversible desertification process.

The risk of current and future analysis of a 4 °C warming represent an important aspect for Brazilian agriculture, as it is an activity directly affected by the increase in CO_2 concentration and more often extreme phenomena. The impacts of climate change on agriculture bring new challenges for the future: where it will be best to plant and what strategies will be employed to ensure that productivity is not lost? Soybean, maize, coffee and beans already show a fall in yield because of climate change. Brazil produces a lot of food and may still keep this position in the world, using techniques like producing seeds more resistant to droughts (like for example, based on the genetic diversity found in the Cerrado) and low carbon agriculture. Climate change impacts lead to the need to plan in order to mitigate the effects of possible alterations to Brazil's agro-climatic risk zoning. Negative impacts resulting from a 4°C increase in temperature outweigh the positive effects in agriculture. It is important that policy makers take into account adapted crops, developed to resist higher temperatures, in order to assist decisions made by farmers. For this sector, the present report points out that: 1) climate impacts on agriculture are highly specific to a place; 2) understanding climate risks correctly, associated to potential impacts on agriculture and food security (vulnerabilities and efficacy of adaptation options) is essential to enable adaptation to climate change; 3) temperature increase equal to or above 4 °C means higher evapotranspiration and demand on the water system.

Climate change is among the most probable threats to public health in the coming years. Analysing current and future consequences of temperature rise above 4 °C points to a great challenge for the sector. Being able to predict the consequences of a temperature rise above the expected rate may improve the health sector's capacity to deal with emergencies and extreme risks associated to climate change. Assessing the effects on human health requires an interdisciplinary approach of healthcare professionals, as well as climatologists, social scientists, biologists, physicists, chemists, epidemiologists, among others in order to understand the relation between social, economic, biological, ecological and physical systems, as well as their relationship with extreme climate change. Intersectorial actions and programmes are necessary to reduce socio-demographic and economic vulnerabilities, as well as access to health services that could diminish the impacts of extreme temperature rise, particularly for diseases transmitted by non-treated water and contaminated food. Extreme climate and environmental conditions, with temperature rising above 4°C may influence human behaviour, due to the direct or indirect need of populations to adapt to these severe risks. Therefore, new lifestyles, daily practices, eating habits, difficulty in relation to access to drinking water may change, resulting in new nosological profiles. In certain situations, some regions not being able to adapt to high temperature rises may favour forced human migration. In addition to social, demographic and economic consequences, an increase in diseases like dengue, malaria, zika and chikungunya might be seen. Indeed, poorer regions in most of Brazil can experience the effects related to the unavailability of treated water, food contamination and lack of personal hygiene, with cases of gastroenteritis and hepatitis A, as well as vector-borne diseases like dengue. For the states of Espírito Santo and Rio de Janeiro, with the average temperature increase predicted for 4.6 and 5.1 °C, a rise in dengue's epidemic potential was seen for 2071–2090.

In relation to biodiversity, the impact of an increase equal to or above 4°C in the average temperature may lead to savannisation and impoverishment of forests; extinction and changes in the distribution pattern of native species with edible and cultural value in the Cerrado are predicted, leading to socio-economic problems in 2080 already; socio-economic impacts resulting from the reduction in the Atlantic Forest native bee species, essential to pollination for agricultural and native species, would already be seen in 2030, getting worse in 2050 and 2080; in 2100, Brazil would lose 200 days a year in suitable plant growing days, causing great impacts on biodiversity, productivity of ecosystems and the economy. In 2100, biodiversity loss on tropical coasts, including Brazil's, will be significant in relation to the impacts on food and the economy. Considering a scenario with a temperature increase of 4 °C or more, biodiversity conservation areas have to start being prioritised now (particularly the western Amazon, Pantanal and Cerrado). In addition, ecosystems have to be recovered (particularly the Caatinga, Cerrado, Atlantic Forest and Pampa) so that these natural resources suffer smaller impacts. At the same time, such actions will allow society to have a higher adaptation capacity in relation to climate change based on ecosystems. Therefore, prioritisation tools for conservation and recovery areas, currently used to guide public policies like the National Conservation Units System and the New Forest Code should incorporate the climate component to their analysis, with scenarios with temperature increase over 4 °C.

In the energy sector, in extreme climate scenarios, a vicious cycle would emerge. An increase in demand for electricity would be seen due to global warming, while energy production would fall short in its supply, as temperature rises would affect renewable sources directly. These results do not depend on future settings of the Brazilian energy system, as renewable sources remain predominant in the medium-long term. Consequently, it is critical to start incorporating the issue of climate vulnerability to Brazil's generation park, as well as the national electricity sector planning. The vulnerability of the Brazilian electricity sector to extreme temperature scenarios will depend on how this system develops in the future. In mitigation scenarios, renewable sources like wind and solar become more important, in addition to fossil fuels coupled with carbon capture and storage. From a qualitative point of view and based on the negative impact on Brazilian hydro plants for the RCP 8.5 scenario (over 4 °C rise in 2100), the electricity supply deficit in the country will be inevitable in an extreme climate scenario until 2040; 5) higher temperatures may also put a stress on the electricity system in terms of use, as demand will increase to deal with hotter room temperatures. Consequently, it is critical to start incorporating the issue of climate vulnerability to Brazil's generation park, as well as the national electricity sector planning. Therefore, studies related to climate change scenarios and their impacts need to be furthered. In addition, policies that help with measures to address the climate change impacts identified in the Brazilian energy sector need to be implemented.

Risk assessments for these sectors, as well as others such as water resources, urban areas, sea level rise and natural disasters are essential to assist in understanding problems and threats consequence of a dangerous climate change due to warming about 4 °C, particularly in regions and sectors of Brazil that are already vulnerable to the extremes of climate variability. These should be conducted systematically and updated regularly, with the aim of reaching the higher government levels. At the same time, minimising climate risks means influencing the developing of policies that prioritise the mitigation of emissions, as well as adaptation to future scenarios. Therefore, policy makers should address climate change as a risk management issue.

In addition to recommendations in relation to public policy action in each sectorial chapter of this book, a set of suggestions for scientific practice has also been provided. As shown by this book, assessing the impacts of extreme climate events due to a warming above 4°C on the energy, agricultural, health and biodiversity sectors is very complex. Hence, scientific investment is recommended in an attempt to find new methodological and technological approaches, such as the inclusion of long data time series. Moreover, it is suggested that events in sentinel areas, representative of Brazil's biomes and urban zones be adopted, so a more refined analysis of particular impact situations may be provided.

At the regional level, while droughts are projected to increase by 2100 and the increase in the likelihood of warming above 4 °C by the middle of this century, it is still uncertain what these rainfall and soil moisture deficits might mean for drought risk in in Northeast Brazil. Even with the possibility of longer and more extensive drought in this region as projected by climate models, and based on the impacts of recent droughts as in 2012–2016 on the population, it seems that adaptation processes to cope the drought are often poorly implemented and losses in the regional economy and ecosystem services induced by climate variability and change are generally high and yet unexplored. For the Amazon region, high emission scenarios project progressively higher warming that may exceed 4 °C in Amazonia in the second half of the century, particularly during the dry season in the region and it is projected a reduction of precipitation year-round. Several negative effects in Amazonia can be anticipated, including short-term hydrological changes similar to the events associated to the extreme 2005, 2010 and 2016 droughts, and longer time-scale modifications of broad scale characteristics such as different biome distribution.

As nations around the world signed a global climate treaty in Paris in December 2015, most leaders and scientists have agreed that limiting the global average temperature rise to 2°C over the next century could yet ward off the worst effects of climate change. But the world remains on a trajectory to experience an increase of 3 °C — even if the national emission reduction pledges to be codified in the Paris Agreement are implemented as manifested by the executive secretary of the UNFCCC. There is a consensus on the scientific community that no doubt that humanity will survive, but it's still a world where there are likely to be massive disruptions.

9.2 Risk Reduction and Assessment

A risk assessment is essential to assist in understanding problems and threats, while at the same time, supplying elements on the nature of solutions. It should be conducted systematically and updated regularly, with the aim of reaching the higher government levels. At the same time, minimising climate risks means influencing the developing of policies that prioritise the mitigation of emissions, as well as adaptation to future scenarios.

Therefore, climate change should be addressed by policy makers as a risk management issue. Currently, sectoral impacts and prospects for Brazil are produced by extrapolating international studies, which may lead to several uncertainties and imprecision. Thus, furthering the analytical and critical study of extreme temperature rise, taking into account different key sectors and biomes in Brazil is a necessity for the country. Consequently, in addition to recommendations in relation to public policy action, a set of suggestions for scientific practice will also be provided. As shown by this report, assessing the impacts of extreme climate events and over 4°C temperature increase on the energy, agricultural, health and biodiversity sectors is very complex. Hence, scientific investment is recommended in an attempt to find new methodological and technological approaches, such as the inclusion of long data time series. In addition, it is suggested that events in sentinel areas, representative of Brazil's biomes and urban zones be adopted, so a more refined analysis of particular impact situations may be provided.

Index

C. A. Nobre et al. (eds.), *Climate Change Risks in Brazil*,
https://doi.org/10.1007/978-3-319-92881-4

MIX
Papier aus verantwortungsvollen Quellen
Paper from responsible sources
FSC® C105338

If you have any concerns about our products,
you can contact us on
ProductSafety@springernature.com

In case Publisher is established outside the EU,
the EU authorized representative is:
Springer Nature Customer Service Center GmbH
Europaplatz 3, 69115 Heidelberg, Germany

Printed by Libri Plureos GmbH
in Hamburg, Germany